Numerical Software –
Needs and Availability

Numerical Software – Needs and Availability

Proceedings of the Conference on Applications of Numerical Software – Needs and Availability held at the University of Sussex, September 19th–22nd, 1977, organized by The Institute of Mathematics and its Applications

Edited by

D. JACOBS

Central Electricity Research Laboratories, Leatherhead, Surrey

1978

ACADEMIC PRESS

London New York San Francisco

A Subsidiary of Harcourt Brace Jovanovich, Publishers

ACADEMIC PRESS INC. (LONDON) LTD.
24/28 Oval Road,
London NW1

United States Edition published by
ACADEMIC PRESS INC.
111 Fifth Avenue
New York, New York 10003

Library of Congress Catalog Card Number: 78–52093
ISBN 0–12–378660–6

Printed in Great Britain by Whitstable Litho Ltd., Whitstable, Kent.

CONTRIBUTORS

C. F. BANFIELD; *Rothamsted Experimental Station, Harpenden, Hertfordshire, AL5 2JQ.*

E. M. L. BEALE; *Scicon Computer Services Limited, Brick Close, Kiln Farm, Milton Keynes, MK11 3EH.*

J. BENTLEY; *NAG Central Office, 7 Banbury Road, Oxford, OX2 6NN.*

T. CHAMBERS; *H.Q. Computing Centre, Central Electricity Generating Board, 85 Park Street, London, SE1.*

A. R. CURTIS; *Computer Science and Systems Division, Atomic Energy Research Establishment Harwell, Oxfordshire, OX11 ORA.*

W. C. DAVIDON; *Department of Physics, Haverford College, Haverford, P.A. 19041, U.S.A.*

L. M. DELVES; *Department of Computational and Statistical Science, University of Liverpool, Brownlow Hill, P. O. Box 147, Liverpool, L69 3BX.*

B. FORD; *NAG Central Office, 7 Banbury Road, Oxford, OX2 6NN.*

S. J. HAGUE; *NAG Central Office, 7 Banbury Road, Oxford, OX2 6NN.*

J. G. HAYES; *Division of Numerical Analysis and Computing, National Physical Laboratory, Teddington, Middlesex, TW11 OLW.*

S. E. HERSOM; *Numerical Optimisation Centre, The Hatfield Polytechnic, 19 St. Albans Road, Hatfield, Hertfordshire.*

D. A. H. JACOBS; *Central Electricity Research Laboratories, Kelvin Avenue, Leatherhead, Surrey, KT22 7SE.*

P. KEMP; *Computing Laboratory, University of Newcastle-upon-Tyne, Newcastle-upon-Tyne, NE1 7RU.*

A. K. MALLIN-JONES; *ICI Limited, Corporate Laboratory, P. O. Box 11, The Heath, Runcorn, Cheshire, WA7 4QE.*

R. W. McINTYRE; *Engineering Computing Centre, Rolls-Royce (1971) Limited, Bristol Engine Division, P. O. Box 3, Filton, Bristol, BS21 7QE.*

J. K. REID; *Building 8.9, Atomic Energy Research Establishment Harwell, Oxfordshire, OX11 ORA.*

H. H. ROBERTSON; *ICI Limited, Petrochemicals and Polymer Laboratory, P. O. Box 11, The Heath, Runcorn, Cheshire, WA7 4QE.*

H. H. ROSENBROCK, FRS; *Control Systems Centre, University of Manchester Institute of Science and Technology, P. O. Box 88, Sackville Street, Manchester, M60 1QD.*

G. J. S. ROSS; *Rothamsted Experimental Station, Harpenden, Hertfordshire, AL5 2JQ.*

B. T. SMITH; *Applied Mathematics Division, Argonne National Laboratory, 9700 South Cass Avenue, Illinois 60439, U.S.A.*

T. TATE; *Engineering Computing Centre, Rolls-Royce (1971) Limited, British Engine Division, P. O. Box 3, Filton, Bristol, BS21 7QE.*

J. WALSH; *Department of Mathematics, University of Manchester, Oxford Road, Manchester.*

J. P. WHELAN; *Philips Research Laboratories, Cross Oak Lane, Redhill, Surrey.*

J. H. WILKINSON, FRS; *Division of Numerical Analysis and Computing, National Physical Laboratory, Teddington, Middlesex, TW11 OLW.*

PREFACE

In September of 1977, the Institute of Mathematics and its Applications sponsored the second of its recent series of conferences concerned with numerical software. Entitled 'Applications of Numerical Software - Needs and Availability', the aim of this conference was to bring together both people who are interested in the development and production of software and also representatives of users and potential users. To fulfil this aim, many of the papers were grouped together with presenters from the two groups. This book contains the written versions of the papers presented at the conference.

A minor regrouping of some of the papers has been made from their order as presented at the conference where this has been judged to assist the reader. The book is divided into two broad sections. The first concerns libraries, their organization, management and upkeep from both the eyes of the library managers and the users. The second section comprises chapters, often in pairs following the theme of the conference, covering many of the facets of numerical analysis and its implementation in computer software. Again, both the details of software currently available, and surveys of application problem areas are included.

Apart from the considerable amount of detailed information presented at the conference and contained in the papers, several important facts emerged. In many cases it is the availability of software libraries with good quality routines which is providing considerable 'professional' assistance to the users. Within each field of numerical analysis tremendous development and specialization has taken place, so much so that in many cases it is only by the communication of algorithms and methods in the form of computer software that many of the new techniques become quickly and widely available. However, it is now widely recognised that for many algorithms, users' requirements are so diverse that it is seldom possible to meet all of them with one piece of software. As a result a suite of different levels of routines may be required for different 'levels' of user. This requirement is discussed in the papers by Bentley and Ford from the library viewpoint and by Kemp from the user viewpoint. At the conference this was amongst the most discussed topics.

The user interface was only one of several valuable topics of lively discussion at the conference. The written papers reflect the views of many of these discussions. It is hoped that with the publication of these proceedings, the communication channels between library managers, algorithm and software developers and users will be further stimulated so that the efforts of contributors can be effectively reaped.

I would like to thank all the conference contributors for preparing their manuscripts so accurately, and Professor Leslie Fox for contributing the foreword. Thanks must also go to all those colleagues of several of the contributors who acted as go-betweens and gave me valuable assistance. The IMA staff, and in particular Catherine Richards and Susan Hockett have both again contributed much to the preparation of this book. I would also like to thank the secretaries of several of the contributors who assisted in the typing of the final manuscript, and of course Academic Press.

February 1978 David Jacobs

FOREWORD

For some years I have been publicising, often stridently and occasionally hysterically, my opinion on the fact that few numerical analysts try to solve real practical problems or help our scientific colleagues to solve them. It is therefore both gratifying and exciting to have attended a conference at which a considerable mixture of numerical analysts and experts in numerical software seem to be moving strongly in this direction. It is also exciting to find an IMA conference at which more than half the participants do not live in the ivory towers of centres of higher education, but inhabit the worlds of industry and research laboratories. In other words those who need these numerical products came not only to listen but also to tell us where we should be going in our future developments of numerical software.

The conference treated three main areas. First, we had talks about numerical libraries themselves, on their value or otherwise in research laboratories, on the details of their construction, and on the current research in the serious problems of portability (or should it be transportability?) in attempts to make less difficult and time-consuming the writing of routines easily adjustable to different kinds of machines and programming languages. Second, we heard talks on the development of better algorithms in various fields for which some algorithms already exist. These included data-fitting and curve-fitting and statistical applications of the work of numerical analysts in the theory and practice of single value decomposition (and how nice to see numerical analysts and statisticians talking together at the same conference!), further advances in routines for sparse matrices, non-linear equations, optimization and mathematical programming, and the ubiquitous problem of stiff differential equations. Third, we heard about both the difficulties and excitements of trying to produce packages of routines, so far attempted only rather spasmodically, for problems like integral equations and partial differential equations in which the number of different problems is very large indeed and in which any reasonably simple classification is therefore very difficult. And there was one extra item, a little gem on the importance of all this work on numerical software in the development of interactive computing. As a stress on an earlier remark, of the 22 talks in the main programme only 9 were given by academics (in any interpretation of this title!), whereas in the York (1976) IMA meeting on the State of the Art in Numerical Analysis no fewer than 15 of the 23 talks were given by academics.

As one who did not contribute to the formal part of this conference I can say without immodesty that the talks were all of high quality, both in content and delivery. I was also impressed with the relevance of the discussions relevant to the talks and the more informal contributions at two evening sessions on "libraries" and "high-level languages" for numerical software.

Several speakers touched on a few important non-technical points peripheral to the main theme of the conference. Some noted that very few users of such software have much numerical expertise, whereas a program library, however good, is used best by such experts, and others noted the lack of numerical analysts able and willing to give advice to users at the computing service level. These are matters of some concern to academics. Others suggested that the writing of library documentation is a very important activity, and that it would be worth training and recruiting real professionals in this field. Some noted the importance of problem formulation and mathematical investigation prior to or perhaps associated with the application of numerical methods and relevant software. And I would like to have seen, and hope to see in the future, more effort given to the admittedly difficult problem of providing more information in the answers about the number of figures which are "meaningful" in relation to data with uncertainties of known sizes which are a feature of most practical problems.

All-in-all this was a very good conference, and it is good to know that many more can now benefit from it by reading these proceedings, published with the well-proven editorship of Dr. Jacobs. The organising committee can congratulate themselves on their work, and we can also congratulate and thank Brian Ford, Jim Wilkinson, Dave Martin and John McDonnell for initiating a conference of such undoubted "timeliness and promise," the IMA for their wisdom in supporting it and the Programme Committee for organising it in such nice detail.

October, 1977

L. Fox
Oxford University

The Institute thanks the authors of the papers, the editor, Dr. D.A.H. Jacobs (Central Electricity Research Laboratories) and also Miss Janet Fulkes and Mrs. Susan Hockett for typing the papers and preparing the diagrams for publication.

CONTENTS

PART I

LIBRARIES OF NUMERICAL SOFTWARE

I.1

A LIBRARY DESIGN FOR ALL PARTIES

B. Ford and J. Bentley

(Numerical Algorithms Group, Oxford)

1. INTRODUCTION TO NUMERICAL SOFTWARE

The preparation of Numerical Software involves three states:

(i) the design of the algorithm

(ii) its realisation as a source-language subprogram

(iii) the testing of the compiled code on a given configuration and its detailed documentation for that configuration.

The end product, numerical software, may have at least three different forms.

(i) Tabulation of Data

The first, a tabulation of data (in the manner of the old tables) may be used in the solution of many, widely differing problems for which almost invariably it has an intermediate role to play in ultimate problem solution.

(ii) An Application Program

An application program is one which is written to provide a solution to a particular problem or class of problems. Two important features of a well-written application program are that both the manner in which the user presents his data, and the form in which the results are produced, are convenient for the particular problem class.

(iii) A Collection of Subprograms

The third, a collection of subprograms, are units available to a programmer to assist in the solution of his problem. However, he must devise or find a method, and write the program, in order to achieve this.

All forms of numerical software can be misused. However collections of subprograms are particularly vulnerable, since their sole use is through inclusion in "foreign" coding where inevitably program control rests with the user. Such collections are therefore open to misuse and abuse by inexperienced programmers. They do nevertheless have one major advantage. A data-base or an application package can only solve that problem or group of problems to which it is directed, with flexibility being achieved only by an increase in program size or significant program sophistication. A collection of subprograms, on the other hand, potentially permits solution of problems from a far broader area, with flexibility derived from the careful selection of units. Needless to say, this freedom is only available to people able to program, whilst use of an applications package requires simply the ability to prepare structured data.

2. THE CONCEPT OF A LIBRARY

The concept of a Library, of scrolls and later of books, has been known for hundreds of years. (The Library of Alexandria was formed in the fourth century B.C.) The extension to a collection of routines was made by Wheeler in Cambridge twenty-six years ago, shortly after the advent of the electronic computer (Wilkes *et al.*(1951)).

Librarianship in general continues to explore structures, forms and modes for optimal libraries of printed information. It is evident therefore why the structures, forms and modes of libraries for use with computing systems have yet to be seriously examined or indeed to be widely understood.

For such libraries of computer software, structure involves questions of classification, codification and presentation; forms are data bases, programs and routines; modes are the source-text, the "pre-compiled" and the "binary" (of a subprogram).

As with a library of books, a collection of routines is prepared according to some principle and purpose. It may be a general library, for example, NAG (1977), IMSL (1974), seeking to cover common requirements over a broad field. It may be a subject library, aiming to cover a particular area in depth (for example the numerical solution of ordinary differential equations, Shampine and Gordon (1975)) or it may be a topic library, addressing the requirements of a particular community (for example quantum chemistry, Q.C.P.E.).

For a library to succeed it must, from the outset, be directed to a particular purpose (for example the solution of numerical and statistical computational problems). It is also of fundamental importance to identify its primary users. The purpose determines which subject areas will be included; the users, the manner and depth in which these areas will be covered and presented.

In considering the analogy of a software library with a library of books, one must exercise great care. It is the responsibility of the reference librarian to cover all aspects of his particular subject area in depth, leaving the reader to absorb the appropriate information and form his own opinions. In the case of a software library such an approach is impractical. The typical user of such a library has a specific problem which he wishes to solve; to present him with every available routine capable of solving his problem would lead not only to confusion on the part of the user but also to gross inefficiency on account of the vast amount of information which would need to be held. In this respect the software library has to be more compact and hence much more selective in satisfying the demands of its users, see Ford (1972) and Cody (1974).

This paper discusses the design considerations in developing a steady state library for numerical software. Such a design strategy may be relevant to the NAG Library (Annual Report, 1977) and indeed to other libraries, for example IMSL (1974) and PORT, Fox (1976), in the next decade, but at present this is not the case. The contents of the current NAG Library consist of material developed to past and indeed present standards. We believe that it would be foolish to force such material into an artificial structure. What we require is a design which is optimal for users, since the library is intended for them, yet natural within the structure of the material involved. Hence we must look to our future software in considering an ideal library design strategy.

3. USER REQUIREMENTS

The main purpose of developing and maintaining a numerical software library is to provide a service to the user community. For this service to be well received and therefore successful it is essential that we begin by attempting to identify that user community. We know, or indeed we hope, that the spectrum of users will be wide, in their knowledge of numerical analysis, of programming and of problem formulation and solution. We therefore require a library design which can satisfy the majority, if not all, of their many requirements.

The user community will, in general terms, range from the "one-off" user, seeking a single easy-to-use routine that solves his mathematical problem, to the numerical analyst who actually contributes material to the library. A library design which attempts to meet their range of requirements is given in Section 4.

3.1 The User and Problem Solution

We have recognised that with a subroutine library the user has to write his own program (primitive as that may be and however much assistance he may receive to complete it!). Every programmer starts with a problem which he wishes to solve. Whatever the academic field from which the problem may derive, it must be presented in terms of mathematics, if the contents of the library are to assist in its solution. The resulting mathematical model may be classified as belonging to one of a limited set of mathematical areas. It is the characteristics of this mathematical model of the problem (irrespective of the discipline from which the problem initially derives) that determine the method of solution. A small percentage of problems may be solved analytically but most must be solved computationally. Some are incapable of solution, at the present time.

3.2 The User and Library Classification

Problem solution is therefore stratified according to the particular numerical area to which a model belongs. Hence the fundamental classification in the library will be best in terms of areas of numerical mathematics, since this approach will most easily segregate the problem types that arise. The major difficulty with this approach is that whilst most library users will be familiar with the vocabulary and thought forms of their own field of interest, they may not enjoy a similar command of computational mathematics. It is essential, therefore to ensure that detailed non-technical documentation is available to users to overcome, or at least to reduce this difficulty.

4. OVERALL DESIGN OF LIBRARY

4.1 Selection of Library Contents

The contents of the Library are selected on the basis of algorithms that solve problems met by users. This selection process is dependent upon users communicating their present and perceived needs to software librarians and upon algorithm developers being motivated by such requirements. Hitherto users have often simply accepted what the analysts have made available. There is however evidence that in an attempt to encourage product utilisation many numerical analysts are seeking to discover user requirement within their area of interest and to satisfy that need. Programmers have at the same time recognised their ability to encourage development of algorithms for specific problem types.

A secondary factor in the selection of contents is the requirement of an algorithm developer in one area to utilise the methods developed in another area. This commonly affects the algorithm ultimately selected for inclusion in the Library and on occasion the interface with which it is made available.

We would prefer each algorithm included in the Library to enjoy five characteristics:

(1) stability

(2) robustness

(3) accuracy

(4) adaptability

(5) speed.

However as the basis of content selection is primarily directed by user need, we may be required (because of the stage of technical development in some area) to provide an algorithm that fails to exhibit any, or indeed all, of these properties (Ford and Sayers, 1976). Our sources and resources will enable us to replace out-moded material at the earliest opportunity.

4.2 Organisation of Contents

It is essential that the contents of the library evolve as research and development permit. We must take care, therefore, to organise the contents in such a way, that all changes are made with the minimum inconvenience to users.

As discussed in Section 3.2 it is convenient in general to divide the contents of the library in accordance with areas of numerical mathematics. Further subdivision will be required (Krogh, 1977) and may follow the natural substructure within each mathematical area. For example, in the area of Linear Algebra we may have the following subdivisions:

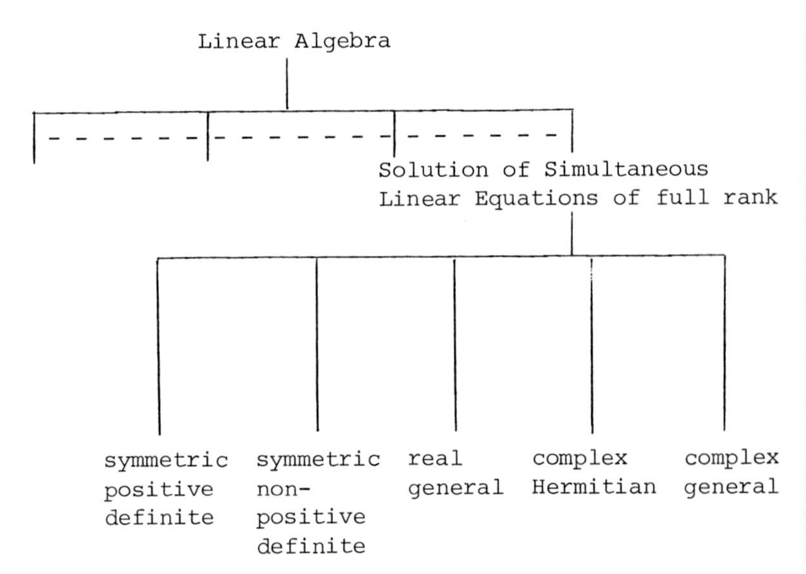

Each mathematical area has its own particular substructure with its own characteristics. To organise the library contents on this basis requires not only that these characteristics be known and explored in each mathematical area (thereby enabling the required detailed classification of problem types) but that the facility to enable the non-specialist user to identify his problem be provided as well. Inevitably the necessary exploration is far more advanced in some areas than in others. And whilst in some areas there may be an accepted mathematical basis for classifying problem types, that basis may not be relevant or apposite for numerical computation.

4.3 Presentation of Contents

As discussed in Section 3, the spectrum of library users, in their knowledge of numerical analysis, of programming and of problem formulation and solution, will be very broad. In general, however, we believe that the majority of their requirements can be satisfied by three types of library software:

(i) Problem solvers - where one routine call is sufficient to solve the problem. An example would be the solution of a set of simultaneous real linear equations.

(ii) Primary routine - where each routine contains one major algorithm. For example an LU factorisation.

(iii) Basic module - a basic numerical utility designed usually by the library contributor (i.e. a numerical analyst who provides routines from his own specialised field) for that particular area and intended primarily for his own and fellow contributors' use. For example, an extended precision inner product routine.

These three types of software provide the three tiers of our steady state library structure.

Each area of numerical mathematics, which we shall refer to as a chapter of the library, will require one or a number of basic modules. These will encode the fundamental, generally distinctive operations required for the basis of the primary algorithms in the chapter. For example, in the ordinary differential equations chapter a single-step integrator for initial-value problems would be provided, amongst other basic modules. Wherever possible each module will consist of a simple algorithm completing a basic operation. The intention is that the basic modules should be the fundamental units upon which the whole library can be constructed.

Each of the primary routines realises a primary algorithm, that is a major operation required in the solution of a specific user problem. Such a routine will generally call at least one chapter basic module, in addition to environmental information and the error mechanism (see Section 5). It offers users the ability to develop their own distinctive subprograms, using the primary routines as building blocks, which receive and return all necessary computational, organisational and control information for the users' requirements. Each routine realises a specific algorithm, which is vital to the program, and is unique in its performance of its intended function.

Each problem solver consists of an ordered set of primary routines collected in a subprogram to solve a specific user problem-set. Provided the user is able to write a program to call the problem solver, to prepare his data for input and to print the computed results, these carefully engineered tools should satisfy all his computational requirements. This means that the user whose interests rest solely at the level of solving a general mathematical problem need at no point concern himself with the algorithmic details and individual steps of the computation.

4.4 Classification and Naming of Contents

As discussed in Section 3, we believe that problem solution and hence library contents should be sub-divided according to the particular areas of numerical mathematics. For purposes of naming, therefore, each numerical area is considered as a chapter within which further sub-division according to problem type and hence library routine is possible.

The names of the individual routines should be

(i) systematic - so that users can easily find their way around the contents of the library,

(ii) easily identified - so that users can recognise the chapter to which a routine belongs,

(iii) "short and snappy" - so that the individual names are easily remembered.

Mnemonic names appear ideal. However, attempting to create unique mnemonic names in algorithmically well developed numerical areas could be a problem. In addition we must remember that each algorithm is likely to be required in different computing languages, and the routines realising these versions should be readily related. Also in some circumstances different precision versions of the same algorithm in each language will be required. Ideally each routine and the document describing its use (Section 4.5) will bear the same name. A classification and naming scheme is therefore preferred which allows a natural place for introductory and for contents documents in each chapter. For ease of use this suggests an alphabetic ordering of names for library contents and hence for library documentation.

Since the major vehicle for the preparation of numerical software is FORTRAN, we must also take into account the fact that the ANSI FORTRAN (1966) standard allows only six character identifiers for subprograms. This is the final body blow to any hope of using a mnemonic naming scheme. The necessary information simply cannot all be provided on a mnemonic basis.

Hence a systematic naming scheme is preferred, based upon an internationally employed classification of numerical mathematics. Whilst in many ways imperfect, the modified Share Classification Index is the best that is generally available at the present time, and will be the basis for the library naming discussed. The name has four fields,

Name of chapter in Share Classification	Major Problem Area	Ordering to Specific Problem Set	Language of Use (and precision of version)
(3 Alphanumeric Characters)	Alphabetic Character	Alphabetic Character	Alphabetic Character
For example, E04	U	A	F
Non-linear Optimisation	Non-linear constraints, no derivatives	First primary routine	For use in a FORTRAN program in the standard local precision

The penultimate character of the name permits a sub-division into primary routines and problem solvers. The basic modules, whilst individually named and documented, are not generally available to library users, and hence exploit other redundancy in the naming structure.

A primitive version of the above has been used within the NAG project for the last seven years and has been widely accepted by programmers. The approach is systematic, the chapter to which a routine belongs is immediately evident and the names are short. Programmers appear to have no difficulty in remembering them, and quickly appreciate the relevance and importance of the individual fields of the names.

4.5 Documentation

In preparing documentation to support the library software we must take care to cater for the different levels of experience we anticipate in users of the library. In addition, many of the problems inherent in the development of library software are relevant also to the documentation that supports that software. For example, we require that the library contents evolve as research and development permit, so too then must the documentation. We require that the library be implemented as wide as user demand dictates, hence the documentation must support the software on the many different machine ranges involved. In short, the documentation is subject to the same design problems as is the software and hence, having determined an optimum structure for the library software it appears desirable that the documentation be designed to reflect that structure.

The documentation should include general information on each mathematical area i.e. chapter of the library. Such information will enable the beginner to learn more of the background to his problem before directing him to a specific routine. The occasional user of the library may choose to ignore the general information and pass directly to the section recommending which routine should be used for his specific problem. The experienced library user, however, will be familiar with the routines available and will require simply the specific information on the routine he chooses to use. We require general chapter information and that specific to each routine.

For ease of use the routine documents, and indeed the chapter introductory documents, should bear the same name as the actual software. They should be machine independent in order that the documentation will be the same for all implementations. Any machine specific information can be handled by providing a brief machine specific supplement with the documentation for a particular implementation. Above all, the documentation should be clear and precise. Its purpose is to assist and encourage use of the library, providing support not only for the experienced numerical analyst but also for the first time user of such software.

5. SPECIFIC DESIGN ASPECTS

5.1 *Design of User Interface (Ford (1972), Smith, Boyle and Cody (1974), Ford and Hague (1973))*

A consistent approach must be employed throughout the library for communication of information, between its constituent parts and in particular to the user. Wherever possible information should only be passed through the calling sequences of basic modules, primary routines and problem solvers. Each calling sequence should maintain the data-structures, the names and the vocabulary of the area of computational mathematics which the subprogram addresses. Such "natural" calling sequences encourage a transparent coding of the algorithm, making the source-text more easily understood and hence more easily maintained. Environmental information is not passed through the calling sequence but is made available by routine call. There is no printing from library routines, except from a few specialised routines with this specific purpose.

We require a systematic approach, throughout the library, to the design of calling sequence layout. A systematic layout within each of the chapters is insufficient. Users will expect the same names to be used for similar variables etc. throughout the library, and should they move to a new machine range then they will expect the calling sequences to be unchanged. We therefore require a consistent subgrouping of parameters within each calling system; for example:

> (input parameters, output parameters, input/output parameters, work space parameters, error flag).

Work space problems and the error mechanism will each be discussed in more detail later in this section.

The length of the calling sequence will in general vary with the type of library software. For the problem solver, we require that the calling sequence be as short as possible. Only vital data-structure parameters and the error flag are included. Because of its completeness, the problem solver neither receives nor returns any intermediate algorithmic control or data information. The error flag is included so that the user can be notified if the data is deficient or if the computation goes wrong.

For the primary routines we accept that the calling sequences will of necessity be longer and often more complicated than those of the problem solvers. The calling sequence of each primary routine must include not only the essential data structure information of the fundamental problem but also the input to and results from the specific algorithm. It must also include any algorithmic control information in addition, of course, to the error mechanism. The intermediate information is required so that the user has the flexibility and control to string the routines together, to his own purpose.

It is incumbent upon the numerical analysts contributing the library software to ensure that their particular basic modules have the flexibility via their calling sequences to satisfy the requirements of other contributors. Due to the frequency of use of the basic modules, within inner loops of primary routines, this requirement should be tempered by consideration of the efficiency overhead of each parameter within the module call.

5.1.1 Working Space Problems

Many primary algorithms and basic modules require temporary arrays whose dimensions depend upon the dimension of the user's problem. This matter is easily overcome in Algol 60 and Algol 68 where local arrays may be included within the individual procedures. However, in FORTRAN we are faced, either with the declaration of the required space outside the subroutine and inclusion of additional parameters in the calling sequence, or with declaring arrays of fixed dimension within the subroutine that are considered sufficiently large for most anticipated problems. The latter proposal is inflexible and often wasteful of storage whereas additional parameters will not seriously affect the calling sequences of the primary routines. The users of such routines will appreciate the necessity of inclusion of the additional parameters and may, on occasion, welcome access to the information they contain.

However, in particular numerical areas (for example ordinary differential equations and non-linear optimisation) there is a need for such a number of local arrays for algorithm work space that the calling sequence is significantly enlarged. A possible solution would be to gather together the individual work arrays into one (or more) large work array(s). This would also enable the algorithm to be changed, even if more work space were required, without the necessity to change the user interface.

This approach runs counter to the attempt to maintain "natural" calling sequences in keeping with the data-structures, names and vocabulary of the particular numerical area. The use of partitioned arrays also renders the source-text opaque even for an experienced programmer, with implications for both direct use and for software maintenance. As there are FORTRAN compilers which limit the number of parameters that may be included in a calling sequence we cannot dismiss the technique completely. However it should only be used with great caution.

The MAP statement, proposed for inclusion in FORTRAN 77 by IFIP Working Group 2.5, (1976), appears to be a suitable means of overcoming such difficulties in future library construction.

5.2 *Modularity*

One of the crucial decisions in designing algorithms and in determining technical features of library structure is the level or levels of modularity to which one should work. Should each module or routine contain one algorithm, which performs one single operation (for example, a double precision computation of an inner product) or may it be a composite of a number of related algorithms (for example with options for single, double or further extended precision computation of an inner product)? Experience has shown that anything but the simplest modularity, with one simple algorithm per subprogram, may at best present unnecessary overheads and at worst render apparently relevant material unusable by other programmers. This is particularly true for algorithms required by other library contributors.

5.3 *Error Mechanism*

A common error mechanism is necessary throughout a library, permitting the user to choose his response when an error is detected during computation in a library routine. Experience to date with the NAG Library suggests that a mechanism offering two options is at least adequate. The user can choose either a "hard" failure or a "soft" failure. If set to "hard", upon encountering an error within a routine, the program control flow is passed to the error module, which outputs a message describing where the error occurred and its nature, then terminates the program. If set to "soft", upon encountering an error the error parameter is reset, to give notice of the error, control is returned to the calling program and the computation continues.

It is the responsibility of the programmer to check for the setting of the error flag and to program his response to this situation. The default setting for the error mechanism in problem solvers and primary routines is "hard". However, the error parameters within the primary routines used in a problem solver are all set to "soft" to ensure that the error message received by a user of the problem solver is intelligible. Such users are likely to be unaware of the routines being used and hence would be disconcerted to receive error messages from these primary routines.

5.4 Transportability

If the library is to be made available as widely as user demand requires, then it is important to consider from the outset what implications this may have, in particular for library implementation. The library will need to be implemented on many different machine ranges (i.e. not just computer families but each relevant combination of hardware and software within the family) and the effects this may have upon contributed software will be of no little consequence.

Much has been said on the subject of portability and transportability (see for example Hague and Ford (1976), Smith (this volume) and the books edited by Brown (1976) and Cowell (1977)) and it is not our intention to repeat such information here. Suffice it to say that effects of the environment will have to be taken into consideration at the contribution stage if the library software is to be implemented on new machine ranges within an economically feasible timescale. A recent project by the NAG Group, instigated in the main by the need for portability, involved updating all existing software to meet new standards. The project is discussed by Ford in the book edited by Brown (1976). As a result of this project, together with the fact that newly contributed software is now written to meet these new standards, the time taken to implement the library on a new machine range has been reduced from 2-3 man years to 3-4 man months (NAG Annual Report, 1977).

As illustrated by the above example, it is vital that we anticipate and hence as far as possible eliminate the need for changes when moving the software to a new environment. If any changes are necessary, and we believe that this may always be the case, then these should be dealt with automatically with the aid of processing tools. Such tools are described in the papers by Du Croz et al (1977), Aird (1977), and Hague (this volume).

6. IMPLICATIONS OF DESIGN PRINCIPLES FOR LIBRARY DEVELOPMENT

The development process of our multi-implementation library may be summarised as follows:

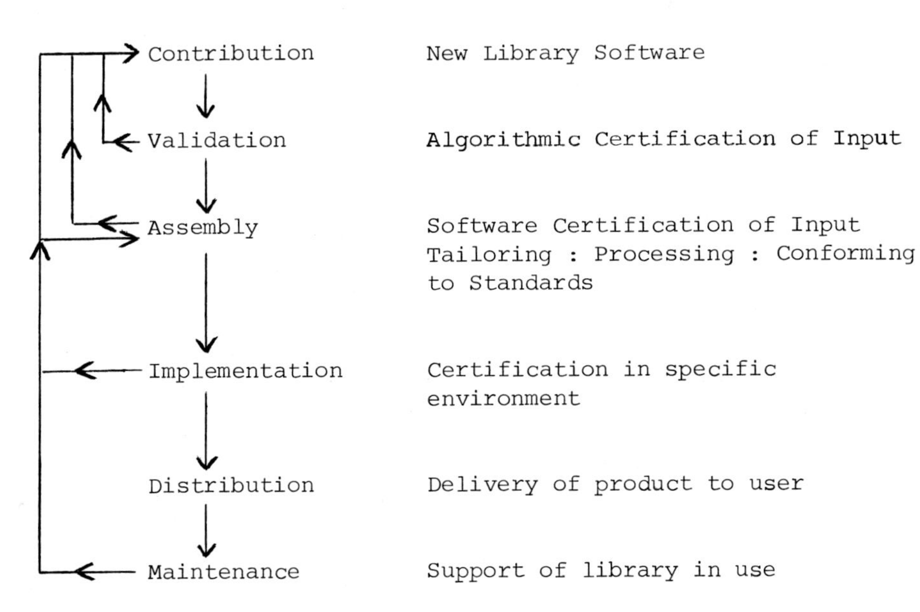

The implications of such a development process are indeed numerous and are discussed in some detail by Ford et al. (to appear). For convenience we shall summarise various aspects below.

If a quality product is to be developed then it is essential that we involve the knowledge and expertise of different technical communities, whose members will inevitably be geographically dispersed. For such a widespread and yet collaborative activity to succeed we require a body of individuals each working to precisely defined standards and each in pursuit of the same specified objectives. We need to monitor and to co-ordinate the work of these individuals. Above all we need to ensure the effectiveness, reliability and usefulness of the product developed.

In pursuit of these objectives, our first requirement is a collection of purpose-built test software for certification of the library. This should include three types of test software:

(i) stringent test programs - to validate each routine and show that it functions correctly and to prescribed standards

(ii) example programs - a relatively simple example, included in the user documentation, which demonstrates the use of the routine

(iii) implementation test programs - to check the correctness of the compiled code, in each implementation, as a realisation of the chosen algorithm.

In the interests of efficiency, consistency and reliability we would prefer, wherever possible, to make use of processing tools in the certification and development process of the library. In particular, tedious tasks such as the checking of results from test software would be best dealt with mechanically. Similarly the checking and where necessary imposition of standards, in the contributed code and indeed in the documentation, should, if possible, be performed automatically. Such mechanisation minimises the cost of development, distribution and maintenance, and is the most reliable method of software processing.

The development process is a continuous one, requiring constant consultation with both user and technical communities; with users to ensure that the library continues to address the problems which they need to solve and with the numerical analysis community in general in order that the library may remain "state-of-the-art" in numerical software.

7. CONCLUDING REMARKS

A numerical subroutine library is a selected collection of numerical software for a user community. Its contents and structure should directly reflect the needs and requirements of these programmers.

A three tier structure for library software has been proposed and shown to satisfy organisational, functional and technical requirements of all the parties.

We have examined the design of the user interface, levels of modularity, an error mechanism and software transportability as specific aspects within this overall structure.

There are implications in these design principles for all aspects of library development and a scheme for preparation and processing of library software has been described.

8. ACKNOWLEDGEMENTS

The authors would like to express their thanks to colleagues at the NAG Central Office, in particular Jeremy Du Croz, Steve Hague and David Sayers, for their valuable assistance in the preparation of this manuscript. We are also indebted to Dr. Brian T. Smith, of Argonne National Laboratory, for the encouragement and guidance he has provided during several years of mutual correspondence.

9. REFERENCES

Aird, T. (1977) "Converter usage in a production environment", *in* "Portability of Numerical Software", Ed. W.R. Cowell, Springer-Verlag.

ANSI FORTRAN X3.9 (1966), American National Standard FORTRAN.

Brown, P.J. (1976) (Ed.) "Software Portability - an Advanced Course", C.U.P.

Cody, W.J. (1974) "The construction of numerical subroutine libraries", SIAM Review, Vol. 16, No. 1, pp. 36-46.

Cowell, W.R. (1977) (Ed.) "Portability of Numerical Software", Springer-Verlag.

Du Croz, J.J., Hague, S.J. and Siemieniuch, J.L. (1977) "Aids to portability within the NAG project", *in* "Portability of Numerical Software", Ed. W.R. Cowell, Springer-Verlag.

Ford, B. (1972) "Developing a numerical algorithms library", Bulletin IMA, Vol. 8, No. 11, pp. 332-336.

Ford, B. and Hague, S.J. (1974) "The organisation of numerical algorithms libraries", *in* "Software for Numerical Mathematics", Ed. D.J. Evans, Academic Press.

Ford, B. and Sayers, D.K. (1976) "Towards a single numerical algorithms library for many machines", ACM TOMS, Vol. 2, No. 2, pp. 115-131.

Ford, B., Bentley, J., Du Croz, J.J. and Hague, S.J. "The NAG Library "Machine" ", To appear.

Fox, P.A., Hall, A.D. and Schryer, N.L. (1976) "The PORT mathematical subroutine library", Computing Science Technical Report #47, Bell Laboratories, Murray Hill, N.J.

Hague, S.J. "Software tools", This volume.

Hague, S.J. and Ford, B. (1976) "Portability - prediction and correction", Software P. and E., Vol. 6, pp. 61-69.

IMSL Library (1974), International Mathematical and Statistical Libraries, Inc., Houston, Texas.

Krogh, F.T. (1977), Private communication.

"MAP statement in FORTRAN to assist in the portability of numerical software" (1976), A proposal to ANS FORTRAN Standards Committee X3J3 from IFIP W.G. 2.5.

Modified Share Classification Index (200 B.C.!)

NAG Annual Report (1977), Oxford.

Q.C.P.E. Quantum Chemistry Program Exchange, Chemistry Department, Indiana University, Bloomington, Indiana.

Shampine, L.F. and Gordon, M.K. (1975) "Computer Solution of Ordinary Differential Equations", W.H. Freeman.

Smith, B.T. "Portability and adaptability - what are the issues", This volume.

Smith, B.T., Boyle, J.M. and Cody, W.J. (1974) "The NATS approach to quality software" *in* "Software for Numerical Mathematics", Ed. D.J. Evans, Academic Press.

Wilkes, M.V., Wheeler, D. and Gill, S. (1951) "The Preparation of Programs for an Electronic Digital Computer with Special Reference to the EDSAC and the use of a Library of Subroutines", Addison-Wesley.

I. 2

PORTABILITY AND ADAPTABILITY -- WHAT ARE THE ISSUES?

B.T. Smith

(*Applied Mathematics Division,*

Argonne National Laboratory,

9700 South Cass Avenue, Illinois 60439, USA)

ABSTRACT

The preparation of numerical software so that it can be easily moved from place to place has been the goal of many software projects over the past fifteen years. Concepts such as portability and transportability of source text, and adaptability of algorithms have been defined. This paper reviews these concepts, and then proposes that the techniques implementing transportability can be applied to algorithmic processes, enhancing their ability to adapt to changing computer environments.

This work was performed under the auspices of the U.S. Energy Research and Development Administration.

1. INTRODUCTION

In this paper, we survey the issues involved in moving software from place to place, and briefly review the techniques to date to solve the problem of moving software about. Much of the experience presented here refers to the restricted class of numerical software that appears in such program libraries as IMSL (IMSL (1975)), NAG (NAG (1977)), and PORT (Fox (1976)) and such program packages as EISPACK, FUNPACK (Cody (1975)), LINPACK (Dongarra et al (1977)), and MINPACK (Hillstrom et al (1976)). However, many of the issues and problems that arise in moving software about are not restricted to this special class of software, but represent more general issues and problems in the development of versatile software.

Section 2 outlines the three aspects of software that contribute to the portability of numerical software. Two of the aspects are discussed in Sections 3 and 4. In Section 5, we briefly outline the compromises that are always made in attempting to prepare portable software. The paper concludes with Section 6 in which a solution to a practical problem in preparing portable software for linear equation solvers is proposed which may avoid a serious limitation of the compromise solutions.

2. THE PROBLEM

We are interested in knowing how to prepare and write our software so that it can be readily moved from place to place. The solution requires techniques addressing three different aspects of the software; first, the source text; next, the algorithm on which the software is based; and last, the documentation. All three aspects impact the ease with which software can be ported about.

The third aspect, although it contributes in significant ways to the ease with which software is moved about, is not discussed here.

3. THE SOURCE TEXT

To describe the ease with which source text (sometimes meaning software as a whole and thus all three aspects listed above) can be moved between computing environments, we have invented the term portability. There are two commonly used definitions of this term. We state both definitions here in order that, by contrasting them, we can explore some of the issues involved in portability.

One definition, used in Ford (1977), applies the term to source text that, without any change, can be moved from place to place without any change and will execute in an equivalent fashion in all places. Thus, portability by this definition is an absolute term; a program either executes within the specifications without change and so is portable, or must be modified (possibly cannot be modified) to obtain the desired behaviour and so is non-portable.

The second definition is quantitative in nature (Brown (1977)). Portability by this definition is a measure of the difficulty involved in moving software from place to place and having it meet specified performance criteria at each computing environment. Thus, source text that is portable by the first definition would be very portable by the second definition, and source text that required modification to execute according to specifications would be portable by this definition, but not as portable as that software that required no changes.

The second definition of portability encompasses more than just the source text. For a measure of the difficulty involved in moving software between environments depends upon the appropriateness of the documentation for the conversion of the software, upon the algorithm on which the software is based, upon the diversity between the computing environments, and upon the experience of the person implementing the conversion to the new environment. The key to this definition is that essentially an expert must be involved in the conversion process in order to obtain reliable conversion of the software.

Whereas the second definition encompasses the practical world where changes are invariably necessary to move software about, the first definition does not. To complement the first definition, the concept of transportable source text (Ford (1977)) is introduced. Source text is considered transportable if the only changes that are needed to move the software from place to place can be automated; that is, such modifications are well understood and are specified well enough that the required changes can be made by program. Such a program analyzes the source text and makes the required changes correctly, but without modifying other parts of the source text unintentionally. Note that source text which is considered portable at an acceptable level by the second definition may not be transportable, for changes may be required that can not be recognized from just analyzing the source text.

Now let us discuss some of the issues concerning absolute portability of the source text. Here we start from the definition of the language in which the source code is written. The first approach, albeit naive, is to write the source text so that it satisfies the "standard for the language". But from experience, we discover that the "Standard for the language" is typically a paper standard. In practice, every compiler for a given language is slightly different, accepting a different input language. There are several reasons for this; the machine architectures are so different that it affects the characters that can be easily represented; the Standards are sometimes ambiguous or inconsistent within themselves so that the designers and implementors of the compilers implement slightly different languages; implementors whose goal is to attain efficiencies in certain cases sometimes create restrictions in the language accepted by the compiler; the tendency to add additional features, often motivated by desires to make the product more saleable, extends the language beyond the Standard; that is, the language accepted by the compiler becomes a superset of the Standard language.

Consequently, to obtain absolute portability, the source text need be written with many definitions of the same language in mind -- the Standard to avoid extensions and the more restrictive subsets to avoid non-Standard sublanguages. In general, this is a formidable task; (cf. Smith (1977) for a survey of the pitfalls in writing portable FORTRAN software). However, some assistance is provided by verifiers such as PFORT (Ryder (1974)) for FORTRAN that check for adherence to a specific sublanguage of the Standard language.

In addition, definitions of portable subsets must be dynamic. For once a subset has been focused upon, there is always another system developed that becomes a counter example to some feature of the subset language. Also the non-standard compilers that contribute to the subset are often later modified or fade into oblivion so that their restrictions no longer contribute to the portable subset. Hence, the concept of absolute portability, though a conceptually useful model for purposes of discussion, is in general, not a practical one.

The concept of transportability is devised to overcome just these ever-occurring changes in computer languages. Here, the idea is to write source text so that it satisfies some Standard, say the formal Standard for the language or possibly some convenient local dialect. Then, when a version of the source text is desired for some specific environment, well-tested program analyzers examine the source text. From the analysis, a new version of the source text is derived, which is expected to compile and execute correctly in the specific environment. The current knowledge about the variety of machine environments is codified in programs that represent the source transformations. Changes in computer environments will cause some expert to add, modify, or delete some of these programs implementing transportability.

Besides the versatility of the transformational approach to move source text about, there is the aspect of reliability. One of the appeals of the absolute portability model is that because there is no need for modification, there is no chance of inadvertent modification of the source text, possibly making it incorrect. With transportability, the changes are made by well tested programs, and provided the transformations are proven correct, and proven correctly implemented, again there is no chance of inadvertent changes. Hence, reliability is an important component in the transportability approach.

Today, transportability is a practical solution only for large packages or libraries of software. In such cases, the number of programs involved in the conversion is large enough to permit economies of scale. Also there is a certain consistency in the way the programs are written permitting simple analyzers to be used for the transformational process. The reason for this restricted application is a practical one; transportability in its general form essentially represents the full language translation problem. Thus, successes with using transportability have occurred in limited areas, and probably will continue to be limited to specialized classes or packages of software. The exception will be if transformational tools become available in compilers and are programable by the average user (cf. Loveman (1977), Standish (1976) and Ganzinger et al (1977) for examples of how source-to-source transformations are being used to produce optimized object code).

A second point is that transportability is attempting to solve the issue of program equivalence. Notice that in the previous discussion, transportability produces alternative versions which are expected to compile and execute correctly in other environments. Today, we must verify that indeed the derived versions work correctly by extensive testing. Some day, we hope this testing stage will be unnecessary. What inhibits a proof of equivalence of the master and derived versions is a valid model for each environment. Also, we need proof techniques that deal with equivalence issues between models. But again, we are dealing with such a general problem that insolvability is at issue here. The practical solution no doubt will restrict classes of software, and restrict the models that can be used. Such a solution is still very useful, however, for the demonstration of program equivalence for even certain model machines increases the reliability of the software.

What does transportability imply for us ordinary users who do not have access to these powerful tools? The key to that question is that the software is to be written in such a consistent and clear-structured fashion that an analyzer, usually a simple one, can make the necessary changes. If a simple analyzer can be written, then the changes required to move the source text about are straightforward enough that the changes can be made manually and reliably. When a complex analyzer is required, the task of manipulating source text manually becomes less reliable; for instance, more of the source may need to be scanned to determine the correct modification. However, the fact that the modification can be made by program implies that the required changes are well-understood and well-specified. Consequently, when great care is taken, the manual modification can lead to reliable manipulation of the source text.

Thus, transportability represents an attitude towards the manner in which software is written -- that is, prepare the source text so that the resulting software can be easily moved between computing environments.

4. ALGORITHMS

The emphasis in the previous discussion on portability has been on the source text, that is, whether the source text can be compiled without severe diagnostics and can be executed without violating the semantic rules of the language. Now, we wish to emphasise the algorithm on which the software is based.

We define the concept of an adaptable algorithm (Ford (1977A), Boyle (1977A)) -- an algorithm is adaptable if it can be expressed in such a way that, relative to some specification, the algorithm performs in an equivalent manner on a wide range of computing environments. The key phrases in this concept are 'in an equivalent manner' and 'relative to some specification'. We consider here a rather special specification of performance of the algorithm in that this specification must be in terms of the environment in which the algorithm is expected to run. In this context, the phrase 'in an equivalent manner' then means that the performance of the algorithm is measured against this environment-dependent specification. For instance, the usual Gaussian elimination algorithm for factorization of a matrix is an adaptable algorithm, because: 1) the specification of its behaviour namely determining the factors of a matrix "near" to the original matrix (where "near" is defined in terms of the precision of the arithmetic unit), is expressed in terms of the environment; and 2), it can be proven that at least using Wilkinson's model (Wilkinson (1963)) for the behaviour of rounding errors, this factorization process satisfies this specification. An example of an algorithm that is not adaptable would be an iteration procedure which terminates when the difference of two iterates is less than 10**(-8) in magnitude. For an environment which has nine or more decimal digits in its arithmetic operations, the algorithm may guarantee (as the result of an error analysis) that the last iterate approximates the solution of the problem to within eight digits. For an environment which has fewer than nine digits, the behaviour of the program may be unpredictable for it may not terminate, or may terminate with no accuracy guaranteed.

Adaptable algorithms in numerical software fall into two categories; 1) non-iterative (direct) processes, as exemplified by Gaussian elimination, which in order to meet their specification, operate on numerical data in a fixed manner independent of the computing environment; and 2) iterative (indirect) processes, which in order to meet their specification, vary the manner in which the numerical data is manipulated depending upon the environment. In order to prepare software, based on such algorithms that can be easily moved about, the dependence of the algorithm on its environment is expressed in terms of machine parameters such as relative machine precision, the base of the machine arithmetic, the largest and smallest positive representable numbers, and several others (cf. (Cody (1977), Ford (1977A), Smith (1977)) and papers referenced by these).

In terms of transportable software, adaptable algorithms take on a new dimension. In the previous discussion, the dependence of the algorithms on the environment were expressed simply in terms of variables or parameters in the program. In the environment of transportable software, the dependence on the environment can be expressed by use of special constructs recognized by program analyzers. A good example of such an approach is given in Schonfelder (1976). There, a package of programs is developed to evaluate special functions. Truncated Chebyshev expansions which yield 30 digits of precision are derived in a machine environment which handles such high precision computations. Then programs for other environments that use fewer digits of precision are generated from this database of 30 digit approximations. The programs so produced are sufficiently accurate and efficient for the applications the author has in mind. The resulting accuracy and efficiency are not as good as could be obtained using other approximations such as minimax rational approximations but are satisfactory for his applications.

Another example of using the transportable approach occurs in the development of LINPACK (Boyle et at (1978)). Here, versions of linear equation solvers are derived from one common form; that is, from single precision complex versions of the programs, single and double precision versions for real input data as well as double precision versions for complex data are derived.

The example given in the last section of this paper suggests how transformations can be used to solve efficiency issues that arise in linear equation solvers in a paged environment. Also, Boyle (1976) gives an extensive survey of several transformational-transportable systems.

5. THE COMPROMISE

The solution of transportable software and adaptable algorithms to the portability problem for software often compromises other important aspects of software such as robustness, efficiency, readability, usability, and sometimes even reliability.

For example, for robust software, we may want to use the error trap facility to gracefully recover from arithmetic exceptions that may occur infrequently on unusual data. On one machine architecture, using special non-standard features of the source language, we may cause a transfer to some part of the program that examines the arithmetic fault and recovers from the fault by some alternative computation that might be less efficient but avoids the arithmetic problems. The software, using this special feature, is now robust, but certainly not portable. It may not be even transportable because the feature used to detect the arithmetic fault and cause certain actions to occur (in this case, transferring to some designated place in the source) may not be available or even possible on other machines.

A compromise is then reached; the program is either less robust (it fails completely when arithmetic exceptions occur), less efficient (it always uses the less efficient but safe evaluation scheme), or even less reliable (the machine recovers from the arithmetic fault in such a way that the subsequent computations are incorrect).

Another major area in which a compromise impinges upon programming is the readability of the source text. The choice of FORTRAN (because it is widely used and readily available) over other languages such as ALGOL, PASCAL, ALGOL 68, PL/1 etc. imposes unwanted restrictions and an awkward programming style on the programmer, and makes the meaning of his source text obscure. The writer is required sometimes to code in a style, imposed by the design of the language, that leads him to use dodges and tricks in place of features that should be in the language (such as dynamic storage allocation, character manipulation, floating point number manipulation, environmental constants etc.).

Finally, there has been a major compromise in favour of portability over efficiency, mainly because the available tools for expressing optimization strategies are not flexible enough. With today's highly complex computer architectures, optimizations that take advantage of paged environments, buffered memories, fast instruction stacks, and pipelined processors, to name just a few features, can improve the efficiency of algorithms by significant factors of 2 or more. Many of these optimizations can be handled at the source text level by clever rearrangements of the source text. Of course, such optimizations are beyond the concept of absolute portability but do come within the scope of transportability. The tools for such optimizations in terms of transportability are beginning to be developed by the designers of highly optimized compilers (cf. Ganzinger et al (1977) and Loveman (1977)).

To illustrate this issue of the efficiency involved in the compromise, we slightly extend an example in Loveman (1977) to exemplify the power of the transformational systems. Consider three forms of the same computational process, written in an ALGOL-like language used in Loveman (1977).

```
(1)

declare  ( A, C)  matrix  ( 1:N,  1:N);
declare  D  diagonal  matrix  (1:N);

comment  PERFORM THE USUAL MATRIX PRODUCT D*A AND
         STORE IN C.

C  :=  PRODUCT( D, A);

(2)

integer  I,J
real  A(N,N), C(N,N), D(N)

loop for  I := 1  to  N;
  loop for  J := 1  to  N;
    C( I, J)  := D(I) * A( I, J);
  repeat;
repeat;
```

(3)

```
integer  I,J
real  Aa(N*N), Ca(N*N), Da(N), T

I := 0;
goto  BEGINI;

label LOOPI;
I := I+1;
T := Da(I);
J := 0;
goto  BEGINJ;

label  LOOPJ;
J := J+1;
Ca(J)  := T*Aa(J);

label  BEGINJ;
if  J-N<0  then  goto  LOOPJ;

label  BEGINI;
if  I-N<0  then  goto  LOOPI;
```

The first version, written in a language invented for the purpose of this illustration, clearly specifies what the program is intended to compute. The operator PRODUCT performs the usual matrix product, but the details of how that is implemented are hidden. In particular, the computational process to form this product may depend on the type of operands supplied to PRODUCT; for example, a specialized program may be used when either operand is a diagonal matrix rather than a full matrix.

The second version is more specific, displaying (and using) the method for storing the full matrix A and multiplying it into the diagonal matrix D. The multiplicatior is by rows, thereby taking advantage in a paged environment of a programming language that stores its two dimensional matrices by rows (Moler (1972)). However, if the full matrix is not stored by rows or the diagonal matrix of version (1) is not represented as a vector, then this version is difficult to adapt to different environments.

The third version is very specific and indeed, although very efficient for special environments, it is difficult to follow, and hence very difficult to modify for other environments.

Of course, the compromise that we all use is version (2); version (3) is much like assembly language in which the details of the implementation are apparent in all statements and the intent of the program is hidden; version (1) is ideal for human communication but the efficiency of the resulting object code is unknown, dependent upon the quality of the compiler. On the other hand, if we had a good transformational system available, the compromise may be avoided, because we could represent our numerical processes in the highest (or most abstract) level and cause the most specific form to be generated by source language transformations. Thus, we obtain the generality of version 1 and the efficiency of version 3, while at the same time preserving the validity of the computational process.

6. FROM ADAPTABLE ALGORITHMIC REPRESENTATIONS TO TRANSPORTABLE SOFTWARE -- A PRACTICAL EXAMPLE.

Using transformational techniques, we can in the future make the compromises more attractive; that is, instead of producing one version that hopefully everyone can accept, we can tailor versions of the software to meet specific requirements from an adaptable representation. To illustrate this point, we briefly consider the process of solving systems of linear equations using Gaussian elimination.

First, we specify the algorithmic process which solves linear equations in an high level ALGOL-like language patterned after the notation in Loveman (1977).

```
comment  CODE FOR THE FACTORIZATION OF A MATRIX A
         INTO FACTORS  L AND  U  USING GAUSSIAN
         ELIMINATION WITH PARTIAL PIVOTING;

real  A(N,N), B(N);
integer  PVT(N);

comment  FACTOR THE MATRIX  A  INTO FACTORS P, L,
         AND  U WHERE P  IS A PERMUTATION MATRIX
         ENCODED IN THE INTEGER VECTOR  PVT(N),  L
         IS A UNIT LOWER TRIANGULAR MATRIX STORED IN
         THE STRICT LOWER TRIANGLE OF  A,  AND U IS
```

```
        AN UPPER TRIANGULAR MATRIX STORED IN THE
        UPPER TRIANGLE OF  A;

loop for  I := 1  to  N-1;

  comment  DETERMINE THE I-TH FACTOR OF THE
           PERMUTATION MATRIX  P, AND ENCODE IN THE
           VECTOR  PVT;

  PVT(I) := MAXIND(A, I);

  comment  PROVIDED AN ELIMINATION STEP IS NEEDED,
           STEP  1: EXCHANGE COMPLETE ROWS TO PLACE
                    ON THE DIAGONAL THE LARGEST
                    ELEMENT IN MAGNITUDE IN THE I-TH
                    COLUMN.
           STEP  2: FORM THE PRODUCT OF THE I-TH
                    ELEMENTARY TRANSFORMATION WITH
                    THE PREVIOUS ELEMENTARY
                    TRANSFORMATIONS.
           STEP  3: APPLY THE I-TH ELEMENTARY
                    TRANSFORMATION TO THE APPROPRIATE
                    SUBMATRIX OF A,
              OVERWRITING THE SAME SUBMATRIX OF A.;

  if  A(PVT(I), PVT(I))¬= 0  then

        call  EXCHANGE_COMPLETE_ROWS(A, I, PVT(I));
        call  FORM_PRODUCT_OF_ELEMENTARY_MATRICES
              (A, I);
        SUBMATRIX(A, I+1) := PRODUCT(
                                INVERSE(
                                  UNIT_LOWER_
                                  ELEMENTARY(A, I)),
                                SUBMATRIX(A, I+1));
  endif;

repeat;

comment  SOLVE THE LINEAR EQUATIONS  P*L*U*X = B
         FOR X;
comment  STEP 1: FORM THE VECTOR  INVERSE(P)*B  IN B;

loop for  I := 1  to  N-1;
  call  EXCHANGE_COMPLETE_ROWS(B, I, PVT(I));
repeat;
```

```
comment STEP 2: FORM THE VECTOR INVERSE(L)*B  IN  B;

B := PRODUCT( INVERSE( UNIT_LOWER_TRIANGLE(A) ), B);

comment STEP 3: FORM THE VECTOR  INVERSE(U)*B  IN  B;

B := PRODUCT( INVERSE( UPPER_TRIANGLE(A)), B);
```

In the above example, we have a high level description of the required computational process, in which the details of the numerical calculations are hidden inside subprocedures such as EXCHANGE_COMPLETE_ROWS, FORM_PRODUCT_OF_ELEMENTARY_MATRICES, etc. and operators such as INVERSE, PRODUCT, SUBMATRIX, etc. We are interested in translating this program into the usual form in which explicit references to the elements of the matrix A and the vector B appear. But there is one additional requirement; we are interested in producing code that runs efficiently in paged environments (Moler (1972)). This implies that we must organize the computational process in a special way; that is, when a page of storage representing part of A say is placed into high speed core, we must use as many of the elements of A in this page as possible before the page is taken from high speed core. But depending on the language (and even in some cases upon the implementation of the language), matrices are stored linearly either by row or by column; for example in FORTRAN, matrices are stored linearly by column so that the last element in the i-th column is adjacent to the first element in the (i+1)-th column of A, whereas in languages such as PL/1 (and sometimes ALGOL), matrices are stored linearly by row.

In this example, the key statements whose efficiency (or inefficiency) contributes most to the computational process are 1) in the factorization process, the product of rank one matrices and submatrices of A; and 2) in the solution process, the products of triangular matrices and vectors. (In both cases, the operator PRODUCT is used to denote these expensive operations.)

We now claim that this is an adaptable algorithmic representation of this numerical process. For in this form, software can readily be produced to suit either orientation for the storage of matrices. There are essentially two approaches; for one, we provide versions of the subprocedures and operators in source code for each of the storage orientations; or second, we can transform the above source text using the techniques of Boyle and Loveman, translating the above representation into a specific language, including the details of the numerical process for the specific orientation.

The central point in the above example is that the algorithm representing the numerical process is specified at a high level without details of specific environments. The details are provided by the transformational process, once the detailed specification of the environment is known. In a sense, the transformations are an encoding of the special properties of the particular machine environments, helping to separate algorithmic considerations from implementation considerations.

The above example illustrates that adaptable algorithmic processes and adaptable representations of them (called abstract forms in TAMPR) complement one another, thus providing a very flexible approach to preparing quality numerical software. The advantages of such an approach are many; the major disadvantage is the inaccessibility of the automated transformational techniques necessary to make this method of preparing numerical software generally useful.

However, the inaccessibility may change. Many of the needed techniques are used in advanced optimizing compilers and some believe that these techniques will become available in specialized compilers (cf. Ganzinger et al (1977) and Loveman (1975)). But despite the lack of availability, we are convinced that this approach to designing software leads to software that is more easily modified to address the needs of present and future users of numerical software.

7. REFERENCES

Boyle, J.M. (1976), "Mathematical Software Transportability Systems -- Have the Variations a Theme?", Portability of Numerical Software, Lecture Notes in Computer Science 57, Oakbrook, Ill., Springer-Verlag, New York, pp 305-360.

Boyle, J.M. (1977), "An Introduction to the Transformational-Assisted Multiple Program Realization (TAMPR) System", Co-operative Development of Mathematical Software, Technical Report, (Ed.) J.R. Bunch, Department of Mathematics, University of California, San Diego.

Boyle, J.M. (1977A), Private Communication, Discussion of Adaptable Algorithms.

Boyle, J.M., Frantz, M.E., and Kerns, B. (1978), "Automated Program Realizations: BLA Replacement and Complex to Real Transformations for LINPACK", Argonne National Laboratory Technical Report (in preparation).

Brown, W.S. (1970), "Software Portability", Report of the 1969 NATO Conference on Software Engineering Techniques, (Eds.) J.N. Buxton and B. Randell, pp 80-84.

Cody, W.J. (1975), "The FUNPACK Package of Special Function Subroutines", ACM Trans. Math. Soft. Vol. 1, No. 1, pp 13-25.

Cody, W.J. (1977), "Machine Parameters for Numerical Analysis", Portability of Numerical Software, Lecture Notes in Computer Science 57, Oakbrook, Ill., Springer-Verlag, Berlin, pp 49-67.

Dongarra, J.J., Bunch, J.R., Moler, C.B., Stewart, G.W. (1977), LINPACK Working Note No. 9: Preliminary User's Guide, Technical Memorandum No. 313, Applied Mathematics Division, Argonne National Laboratory.

Ford, B. (1977), "The Evolving NAG Approach to Software Portability", Software Portability, An Advanced Course, Cambridge University Press, London.

Ford, B. (1977A),"Preparing Conventions for Parameters for Transportable Numerical Software", Portability of Numerical Software, Lecture Notes in Computer Science 57, Oakbrook, Ill., Springer-Verlag, Berlin, pp 68-91.

Fox, P.A. (1976), "The PORT Mathematical Subroutine Library User's Manual", Bell Laboratories, Murray Hill, New Jersey.

Ganzinger, H., Ripken, K., and Wilhelm, R. (1977), "Automatic Generation of Optimizing Multipass Compilers", Proceedings of IFIP Congress 77, Toronto, Canada, North Holland, Amsterdam, pp 535-540.

Hillstrom, K., Nazareth, L., Minkoff, M., More, J., and Smith, B. (1976), Progress and Planning Report, MINPACK Project, Internal Memo, Applied Mathematics Division, Argonne National Laboratory.

IMSL Library Manual (1975), International Mathematical and Statistical Libraries, Volumes 1-2, Edition 4, Houston, Texas.

Loveman, D.B. (1977), "Program Improvement by Source-to-Source Transformation", JACM Vol. 24, No. 1, pp 121-145.

Moler, C.B. (1972), "Matrix Computations with FORTRAN and Paging", CACM Vol. 15, No. 4, pp 268-270

NAG FORTRAN Library Manual, Mark 6 (1977), Numerical Algorithms Group, Oxford.

Ryder, B.G. (1974), "The PFORT Verifier", Software Practice and Experience, Vol. 4, No. 4, pp 359-377.

Schonfelder, J.L. (1976), "The Production of Special Functions for a Multi-Machine Library", Software Practice and Experience Vol. 6, pp 71-82.

Smith, B.T., Boyle, J.M., Garbow, B.S., Ikebe, Y., Klema, V.C., and Moler, C.B. (1974), "Matrix Eigensystem Routines, EISPACK Guide", Lecture Notes in Computer Science 6, Springer-Verlag, Berlin. (Cf. 2nd edition 1976, and guide for the 2nd release of EISPACK, EISPACK Guide Extension, Lecture Notes in Computer Science 51, 1977.)

Smith, B.T. (1977), "Fortran Poisoning and Anecdotes", Portability of Numerical Software, Lecture Notes in Computer Science 57, Springer-Verlag, Berlin, pp 178-256.

Standish, T.A., Kibler, D.F., and Neighbors, J.A. (1976), "Improving and Refining Programs by Manipulation", Proceedings of the ACM Annual Conference 1976, Houston, Texas, pp 509-516.

Wilkinson, J.H. (1963), "Rounding Errors in Algebraic Processes", Notes on Applied Science No. 32, Her Majesty's Stationery Office, London.

I. 3

LIBRARIES: THE USER INTERFACE

P. Kemp

(NUMAC Computing Service, University of Newcastle-upon-Tyne)

1. INTRODUCTION

It is well known that a piece of software will not be used voluntarily unless it is easy to use, however good it may be internally. Examples range from the way users will desert OS Job Control Language on IBM machines, given half a chance even if it means a restriction of facilities, to the widespread use of the Statistical Package for the Social Sciences despite the availability of packages with superior statistical capabilities. In many areas of numerical analysis, we are now able to write reliable programs to solve most problems, or at least to give an indication of the precision of a computed result. It is therefore vital that the user interface should be carefully considered by writers of numerical software, to ensure that their efforts are not wasted by not being used.

First we must identify potential users of numerical software and consider their needs. Only then can we attempt to evaluate the utility of a particular piece of software. I would identify three distinct classes of user.

The first, and most important, is the end user who will use the software to solve his problem. This group spans an enormously wide spectrum. At one end is the user, perhaps in a biological science, who possesses so little mathematical knowledge that he cannot even state his problem mathematically; for example he will be looking for a curve to fit his data so that he can approximate intermediate values. He has a problem that a library ought to be able to solve but he does not care how it does it provided that the results are usable. In particular he does not understand terms like "least-squares" and "minimax". It is vital to remember that, however naive he is mathematically, he is an expert in his own field. At the other end is the sophisticated user who has a difficult mathematical problem to solve and wants to take time tuning a method to suit his particular problem so that he gets the best results. He cares passionately what goes on inside the library routine and needs a large number of fine tuning aids. An example is the fluid dynamicist who knows he has a small, difficult boundary layer which can be handled analytically and does not want an ordinary differential equation solver from a library to waste time trying to handle it.

The second class consists of the staff of the computing centre. They are mainly interested in the ease of supporting the library. They are less important than the end user but their requirements must be considered seriously by library distributors for two reasons. A piece of software which is difficult to support is likely to be implemented badly, to the detriment both of the reputation of the software and of the ease with which the user can solve his problems. Second, it is frequently the case that software becomes widely used only because the computing centre staff recommend it; they are unlikely to recommend software which they cannot maintain effectively. There are several software packages which have been requested by NUMAC users but which I cannot recommend because of the difficulty of support.

The third class comprises algorithm developers. For example a stiff ordinary differential equation solver needs access to routines for matrix decomposition. It would be nonsensical not to use the routine provided for this purpose in the library. They can be regarded as an extreme case of the first group but are better considered separately. Since their requirements have been discussed elsewhere in this volume (by Ford), this paper will not consider them further.

In Sections 2 and 3 of the paper, we will discuss the special needs of the first two groups of users and then in Sections 4 to 8 will consider ways in which these needs can be met. Finally, in Section 9, we will consider how the users' needs have been met by a variety of libraries, past and present.

2. THE REQUIREMENTS OF THE END USER

As previously noted, the problems of prospective users of a numerical subroutine library are diverse. This is especially true in a university environment and may not apply in an industrial one. However, their requirements are all similar although they may need to be met in different ways.

First users need the library to have the capability of solving their problem. This is a matter largely of judicious choice of library contents but it also means that the library must be documented in such a way that users in a wide range of disciplines can find what they need described in a language they can understand. It is for this sort of reason that I believe that libraries should contain such routines as Runge-Kutta ordinary differential equation solvers, whatever opinions numerical analysts may have as to their suitability as general methods. Most users with a little mathematical knowledge think they understand the Runge-Kutta method and they are thus given a name to whet their appetite.

Users' second requirement is to be given confidence that library routines will behave as claimed. There is not much that the library vendor can do directly about this except to ensure that his software is of top quality and to make moderate claims only. The user himself will only by convinced by the attitude of the computer centre (*i.e.* does it make the software easily available and recommend it without hesitation?) or his friends who have already used it. If they show that they have confidence in the library, our potential user will be encouraged to try it, hopefully will get the results he expects and then proceed to use other parts of the library to solve other problems.

The third, and most important requirement, arises because users of computers are always busy. Therefore use of the library must save time. To return to the ordinary differential equation example, any user can copy the FORTRAN statements from a book to implement a simple Runge-Kutta method in about half an hour. The resulting program may run slowly but this is of little concern to him unless his computing time is costing him real money. Unless it takes less than half an hour to get the right library routine integrated into his program, he may well not use the library.

3. THE REQUIREMENTS OF THE COMPUTING CENTRE

The basic computing centre requirement is to be able to offer a good service to users, allied if possible to a quiet life. This is only possible if the software and documentation of the library satisfy certain criteria.

The first impression a centre gains of a piece of software is when it tries to mount it. On an IBM system, for example, it is perfectly possible to have a new piece of software running in a morning, provided the distribution tape is structured reasonably and adequately documented. So wrong decisions about distribution formats can create bad impressions which will take time to dispel.

A good initial impression will be enhanced if the software turns out to be easy to maintain. In the case of a subroutine library this means that the code must be structured so that individual items can be extracted easily using standard utilities or ones provided by the library supplier. Corrections must be supplied in such a way that their incorporation and verification is easy. On the assumption that the computing centre will try to investigate alleged or actual malfunctions itself rather than passing them direct to the supplier, adequate technical documentation must be provided to describe not only the mathematical methods used but also how they are implemented in the routine. Good, well-structured code is of great benefit in this task. The centre also needs to be confident that any fix it applies, either supplied by the library vendor or in response to a user problem, will not impair the general performance of the routine. Therefore adequate test software must be provided.

Once the software is available to users, the centre requires that it shall be easy to use so that valuable staff time is not wasted on trivial enquiries. This imposes an obligation on the centre to ensure that the library is mounted in such a way that it can be accessed easily and that it is documented. There is a much larger obligation on the supplier to ensure that the basic documentation is clear and that the structure of the library is helpful.

If all the above conditions are met, the staff of the centre will have some spare time to fulfil another rôle, that of helping users with genuinely difficult problems. A good consultancy service, based on the library, provided by the computing centre will enhance the reputation of the library. To be possible, it requires that the library contain a wide range of good algorithms, documented so that a specialist is able to use them and where necessary adapt them to solve a difficult user problem.

4. ASPECTS OF THE USER INTERFACE

Having identified the user population which a subroutine libr
is trying to serve and described in general terms the requirements

of the various sections, we now turn to the details of the methods by which the needs may be met. It is clear that we must consider documentation very carefully since this affects users at all levels; a number of requirements have become clear already. Library structure has been seen to have a bearing on the ease with which a library can be used and maintained. The quality of advice available on the use of the library is of great importance to the end user and the algorithm selection policy affects both the sophisticated user who needs state-of-the-art routines to solve hard problems and the centre staff who have an added maintenance burden if new routines appear frequently. These aspects will be considered in the next four sections.

5. DOCUMENTATION

Good documentation is vital to the success of a software project since, without it, the software will not be used to its full potential. This is almost universally recognised and the problem becomes one of identifying what constitutes good documentation for a particular piece of software and then establishing mechanisms for ensuring that it exists. Lill (1974) described the two-level scheme used by NAG which consists of a main manual giving full details of the library and a mini-manual giving just a summary, together with advice on the choice of routine. She also drew attention to the need for standards in documentation, particularly in a project involving several collaborators. Newman and Lang (1976) made a number of observations, drawing attention particularly to the wide range of documentation required, both in terms of medium and coverage, and to the degree of professionalism involved. Their remarks are not addressed to any one type of software and it is worth considering the implications of their conclusions for libraries.

The user with a problem to solve needs convincing first that the library can solve it. The documentation required at this stage must be written in such a way that the potential user can understand it, whatever his discipline, and should be aimed at selling the library. The ideal would be a series of general information manuals aimed at different disciplines. Given the limited resources available, this will rarely be achieved and the tacit assumption is usually made that all users are able to describe their problem accurately in terms of mathematics. This is not always a safe assumption. I have several times been approached by a user requesting a multi-dimensional quadrature routine only to discover that his original problem was a high order linear differential equation which he had integrated by factorizing the operator. Whether this is a failure of the teaching of mathematics or of library documentation is a moot point.

Once the user has decided to try the library, he needs detailed information on how to set about it. This must include

(i) advice concerning the exact routine to use for his problem. This must be well written in language which is as non-technical as possible since few users will be expert mathematicians. For example we must say exactly what we mean by a sparse matrix. If it is not well written, a non-expert user will simply look for keywords remembered from an elementary mathematics course in the distant past and probably choose an inappropriate routine.

(ii) full details on the usage of the chosen routine. This must cover calling sequence, error conditions and an example of how it is used.

(iii) details of how to pick up the library on his computer.

It will be clear that this manual will be a large one; therefore it will be expensive and the user is unlikely to have his own copy so will be using a shared one. The manual must be trying to boost the user's confidence in the software and so must be well produced in a format which is standard throughout. For these reasons, and also because it will contain many mathematical symbols and diagrams, a computer produced manual is unsuitable, although microfiche may be acceptable.

Since the user is likely to be consulting the manual somewhere other than the place where he is writing his program he will have to copy extracts from the manual. He will probably get these wrong or he will have failed to extract some information which turns out to be necessary. In either case his program will fail and he needs more help. The least that can be expected is the availability of a manual small and cheap enough that all users might be expected to have their own copy it should contain brief details of the purpose of the routines a check list of parameters (without full specifications) and a list of the error conditions. If the computing system being used is a terminal based one, it would be highly desirable if this information were kept on-line so that a user at a termina with an error can quickly find out possible causes. Some cent have gone a step further and keep example programs using each routine in the library on-line; their users can thus start from a program which uses the routine correctly on a simple problem and build up confidence that way.

In most cases the documentation described above should be sufficient for the user to obtain the results he requires. However there remain the possibilities that his problem is so difficult that it requires more sophisticated use of the library than he can manage unaided, or that he requires a routine to be modified or even that his data uncover a bug in the routine. In each case he must approach staff in his local computing centre; the rate at which he can proceed then depends on their expertise and on the documentation available to them. If his problem is merely one requiring more sophisticated use, the normal user documentation should be sufficient. Many libraries have a two level routine structure so that there are general routines with simple calling sequences which will solve most problems and lower level routines with large numbers of parameters that can be tuned to solve particular problems, given enough expertise on the part of the user. However, if the requirement is for the routine to be modified (for example to allow a quadrature routine to call a copy of itself in order to get a two dimensional routine) or to find a suspected bug, the computing centre adviser must be able to discover precisely how a routine in the library works. It is possible for a routine to be self documenting but few programmers are sufficiently disciplined to work in this way. Additionally the library routine will probably require a work array to be passed to it via the argument list. The elements will be used for diverse purposes within the routine and, since EQUIVALENCE cannot be used in these circumstances, readability is impaired. Hence the ideal solution is a document describing precisely how each routine works, on the pattern of the Program Logic Manuals which IBM produce for much of their software. Armed with this, and a knowledge of how the mathematics works, the adviser should be able to deal with the problem much more quickly than the alternative solution, which is to refer the problem back to the library supplier.

A final item of documentation, not mentioned above but which must not be forgotten, is the implementation guide. It tells the computing centre staff how to mount and modify the library on their machine and is the first direct experience the centre has with the library. Thus a poor one, or even worse none at all, gives a bad initial impression which will certainly persist.

6. LIBRARY STRUCTURE

Most previous discussions on library structure (e.g. Ford and Hague (1974)) have been from the library supplier's viewpoint. Many aspects have significant implications for the ease of use of the library also; what is best for the library developer is not always best for the user. For example, a library which is intended to run on several machines could be written virtually machine independently without regard to efficiency or accuracy. This might help the short term aims of the developer but is unlikely to appeal to the end user. The aspects of library structure which impact on the user are routine naming conventions, calling sequences, choice of routine types and error handling mechanisms.

6.1 Routine Naming Conventions

There is little doubt that most users would like a mnemonic name for routines, for example EIGEN for a routine to extract eigenvalues. They usually use meaningful names for routines they write for themselves and would be helped if library routine names were similar.

Unfortunately this is not possible in a general purpose library because there are insufficient distinct meaningful names. A library is likely to have about six eigenvalue routines and the naming problem is immediately severe. The closest one could be to mnemonic names would be something like EIGSYM, EIGUS, EIGCOM,...; this is not significantly easier than FO2AAF, ... Hence the need for more structured names is readily accepted by most users. The reason which is usually advanced, namely that a structured naming convention eases library management, is relevant to the library producer but not to the end user.

The major drawback of a structured scheme is that it is not always immediately obvious if a name is typed wrongly. This is particularly true of naming conventions based on the SHARE classification because it results in names which contain both letters and digits and few output devices distinguish adequately between "zero" and "oh". One of the most common errors made by NUMAC users meeting the NAG library for the first time arise from this.

6.2 Calling Sequence

Library designers, particularly if they use FORTRAN with its lack of dynamic local arrays, are faced with an insoluble problem when deciding how routines should be called. To a user, the obvious call to a function minimization routine is:

```
CALL MINFUN(F,FMIN,X,N)
```

In this F is a function name which, when called evaluates the function and possibly its partial derivatives, FMIN is the minimum function value found and X is an array in which the values of the N variables at the minimum are returned. There are no other parameters which are relevant to the mathematical statement of the problem and if the routine requires more, there should be a very good reason. In comparison, one of the NAG minimization routines needs 23 parameters. Its authors would explain, with considerable justice, that this is a routine which is to be used by experienced users only who require the extra parameters for tuning purposes. The ideal must lie somewhere in between and most users would place the ideal as close as possible to the four used above.

In FORTRAN it is necessary to provide more parameters for work space arrays and their sizes. Other languages allow dynamic array declarations and remove this necessity; however it may often be useful to include them because important information may be available in the work space arrays on exit. In the minimization example, an estimate of the Hessian matrix will be returned somewhere and this may be useful if its position is documented.

It is often necessary for the routine to use a number of parameters whose value is retained across several calls (for example the current step size in an ordinary differential equation solver). In ALGOL 60 this can be achieved by the use of own variables but in FORTRAN the only solutions are COMMON blocks or extra arguments. The former are not particularly easy to use and may interfere with the user's own organization of the COMMON blocks. Thus the latter choice should be made and the user will find it easier and hence less error prone if the work space variables are packed into arrays. Thus he only has two extra (useless to him) variables to define (the array and its size). The algorithm writer has the problem of whether to use the array elements (making his code less efficient and less readable) or to unpack the array into local variables. This need not be the concern of the user.

The routine will also be easier to use if the different types of arguments are grouped in a consistent manner throughout the library. Such an order might be:

INTEGER variables and arrays } arising from the
REAL variables and arrays } mathematical formulation
Routine names } of the problem

INTEGER work array and its declared size

REAL work array and its declared size

Routine names which might be called for monitoring progress etc.

If such an order is consistently used, it becomes much easier for the user to ensure that his argument list is in fact correct and also for the advisers in the computer centre to check it.

6.3 Choice of Routine Types

In view of the large potential audience for a subroutine library, routines must be available at different levels. In the previous section we saw that a minimization routine for a naive user need only have a few arguments, whereas a more sophisticated user (perhaps an algorithm developer) needs many more for finer control. If a library, perhaps in the interests of making the results of advanced research available, contains only sophisticated routines, a naive user will either be frightened off or will get the calling sequence hopelessly wrong and become an unnecessary burden on the computer centre advisory service. On the other hand, if the library policy is to provide only simple routines aimed at the naive user, the more sophisticated user will want to modify routines himself to provide extra control and the algorithm developer will have to incorporate modified copies of other routines in the library. The NAG library has been guilty of both policies in different sections in the past.

A solution is to build up from a series of basic building blocks each of which performs some trivial operation towards the complicated routines with simple calls required by the naive user. Part way between will be the sophisticated routines required by the algorithm developer. A possible hierarchy for the solution of stiff differential equations is:

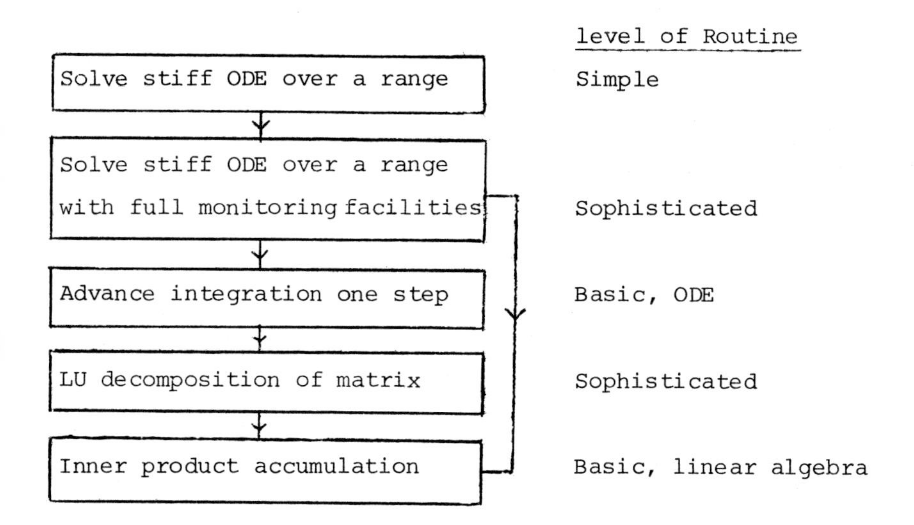

If the library has this type of structure, the average end user would use the full sophistication of the library through the simple routines only. However, if his needs became more demanding, he could advance naturally to the more sophisticated routines where necessary. The algorithm developer would have the ability to write modular programs by using the basic and sophisticated routines as appropriate, leading to easier debugging, more reliability and less duplication in the library.

6.4 Error Handling

Any piece of software requires a mechanism for reporting on, and if possible recovering from errors. It is rarely satisfactory to allow an underlying system, be it the operating system or the FORTRAN run-time system, to trap errors, if only because the message produced will not be in terms of the user's application.

In the case of library software, there are a number of distinct possible sources of error, which need different treatment. These include algorithmic failure, in which the program terminates but produces incorrect results, error conditions which arise during the computation and which might be trapped by the hardware or operating system, and blunders which usually result from an incorrect calling sequence. It should be an act of faith on the part of library designers that all errors should be found by the library software itself, unless there is a very good excuse for not doing so.

The case of algorithmic failure is most difficult to detect but the easiest to know how to deal with if detection is possible. In many areas of numerical analysis it is possible to devise algorithms which return a value expressing a confidence level in the result. One of the earliest to do this was de Boor's CADRE (1971) and the idea is carried through many of the NAG routines, particularly in the optimization chapter. In other areas, for example ordinary differential equations, it is almost impossible to detect malfunction of the algorithm without an unacceptable overhead. In such circumstances it is vital that the documentation should contain clear advice on how to check that results are correct.

Error conditions arising during the computation include overflows and underflows. All library software should check for these problems before they occur. An attempt to perform a Choleski decomposition on a matrix which is not positive definite should result in an error message in terms of the matrix and not one saying something like "ATTEMPT TO TAKE SQRT OF NEGATIVE NUMBER AT OFFSET 1BE8". The latter will bring the user running to his advisory service complaining that the library routine has broken and his confidence in the library will be reduced.

The question of how blunders can be caught is a delicate one. They most frequently arise because the user has an error in the calling sequence or has failed to initialize his variable Left to themselves, on IBM equipment, such errors usually result in an exception error of some sort being reported by the operating system, much to the confusion of the user. It is arguable that library routines should perform extensive argument checking but this imposes a large penalty on those production programs which are correct and in any case is not always possible (e.g. it is impossible to check for an unassigned variable in FORTRAN). Most systems have a debugging compiler (e.g. WATFIV on IBM machines) and, if library software can be fitted into this, most blunders are dealt with gently. The WATFIV compiler will detect incorrect types of argument, wrong numbers of arguments and unassigned variables. The problem is that, along with other debugging compilers, there are restrictions on the language which it will accept and library developers oft do not write in the required subset. Neither SSP nor the Harwe library can be used in WATFIV although the NAG library can be interfaced with a little effort.

Hence one of the keys to a successful library is that it should handle any errors itself. Users are much happier with software which reports errors in a language which they can understand. In the case of a numerical subroutine library this is either in terms of the underlying mathematics or in terms of the program statements they have written to call the routine. Any message reporting errorsin operating system jargon causes confusion and lack of confidence.

A user may, of course, discover a real error in a library routine. If so, he needs a fix rapidly and the support system for the library must be able to respond quickly. A query entering the NAG error handling system inevitably takes a long while for solution since it must be checked at the levels of machine implementation, Central Office and contributor to discover where it originated. This is not a particular criticism of NAG; in fact it responds to errors more rapidly than most other software suppliers. It does mean, though, at NUMAC we expect to investigate errors ourselves and provide temporary remedies.

7. ADVICE ON LIBRARY USE

We have seen in several places in the preceding sections that there is a need for an advisory service to provide help for a user beyond the contents of the documentation. In a large university or research organization it will usually be provided by a computer centre but a smaller organization will probably rely on the library supplier. Largely because of this latter problem, it is essential that the library documentation should give comprehensive advice on the use of routines. For the purpose of this section, we will assume that the advisory service is staffed by experts who are capable of offering advice at all levels so that the user will get a satisfactory answer. We must, though, consider the tools needed by the adviser so that he can do his job properly.

The key must be to give the adviser job satisfaction. At NUMAC, one of the most frequent queries on the NAG library arises because the user has forgotten that variables must be double precision in the IBM implementation and he usually gets an exception error eventually. This is frustrating both to the user and to the adviser, whose talents are being wasted. The ideal would be a situation in which the user documentation would be so perfect that the adviser would only be called in to solve intellectually stimulating numerical problems. There seems to be no way of achieving that state at present but it should remain the aim of library documentation designers.

Fletcher and Hebden (1974) have described their experienc at Harwell with a numerical methods advisory service. Their chief finding was that the major tool required was a comprehensive, well-documented, reliable subroutine library. Armed with this, the adviser can confidently give advice on the various methods a user might employ to solve his problem, knowing that they are reliably implemented. It is essential that the advise should know about the underlying mathematics of the routines in the library and documentation at this level must be available, either in the open literature or provided by the algorithm developer.

At NUMAC, we run a two-level advisory system. A duty officer is available daily and this service is staffed on a rota basis by the centre staff. Only one person on this rota would claim to be a numerical analyst but all are able to advis on the basic use of most of the software available and thus are able to solve the common problems which arise from inadequacies in communication. Problems which cannot be solved immediately, the vast majority of which are genuine numerical difficulties, are referred to one of the two numerical analysts on the servic staff or to one of the numerical analysis lecturers in the computer science department. Using our pooled knowledge of numerical analysis and of the library we would expect to get the use moving again quickly. The two level scheme has the particular merit that the frustrating, trivial queries are spread over a large number of people and the expert numerical analysts can concentrate on solving intellectually demanding problems. The advantage to the users is that they can get straightforward queries answered quickly, without having to wait until a numerical analyst is available.

8. LIBRARY CONTENTS

Most users are attracted to a library because it will solve their problem more quickly than they could themselves. The typical naive user is interested above all else in reliability, stability and ease of use. However the more sophisticate user needs routines embodying the latest research. This argues in favour of a large library, part of which presents a stable interface providing reliable routines for the naive user, and the other part is rapidly changing to reflect the latest ideas for the sophisticated user. Most generally available libraries simply cater for the former need because constant updating thrc a strain both on the library vendor and the computer centre supporting it locally.

Stability and reliability mean that the library changes slowly and that algorithms are selected only after careful certification. There must be a mechanism for removing outdated routines; in an ideal world this would be by leaving the user interface upward compatible and changing the underlying routine. This is what manufacturers attempt when updating operating systems. However, this is not possible in a subroutine library since it is extremely unlikely that requirements such as work space will be identical. So generally routines are completely replaced at intervals. A conservative algorithm selection policy is essential to ensure that these intervals are long.

The foregoing means that a general purpose library is unlikely to embody research less than 18 months old. Could this research be made available to the end user? Library suppliers usually have new algorithms available soon after they are coded so that they can consider them for possible inclusion. It ought to be possible for a library vendor to offer an uncertified library service to make those routines available.

A further feature of algorithm selection has nothing to do with reliability, but rather with gaining the user's confidence. Many users feel they know a little about mathematics and have learnt some keywords, for example Runge-Kutta or Gaussian elimination. It is arguable that a library should contain routines implementing these methods just to encourage this kind of user - once he has started to use the library, he can be encouraged to move on to the better algorithms contained in it.

9. A LOOK AT SOME LIBRARIES

In order to illustrate some of the points made, it is interesting to consider the user interfaces of a number of libraries. We will consider three, namely the EDSAC 1 library (Wilkes, Wheeler and Gill (1951)), the IBM Scientific Subroutine Package (IBM, 1969) and the NAG library. These are chosen not because they are the best available but because they illustrate the evolution of library design from the early 1950's to the present day.

The EDSAC subroutine library existed almost as an extension to the hardware. It was used by copying a physical paper tape on to a user's program tape and writing the necessary instructions to pack round it. It contained routines for quadrature (Simpson's rule), trigonometric functions and ordinary differential equations (Runge-Kutta-Gill), all superbly documented in the book. The book assumes that all users are experts, a reasonable assumption for 1951, and promised routines for things like the solution of algebraic equations at some stage in the future. The library was thus well documented, easy-to-use by the standards of the day and advice readily available (since users, including the routine author, stood in a physical queue to use the machine). Miller (1966) has suggested that one of the reasons that research into ordinary differential equations progressed so slowly in the 1950's was the existence of such an easy-to-use routine so early in the history of computing - perhaps an example of a user interface which was too good!

IBM's SSP was one of the first comprehensive subroutine libraries. Its main virtue was its width of coverage although the quality of algorithms did not always reflect the state-of-the-art. The documentation was clear, although expressed entirely mathematically, but IBM never produced a Program Logic Manual so it was difficult to support. The user interface was extremely stable since there was no updating policy. Its success stemmed from the fact that when it was produced, it was by far the most comprehensive library of its day. By the standards of the 1970's, it looks inferior but many users still make use of it.

The NAG library was born out of the need for a library for the ICL 1906A and has only recently rid itself completely of that image. It did however introduce at the start the ideas of a consistent user interface thoughout the library (even though that was not achieved at first) and of an algorithm replacement policy so that the library would incorporate the latest ideas. Early in the life of the project, it began to be transferred to other machines so that it became a transportable library and the user interface is not only consistent within the library but also across machine ranges. That it is so wide used is a tribute to its quality.

Table I compares the libraries on some of the aspects mentioned so far.

Table I

Feature	EDSAC 1	SSP	NAG
1. Range of Capabilities	Limited	Wide	Wide
2. Quality	Good (by 1950 standards)	Variable	Generally good
3. Ease of use	Good	Variable	Good
4. Ease of maintanance	Good - author on site	Poor - user is not assisted by originator	Good - much backup support
5. Documentation	Good - for the expert	Fair	Good
6. Structure	n/a	Little	Good
7. Advice on use	Plenty - not written down	Little	Widely available
8. Contents	Limited - not updated	Good at first, becoming poor because of no updates	Good
9. Availability	Cambridge	IBM machines	Almost universal

10. CONCLUSIONS

No existing library is perfect, and nor will any future library be. There are too many conflicts of choice involved for this not to be so. This paper has tried to put forward the users' requirements in the areas of documentation, library design, availability of advice, algorithm selection and supportability. The user population is so varied that this in itself makes it difficult to satisfy all users; when the addition of the constraints of the library developer are added it probably becomes impossible. We have seen that the main requirements are:

- ease of use, so that a user is tempted to use the library rather than his own program
- good documentation, so that all those in contact with the library are kept informed

- good algorithms, so that as many problems as possible can be solved
- reliability, and that the user retains confidence
- good backup services, so that if the user runs into trouble he can extricate himself

These could probably be summarised as comprising a friendl library and if library developers kept this in mind they would not go far wrong.

11. REFERENCES

de Boor, C. (1971) "CADRE: An algorithm for numerical quadrature in "Mathematical Software", Ed. J. Rice, pp. 417-449, Academic Press, New York and London.

Fletcher, R. and Hebden, M.D. (1974) "Setting up a numerical advisory service", in "Software for Numerical Mathematics", Ed. D. J. Evans, pp. 413-421, Academic Press, London and New Yo

Ford, B. and Hague, S.J. (1974) "The organization of numerical algorithms libraries", in "Software for Numerical Mathematics", Ed. D. J. Evans, pp. 357-372, Academic Press, London and New York.

IBM (1969) System/360 Scientific Subroutine Package, Version II Programmers Manual.

Lill, S.A. (1974) "User documentation for a general numerical library: The NAG Approach", in "Software for Numerical Mathemati Ed. D. J. Evans, pp. 423-432. Academic Press, London and New York.

Miller, J.C.P. (1966) in "Numerical Analysis - An Introduction" Ed. J. Walsh, pp. 63-98. Academic Press, London and New York.

Miller, J.C.P. (1966) "The Numerical Solution of Ordinary Differential Equations", "Numerical Analysis - An Introduction", Ed. J. Walsh, pp. 63-98. Academic Press, London and New York.

Newman, N. and Lang, T. (1976) "Documentation for Computer user *Software-Practice and Experience*, **6**, 321-326.

Wilkes, M.V., Wheeler, D.J. and Gill, S. (1951) "The Preparatio of Programs for an Electronic Digital Computer". Addison-Wesle Cambridge, Mass.

I. 4

SOFTWARE TOOLS

S.J. Hague

(*NAG Central Office*

7 Banbury Road, Oxford OX2 6NN)

1. INTRODUCTION

This paper might well seem as the 'odd-man-out' in a series of papers devoted to mathematical software. The terms used in Section 4 particularly might suggest that the paper ought to have appeared in the proceedings of a computer science conference. Anyone who does suggest that is, however, quite wrong. Though by no means exclusively devoted to NAG, this paper has been written against the backcloth of the NAG Library activity. Our collective experience over several years is that making available an increasing amount of high quality numerical software on a wide variety of computing systems is far from being the perfunctory activity that some might think. Indeed it can be so difficult, tedious and time-consuming that the whole activity is in danger of collapse unless some degree of mechanisation is introduced. Other major mathematical software groups throughout the world share this view. Mechanisation (i.e. the development and application of software tools) is therefore an appropriate topic for discussion at this conference since it can have an important role to play in the development and distribution of mathematical software.

The term software tools refers to those programs which are used to assist in the development, maintenance or distribution of other programs. Alternative terms used to refer to such programs include programming aids and mechanical aids. In this paper we will concentrate on those tools which analyse or manipulate programs at the source text level. This means that we will not be considering compilers such as WATFIV whose diagnostic abilities make it a most useful tool nor shall we refer to optimising compilers which can sometimes significantly improve program performance. Such compilers which map a high-level program to a lower-level form are effectively 'black boxes'; their effects though perhaps well documented or reflected in the behaviour of the compiled program cannot be explicitly inspected by the author of the high-level program. Here we are concerned with tools whose effects are demonstrably visible at the source text level (assuming that the mathematical software is written in languages such as FORTRAN, Algol 60 , Algol 68, PL/I etc.).

Software tools in the category of analysis include verifiers, text comparison programs, tools for gathering statistics, flow-charting tools and table-generating tools. The category of program manipulators embraces software tools for variant extraction, precision changes, adjustment of constant values, and other aspects of the portability problems posed by mathematical software.

For a discussion of the issues of portability and transportability see Hague et al (1976) for instance. Other program manipulation devices include standard text editors, string processors, standards imposers, testing aids such as data flow and control flow analysers, program generation systems, preprocessors for language extensions (these might loosely be described as higher-to-high level language compilers), program verification systems and general-purpose transformational systems.

It should be clear from the foregoing that the range of tools currently in use is extremely broad. Most of the mathematical software institutions and groups throughout the world have been using or developing tools of various kinds to assist their work. Much of this interest in mechanisation is fairly recent, and this may in part explain why, despite the large number of tools in use, there appears to be widespread ignorance about them. The development of these tools has taken place largely in an isolated and unco-ordinated fashion and so the adoption by one group of a programming aid developed by another is often simply a matter of chance. One way to improve this state of affairs might be to compile a comprehensive list of as many of the current software tools as could be identified. In specific areas, e.g. the cataloging of

FORTRAN preprocessors, (Meissner (1976)) this has already been done. The general task, however, is beyond the scope of this paper so we must be content with giving a sufficient number of examples of tools to illustrate the wide spectrum of application (and of sophistication). The reader must be cautioned against assuming that any of the aids mentioned in this paper are necessarily available for general distribution.

In Section 2 we discuss the issues of manipulation and mechanisation. Section 3 gives outline sketches of a number of software tools, several of which are either developed or used by NAG. The fourth section is devoted to a condensed description of the TAMPR transformational system and its uses, and in Section 5 we speculate upon future developments in the field of software tools.

2. MANIPULATION AND MECHANISATION

Before discussing software tools further, we must provide answers to two basic questions which might be posed by the 'lay user' or perhaps by the numerical analyst who suspects that his more software-orientated colleagues are attempting to create a major new branch of computer science with no real need for it. Those basic questions are:

(i) why is it necessary to manipulate mathematical software?

(ii) does the mechanisation of that manipulation process bring significant advantages?

2.1 *Why May Changes Be Required?*

Answering this question poses little difficulty particularly if one has witnessed the development of NAG from a single machine range library project to its present state in which there are implementations of the NAG FORTRAN Library on 19 distinct machine ranges,(9 for the Algol 60 version). Within each implementation there may be sub-implementations for particular computing systems, e.g. on the CDC 7600, there are source text differences between the Small Core Memory and Large Core Memory implementations. Similar problems are faced by other groups who are also interested in developing and maintaining high-quality software on many machines. Whether for reasons of uniformity, portability or refinement, it is frequently necessary to alter a body of source text either on a small scale or throughout an entire suite of programs. Below are summarised a number of reasons which might precipitate such changes:

- correcting a coding error either of an algorithmic or linguistic nature.
- altering the structural property of the text e.g. imposing a certain order on non-executable statements in FORTRAN.
- standardising the appearance of the text.
- standardising nomenclature used e.g. giving the same name to variables having the same function in different program units.
- conducting dynamic analysis of the text (e.g. by planting tracing calls).
- ensuring adherence to declared language standards or subsets thereof.
- changing an operational property of the text (e.g. changing the mode of arithmetic precision).
- coping with arithmetic, dialect and other differences between computing systems.
- altering similar algorithmic processes and similar software constructs in large collections of programs.

2.2. *What are the Benefits of Mechanisation?*

Several published papers have eloquently argued the case for the mechanised approach to program manipulation. Standish et al (1976) discuss the merits of improving and refining programs by means of an interactive program manipulation system. Perhaps more immediately relevant to mathematical software, a paper by Boyle et al (1976) discussed the advantages of automating multiple program realisations; that is, deriving by mechanical means, several members (realizations) of a family of related programs from a prototype or generalised program. (This topic is discussed further in Section 4).

The main arguments presented for mechanisation usually concern two factors; economy and reliability. If numerous changes are to be repeatedly made to a large body of software, then the use of a mechanical aid offers the prospect of considerable savings in time and effort. Presumably some poor programmer is relieved of performing what would be a tedious and slow manual task and can be employed on some activity with a greater intellectual stimulus. The fact that the changes are made mechanically means that we can at least expect consistency. It may also be that the mechanical nature of the alterations are amenable to at least an informal (if not a formal) proof of correctness for the transformed program. The study of correctness-preserving transformations is an active field of research, on the basis of which, some developed form of a TAMPR-like system (see Section 4) may eventually lead to software tools whose operations are demonstrably reliable.

In the light of the experience of the NAG Central Office in using automated aids (Du Croz et al (1977)), our overall view would be that the use of such aids can indeed lead to greater economy and enhanced reliability. We would add two notes of caution, however. The first is that programming projects in general are prone to take longer than expected. In an organisation of limited resources, practical aims and subject to the day-to-day pressures of both academic and commercial life, the decision to undertake the design and implementation of a new software tool should be taken with perhaps more caution than in, say, a research establishment. A second point of concern is that the use of a software tool may lead to over-reliance upon its effectiveness and so to a temptation not to check the output software closely. After a tool has been successfully operational for sometime, complacency can arise. If the tool is applied to data (i.e. programs) which contravene some un-documented assumption made by its developers, then what we must hope for is that the output is unmistakeably wrong even at a casual glance. If such a contravention caused a somewhat obscure malfunction to occur, however, incorrect coding may be generated without it being noticed.

3. VARIOUS TOOLS USED IN MATHEMATICAL SOFTWARE DEVELOPMENT

To illustrate the wide-range of tools currently used by mathematical software groups, we outline here the principal features of a number of programs from different areas of application. The interested reader should refer to the references given for further details. Unless otherwise stated these programs are written in FORTRAN and process FORTRAN software.

3.1 *Transportability Aids*

A number of tools could be discussed under this heading. The particular tool described here is the Converter used by IMSL. A similar facility (APT - see Du Croz et al (1977)) is used by NAG. A less elaborate program, the Rothamsted Conversion System, is used to assist in the handling of multi-machine statistical packages such as GLIM and GENSTAT. The Krogh Specializer (Krogh (1972)), is one of the more sophisticated tools in this class. Each of the tools mentioned above has been designed to meet the particular requirements and constraints of its intended application, hence the variation in the level of sophistication. They all tend to have, however, facilities for

- precision changing
- value substitution
- variant embedding and extraction

and so in that sense, the IMSL Converter can be regarded as a typical aid to transportability.

<u>IMSL Converter</u> (Aird et al (1977))

This FORTRAN portability aid operates on a program which is compilable in one environment by making changes which convert the program to a form suitable for a new (target) environment. Input to the converter is a basis deck (the program plus converter instructions) and output is another basis deck or a distribution deck, the latter being produced by deleting all converter instructions from the basis deck.

According to the chosen target environment, the converter ensures that machine and mathematical constants are adjusted to the appropriate values. Precision changes such as SQRT→DSQRT are made if requested in the options to the converter. Certain statements are selected or rejected on the basis of proceeding converters instructions, e.g.

```
C$    IF (SS1G10.LT14) 4 LINES
```

causes the next four records to be 'activated' (by placing a blank in column 1) if the number of significant digits in a single precision variable for the target environment is less than 14.

3.2 *Language Dialect Verifiers*

To overcome the problems of different dialects on different computing systems, several mathematical software groups have resorted to the intersection approach. This means deciding upon a set of permissible language features which lie in the intersection of most if not all the dialects involved. If such a set can be found (the task may be complicated by efficiency considerations) it is obviously desirable that adherence to the chosen rules and restrictions can be mechanically verified. The PFORT verifier is perhaps the prime example of such a mechanical verification aid. Another example is the syntax-checker of Schaumann and Strudthoff developed at Munich.

<u>The PFORT Verifier</u> (Ryder and Hall (1975))

Written at Bell Telephones Laboratory, New Jersey, the PFORT verifier checks that submitted programs conform to the PFORT dialect of FORTRAN. This dialect is a slightly more restrictive form of the ANSI 1966 FORTRAN language. The PFORT designers claim that any coding written in their dialect is portable; that is, it will run correctly without modification under almost all compiling systems which implements the 1966 standard.

The verifier itself is written in PFORT. It is particularly effective in detecting inter-program unit inconsistencies, the type of error which most conventional compilers overlook. A useful secondary function of the PFORT verifier is that of a table generator. The tables which record the use of variables or labels can be utilised by other software tools (see 3.3 (DECS) for example).

3.3 *Standardisers*

A wide range of software tools could be classed in the category of standardisers. They include formatting aids for improving the appearance of programs, tools for making programs more consistent and special-purpose utilities for the imposition of particular programming standards. As examples, we describe the principal features of the POLISH, SOAP and DECS programs.

POLISH (Dorrenbacher et al (1974))

This program, from the University of Colorado at Boulder, is described as a FORTRAN Editor, implying that its function is not merely to improve the layout of a FORTRAN program but to impose a greater consistency on the use of labels and the termination of DO loops. It is therefore both a formatter and a standards-imposer. Its effects are partially parameterised.

The principal features of POLISH are:

- recalculation of statement and format labels, the former into regular ascending order, the latter into regular descending order.
- layout adjustment of lists, comments, expressions, and indentation of DO loops bodies.
- control of record-length (both start- and end-columns).
- termination of each DO loop by its own CONTINUE.
- migration of FORMAT statements to the end of the program.

SOAP (Scowen et al (1971))

SOAP is an Algol 60 program, the function of which is to Simply Obscure Algol 60 Programs. It was written at the National Physical Laboratory, Teddington.

Its main functions are to adjust indentation levels to reflect program structure, to impose regular spacing of lists and expressions and to limit the number of basic symbols per physical record. It can also produce cross-reference tables for variables, constants and labels.

DECS (Sayers (1977))

This is an example of a tool which imposes a standard on the software that it processes. Produced by the NAG Central Office, its function is to ensure that all variables are explicitly declared. It derives much of the variable-type information it needs from the tables output by the PFORT verifier and so is a table-driven tool.

The transformed program, now with all variables explicitly declared, has one set of declarations for dummy arguments (if any), a second set for variables in COMMON (if any) and a third set for local variables. Each set is preceeded by a standard comment. Within each set, type statements appear in a specified order.

A secondary role for DECS is the reordering of all leading, non-executable statements.

3.4 *Language Extensions*

One branch of the program manipulation activity that has been particularly active recently is that of mapping some form of extended FORTRAN into 'ordinary' FORTRAN. The programs which perform the mapping are usually called FORTRAN preprocessors. By implementing such tools, their developers hope, at the cost of a preprocessing pass before conventional compilation, to overcome some of the linguistic inadequacies of FORTRAN but still be in a position to exploit the near-universal availability of the language. (The impact of the new FORTRAN 77 standard on the use of these preprocessors is awaited with interest.). As an example of an extension-cum-preprocessor, we describe RATFOR below. Another system worth mentioning is MORTRAN (see Meissner (1976)) from Stanford Linear Accelerator Centre. As well as providing its own brand of 'structured' FORTRAN, MORTRAN has a macro facility for text manipulation.

RATFOR (Kernighan et al (1976))

RATFOR ("Rational Fortran") is a simple extension of FORTRAN. It can also be readily translated into PL/I. The primary purpose of RATFOR is to make FORTRAN a better programming language, for both writing and explaining, by permitting and encouraging readable and well structured programs. This is done by providing the control structures that are unavailable in bare FORTRAN, and by improving the "cosmetics" of the language.

The language is free-form; statements may appear anywhere on an input line. The end of a line generally marks the end of a statement but there are continuation conventions. A # character anywhere in a line signals the beginning of a comment.

In RATFOR's grammar, these are some of the extensions to ordinary FORTRAN:

if (condition) statement

if (condition) statement else statement

while (condition) statement

for (initialize; condition; reinitialise) statement

repeat statement

repeat statement until (condition)

do limits statement

RATFOR is block-structured in the sense that a single statement can be a group of statements in braces. The RATFOR-to-FORTRAN translator can be coded in RATFOR itself and hence in FORTRAN. For further details, see the above reference, which is most relevant reading because it is also devoted to the theme of software tools.

3.5 *Testing Aids*

Increasing emphasis is being placed on software verification by mechanical means. Proving the correctness of a program mechanically must, however, be regarded as a long term goal. The difficulties in formally verifying non-trivial programs, particularly those involving real arithmetic, are formidable, so in practice we must at present be content with establishing other properties of our programs. Two software tools which are designed to assist in such validation attempts are BRNANL and DAVE, both from the University of Colorado at Boulder. A brief description of the former is given below. The latter is concerned primarily with data flow analysis. For further details, see Osterweil et al (1976).

BRNANL (Fosdick (1974))

This utility enables dynamic flow analysis to be performed during testing of FORTRAN software. It can be used, for example, to establish that a specific set of test problems force control to pass at least once through every statement of the program under test.

Its principal function is to identify each basic block (which has the property that if any one statement in such a block is executed then all the others are also). Having located a basic block BRNANL plants a tracing call at its head thus providing a means of determining the number of times that each block is entered during execution. It also transforms logical IF statements into two separate statements to establish proper block boundaries.

4. TAMPR - A TRANSFORMATIONAL SYSTEM (Boyle et al (1977))

The TAMPR (Transformation-Assisted Multiple Program Realization) system developed at Argonne National Laborator by Boyle, Dritz et al is a general purpose tool for the transformation and formatting of software. Developed primarily as a research tool, it has been used internally a Argonne in a production capacity (see 4.2). Surprisingly few people in the mathematical software community appear to know much about TAMPR, and the number of people who have actually used it is extremely limited. One purpose in devoting a section to this particular tool is to help to remedy the relative lack of publicity about TAMPR. A more important reason for discussing this system is that it coul be viewed as some form of pioneering attempt; the fore-runner of a range of transformational systems of varying degrees of generality and sophistication, which employ syntax-directed transformational techniques in a manner similar to TAMPR. This prospect is discussed further in Section 5. We now describe in a somewhat condensed form, the major elements of the TAMPR system and then give several examples of its application.

4.1 *Principal Features of the TAMPR System*

The flow of information in the system can be summarised by this diagram:

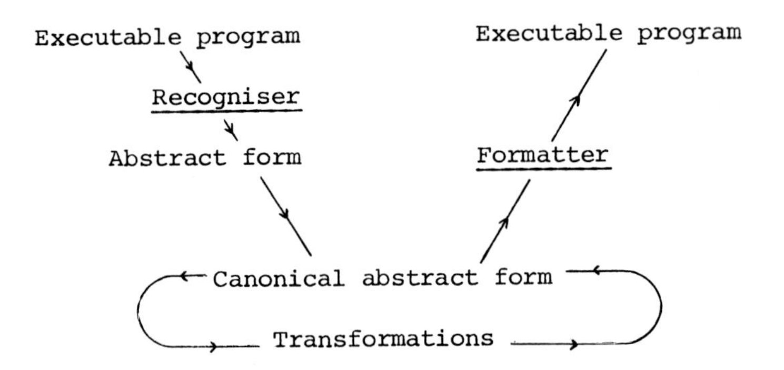

The abstract form is produced by the recogniser from the executable program according to the supplied BNF grammar (in principle, TAMPR can be used with any grammar). This form is essentially the parse tree of the program; that is, an explicit syntactical representation of the program in tree form. Suppose for instance we are recognising a FORTRAN program which contains this assignment statement:

```
X = 0.0E0
```

The corresponding fragment of the parse tree might under some appropriate grammar (see Boyle et al (1976) for instance) be:

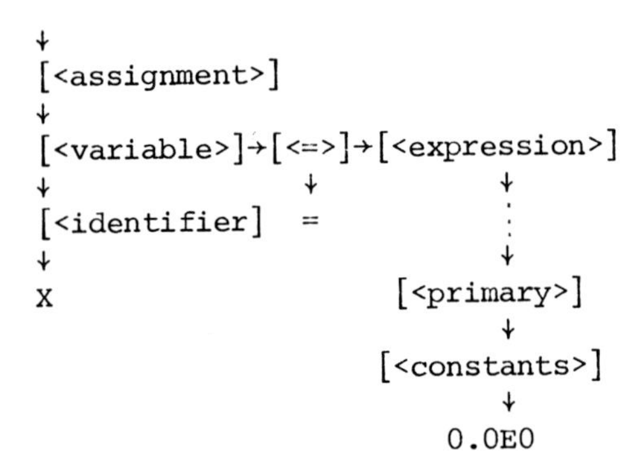

where [<x>] is a parse tree node of gramatical type <x>.

Once formed, the abstract form is immediately processed using a set of structure-clarification transformations. This produces the canonical abstract form. The program can now be manipulated none or more times by user-written transformations. The final canonical abstract form is then processed by the formatter and reconstructed in executable form.

Both the transformational and formatting components are programmable. In the formatting phase, the Formatting Control Language allows fine control to be exercised over layout (indentation, spacing, conditional line breaks). It also permits actions such as the renaming of variables. The transformational component consists of Intra-Grammatical Transformations (IGTs), a brief outline of which is given below (see Boyle et al (1976) for further details).

An IGT comprises a pair of syntactical patterns which are sequences of terminal and non-terminal symbols in the grammar being used. The first of the pair is called the Structural Description. If the SD is found to match a subtree in the parse tree, the second of the pair (the Structural Change) is used to reconstruct that subtree. Both members of the pair have the same dominating symbol <x>, e.g. <variable> or <specification part>. The attempt to match the SD is applied only to the subtrees of parse tree nodes of type <x>. Both the SD and SC patterns are constrained to be syntactically-valid instances (productions) of the dominating symbol <x>. Thus the grammatical legality of the program is preserved, whence the term intra-grammatical transformations.

An elementary example of an IGT is given below:

```
     dominating symbol →  |  <primary>
                          |
                          |  { .SD.
                          |
               comment →  |     /* maps a FORTRAN comple:
                          |        constant into a CMPLX
                          |        function call */
                          |
Structural Description →  |     (<const> "1",
                          |      <const> "2")
                          |
          is mapped to →  |  ==>
                          |
     Structural Change →  |     CMPLX (<const> "1",
                          |            <const> "2")
                          |
                          |     .SC. }
```

For the simple example above, the appropriate fragments of a suitable grammar could be:

```
<primary>  :: =  <variable>  |  <constant>  | ..

<variable> :: =  <identifier>(<expression list>) | ..

<constant> :: =  (<const>,<const>)  |  ..
```

so indeed both the SD and the SC are valid instances of a <primary>, since the former is a <constant> and the latter, a <variable>.

The example gives an illustration of the indexing facility; the ability to label any symbol in the SD (e.g. <const> "1") and to refer to it by index in the re-construction part. Other powerful features not illustrated are the use of indefinites for the matching (under certain conditions) of arbitrary tree fragments, and the use of subtransformations within the SC. There is also a marking facility for <identifier> and <statement> nodes.

At the top level, IGTs are organized into sequences of transformation sets; each set operating on the results of its predecessor. Sets, in turn, are comprised of lists of IGTs, all the IGTs in a particular list having the same dominating symbol. The IGTs in a list are applied in a Markovian order; if one applies, the list if considered again from the beginning and applied to the result. If none apply, the application of the list is finished and application of the containing set moves to the next node of the parse tree in a bottom-up, left-to-right sense.

4.2 *Some of TAMPR's Applications*

To give an indication of the potential range of TAMPR's application we describe here several instances of its use.

4.2.1 *Linpack Package Development*

TAMPR has been utilised as a production tool (see Boyle et al (1976A)) in the LINPACK linear equation package project based at Argonne. Transformations have been applied to complex double precision algorithms to produce realizations in complex single precision, real double precision and real single precision. The package developers believe that these alternative forms derived by mechanical means have helped in reducing development and testing costs and will assist in future maintenance. Another role for TAMPR in the same project has been that of source level optimiser. It is used to replace calls to certain key innerproduct routines by suitably optimised in-line coding. On some computing systems and for some ranges of problem size, the in-line code version performs significantly faster than the modular version. By automating the mapping so that either form can be reliably produced, TAMPR has helped to minimise a conflict often faced by mathematical software specialists; modularity versus run-time efficiency. Finally, TAMPR is also used in the LINPACK project to format the codes after these various transformations. In addition to using the FCL component for the standard tasks of layout adjustment, an attempt is being made to alter certain details of the encoded documentation so that it reflects the transformed state of the program. Thus in this single project TAMPR is being used to

- map an algorithm in the complex domain into one in the real domain.

- change from one precision to another.
- optimise innerproduct calculations.
- adjust the layout of the transformed code.

(- update elements of the embedded documentation.)

4.2.2 *Automatic Theorem Proving*

In another application of TAMPR, it is being used to facilitate automatic program proving. Assertions are placed in a FORTRAN program. The usual grammar for FORTRAN is extended to include the syntax of these assertions. TAMPR is then used to transform the augmented program into a series of clauses in the first order predicate calculus. This series is now in a suitable form for input to a theorem proving program.

4.2.3 *The CARITH Exercise*

As a third example of the use of TAMPR, we briefly refer to a recent exercise to convert FORTRAN programs using COMPLEX facilities into a COMPLEX-free software. That is, the algorithms which the programs implement remain in the complex domain but COMPLEX entities or expressions are replaced by a pair of REAL entities or expressions, the first of the pair being the real part, the second being the imaginary. Whereas the LINPACK application could be regarded as transforming a general form to a more specific form (e.g. complex double precision to real single precision) this exercise would be seen by some people, pure mathematicians perhaps, as no more than a change of notation within the complex domain. The attempt to mechanise the process revealed, however, several unexpected difficulties inherent in the transition whether by manual or mechanical means. A note describing these difficulties is in preparation.

To give an indication of the overall effect of the above transformation exercise, consider the following code:

```
COMPLEX X,Y
REAL A
READ (...,...) X,Y
A = AIMAG (X+Y)
X = Y*Y - Y
```

The final transformed code would be

```
REAL XR,XI,YR,YI
REAL A
READ (...,...) XR,XI,YR,YI
A = XI + YI
XR = YR*YR - YI*YI - YR
XI = YR*YI + YI*YR - YI
```

where COMPLEX X is replaced by REAL XR, XI etc.

To achieve this effect, the original code is subjected to several TAMPR transformation sets known as the CARITH sets. The first of these is primarily concerned with the identification of COMPLEX entities throughout the program, and the last with tidying up the cumulative effect of the preceeding sets. The most significant role of the intermediate transformation sets is the expansion of arbitrary COMPLEX expressions into explicit real and imaginary expressions.

After the application of the first TAMPR set, all COMPLEX variables such as X are temporarily replaced by CMPLX(XR,XI). To show how the expansion process relates to the previous discussion of IGTs, a representative member of the intermediate transformation sets might be

```
<a e>

{  .SD.

      /* transforms a complex addition into an
         equivalent real & imaginary addition */

      CMPLX (<a e> "1", <a e> "2") +
      CMPLX (<a e> "3", <a e> "4")

==>

      CMPLX (<a e> "1" + (<a e> "3"),
             <a e> "2" + (<a e> "4") )

   .SC. }
```

where <a e> is an abbreviation for <arithmetic expression>.

Since XR is <variable> which is a <primary> which is a <factor> .... which is an <arithmetic expression>, the above IGT would map

```
CMPLX (XR,XI) + CMPLX (YR,YI)
```

into

```
CMPLX (XR + (YR), XI + (YI))
```

When this transformation has applied, it may then permit others to apply to continue the expansion (or accumulation process). If, as in the original code, X + Y was the actual argument of an AIMAG call, then the specific transformation for AIMAG, namely

```
      AIMAG (CMPLX (<a e> "1", <a e> "2"))

==>

      <a e> "2"
```

can now apply because the IGT for addition has transformed the argument of AIMAG into the required form. Having applied this mapping, the original statement

```
A = AIMAG (X + Y)
```

is now in the form

```
A = XI + (YI)
```

The final transformation set would remove the parentheses (unnecessary in this case) around YI.

To reiterate the point made at the end of Section 4.1, the order of application of transformations is not controlled interactively by the user; for a given program parse tree and traversal strategy, it is essentially determined by the grammar being used. TAMPR is therefore truly a syntax-driven system.

5. PRESENT STATE AND FUTURE DEVELOPMENTS

It is evident from the previous section that TAMPR is a versatile and powerful transformational system. This does not mean, however, that it provides a complete answer to all program manipulation problems. There will always be a need for simple tools for specific, undemanding tasks. TAMPR is, or with modest extensions could be, capable of performing the roles of a variety of the tools currently in use. Unfortunately, at present this capability is more conceptual than actual for at least three reasons:

- TAMPR is still a research project undergoing development.
- the current implementation is written in XPL, LISP and PL/I and so could not be mounted on systems without those compilers.
- its core size and run-time requirements would be considered prohibitive by many sites.

TAMPR cannot therefore provide an immediate answer to the current range of problems facing mathematical software groups. These groups must therefore resort to using such tools as are available. Each tool is usually designed for just one or two purposes with perhaps a few secondary roles.

It may therefore be that several such tools are required in order to achieve the desired effect. Using a collection of independently - developed tools is fraught with problems such as:

- the software tool developed on one system might be difficult to mount on another.
- the tool from an outside source may not be properly maintained.
- such a tool is likely to have a different user-interface from that of others from different sources.
- its effect may not be easily adjustable.
- its actions may be incompatible with other tools.
- it may rely on undocumented assumptions about the input program.
- each tool of any sophistication must perform some analysis of the program being processed. Much of that analysis is common to many tools but each performs it separately.

The present state of the software tools activity cannot therefore be described as satisfactory. It is our belief that until some experience is shared, resources pooled, standards devised and requirements established, that this state of affairs will continue. Perhaps mathematical software itself sets a good example. It was not until active collaboration between algorithm developers, software experts and distributors began, that a significant proportion of the computer user community started to enjoy the benefits of the many theoretical and practical advances made in the last decades. One step towards achieving the same effect for software tools is to ascertain what the experiences and requirements of the users are with regard to such tools. In this case, the 'users' are the numerical analysts and software developers themselves. An attempt to elicit such information has recently been conducted in the form of an informal survey (Hague (1977)). If the responses to that survey indicate a sufficiently broad area of common interest, practice and needs, some progress towards developing a set of widely-available, well-documented, useful tools might be made.

What might these tools be like in terms of design? Hopefully this breed of 'second generation' tools will be more modular, more uniform, more portable than their predecessors. They could have several standard modules in common. One can envisage a number of table-driven tools in the manner of DECS for instance. These tools would utilise tables produced by a common set of parsing and table-generating modules. They might be constructed in part from a set of lower-level utility modules designed to facilitate the process of program text manipulation. It seems probable that these tools of moderate to medium sophistication would have to be language-dependent, i.e. they could only operate on programs in one particular language. Looking towards the more sophisticated end of the spectrum of tools, one would hope to see programming aids which achieve in practice the language-independence that TAMPR conceptually offers. These tools might not only include tree manipulation facilities in a TAMPR-like manner but also cater for the procedural type of change; namely those changes which are more easily specified by a procedure of several steps than by syntax pattern matching and replacement.

The software tool activity seems likely to continue its rapid growth. Whether the mathematical software community as a whole will have the chance to benefit from this growth depends on many factors including political, commercial, proprietorial and managerial issues. Perhaps the most important factor, though, depends upon the attitude of the developer of the software tool. So often, it seems, tools have been produced which, with little extra effort, could have been made more widely available or more generally useful. Unless the software tool developer is prepared at least to consider the possibility that what he is doing might be of interest or use to someone else, then many of the potential benefits of mechanisation must indeed remain as potential rather than actual.

6. ACKNOWLEDGEMENTS

The author wishes to thank his colleagues, Mr. J.J. Du Croz, Miss J. Bentley and Dr. B. Ford, of the NAG Central Office for their helpful criticism.

7. REFERENCES

Aird, T.J., Battiste, E.L. and Gregory, W.C. (1977) "Portability of mathematical software coded in FORTRAN", ACM Trans. Math. Soft., Vol.3, No.2, pp 113-127.

Boyle, J.M., Dritz, K.W., Arushanian, O.B. and Kucherskiy, Y.V. (1977) "Program Generation and Transformation", Proceedings of IFIP '77 Congress in Toronto, North Holland, Amsterdam.

Boyle, J.M. and Frantz, M.E. (1976A) "Automating multiple program realization II. BLA Replacement and complex to real transformations for LINPACK",
Report ANL-7655, Argonne National Laboratory, Illinois.

Boyle, J.M. and Matz, M. (1976) "Automating multiple program realization",
Proceedings of Computer Software Engineering Symposium, New York.

Dorrenbacher, J., Paddock, D., Wisneski, D. and Fosdick, L.D. (1974) "POLISH - A FORTRAN program to edit FORTRAN programs", Department of Computer Science Report#CU-CS-050-74, University of Colorado at Boulder.

Du Croz, J.J., Hague, S.J. and Siemieniuich, J.L. (1977) "Automated aids in the NAG project", in Proceedings of Argonne Workshop on Automated Aids, (Ed.) W.R. Cowell (To be published by Springer-Verlag).

Fosdick, L.D. (1974) "A FORTRAN program to identify basic blocks in FORTRAN programs",
Department of Computer Science Report#CM-CS-040-74, University of Colorado at Boulder.

Hague, S.J. (1977) "Automated source-to-source transformation - a request for information", NAG Central Office Report, Oxford.

Hague, S.J. and Ford, B. (1976) "Portability - prediction and correction", Software Practice and Experience, Vol.6, No.1, pp 61-69.

Kernighan, B.W. and Plauger, P.J. (1976) "Software tools", Addison - Wesley, Reading, Massachusetts.

Krogh, F.T. (1972) "A language to simplify the maintenance of software which has many versions", Section 914, Technical Memorandum 314, Jet Propulsion Laboratory, Pasedena, California.

Meissner, L.P. (Ed.) (1976) "FOR-WORD", FORTRAN Development Newsletter, Lawrence Berkley Laboratory, California, Vol.2, No.1, pp 4-5.

Osterweil, L.J. and Fosdick, L.D. (1976) "Data flow analysis in software reliability", Department of Computer Science Report#CU-CS-087-76 (Revised), University of Colorado at Boulder.

Ryder, B.G. and Hall, A.D. (1975) "The PFORT Verifier", Computing Science Technical Report#12, Bell Telephones, Murray Hill, New Jersey.

Sayers, D.K. (1977) "The DECS declaration-of-variables program", NAG Central Office Report, Oxford.

Scowen, R.S., Aillin, D., Hillman, A.L. and Shimell, M. (1971) "SOAP" - A program which documents and edits ALGOL 60 programs, Comp. J. Vol.14, No.2, pp 133-135.

Standish, T.A., Kibler, D.F., Neighbors, J.M. (1976) "Improving and refining programs by program manipulation", Proceedings of the ACM annual conference, Houston, Texas, October, pp 509-516.

Woodward, P.M. (1975) "Skeleton analyser and reader for Algol-68R ('SARA')", Internal Note, R.S.R.E., Malvern.

I. 5

THE VALUE OF A NUMERICAL LIBRARY TO A LARGE INDUSTRIAL ORGANIZATION

R.W. McIntyre and T. Tate

(Rolls-Royce Ltd, Bristol)

1. INTRODUCTION

Rolls-Royce is now the sole large scale manufacturer of gas turbines in the United Kingdom. It has some fifty thousand employees based at some ten sites across the country. It has international trading and manufacturing links with a number of other countries. In these and other ways, it is a typical, large industrial organisation.

Early in 1975, the Company took out a subscription for the IMSL Library. IMSL stands for International Mathematical and Statistical Libraries, Inc., of Houston, Texas. For those who are better acquainted with the NAG library, the libraries of these two organisations are broadly similar in scope and concept.

The IMSL Library was leased mainly for engineering, although it is realized that it contains much that is useful also for management sciences. Section 2 outlines the way Rolls-Royce uses engineering computing in support of its business. Aspects of internal and external transportability are discussed.

Section 3 summarizes the way engineering software is produced internally, and lists the main external sources. With this background, Section 4 describes the main numerical needs of engineering, and the steps taken to keep pace with growth in demand.

Section 5 deals with the stages leading to the decision to lease a library, the factors governing choice, and the way IMSL was introduced into the computing systems operation in the Company.

Finally, Section 6 discusses IMSL as an integral part of the engineering computing service.

Some of the discussion is specific to the services supported from the Engineering Computer Centre at the Company's Bristol site, where the co-authors work.

2. ENGINEERING COMPUTING

Engineers at Rolls-Royce obtain their main service from the Company's own Computing Centres, which are located at the two largest sites Derby and Bristol. Large aero-engine projects are associated with either the Derby or Bristol Engine Groups, and so each Centre offers full computing support for gas-turbine engineering. Like many others of a similar size, the Company is virtually self-sufficient in hardware, and, in particular, has several large mainframes. At Derby, the Computing Centre currently has IBM 370 machines, whereas at Bristol, mainstream engineering calculations are now on a CDC CYBER 74. It is not uncommon for a company of this size to hav more than one mainframe type.

Engine projects provide the main stimulus for computing development, and this accounts for the considerable degree of independence retained by the Derby and Bristol Centres. The respective groups of engineers get most of their computing support from their local Centres, but this does not imply an absence of technical dialogue between the sites. Wherever appropriate, computing methods are developed jointly. In this context, libraries such as IMSL and NAG, that are available on all the mainframes used by the Company, are of particular benef

Rolls-Royce has important engineering interfaces with other companies. For example, several projects have been set up jointly by either the Bristol or Derby Engine Groups and engineering teams belonging to other large European manufacture of gas-turbines.

With a joint project, throughout design and development, it is necessary to establish and maintain consistency however the computing support is provided. To this end, key programs, such as those for the analysis of component and engine tests, need to be exchanged amongst the partners in the consortium.

A second, and equally vital, interface is with the custom The aircraft manufacturer and the airline need to evaluate riva bids from competing aero-engine companies. Such an evaluation depends heavily on large computer programs supplied to the customer to simulate engine behaviour under his operational conditions. Having chosen an engine, the customer is given further support in the form of computer programs that can monitor maintenance needs.

These internal and external interfaces give a clue to the sort of transportability that most concerns a company such as Rolls-Royce. The majority of programs have been written for running on one machine at one site. A growing number must be able to run at both of the Company's Computing Centres. A smaller but growing number of key programs are required for running at sites external to the Company, mainly outside the United Kingdom. The other sites and machines are known in advance for each program that needs to be transported.

In dealing with external companies, it is not generally possible to take advantage of common libraries in the way that applies between the Company's own Centres. The benefits in having a numerical library may be less tangible, but engineering companies should feel more confident if they know that programs they exchange embody good quality numerical routines.

At this stage, it should be noted that the value of a library to an engineering company would be diminished if undue restrictions were placed on its use for programs that need to be transported.

3. ENGINEERING SOFTWARE

3.1 Software Produced in the Company

Out of a total workforce of fifty thousand, some seven thousand are engaged in engineering. Of these, about one hundred, mainly graduate mathematicians and computer scientists, could be classed as programming on a full-time basis. In addition, there are several hundred (it is difficult to be more precise) who write programs from time to time, as part of their over-all engineering tasks. In terms of output, this additional group, made up mainly of engineers, would be equivalent to an additional full-time programming staff of perhaps fifty.

The most fundamental needs were satisfied very soon after computers were introduced into the aircraft industry some twenty years ago. It says much for the quality of the early systems analysis that many features of the earliest programs have been retained despite the various changes of machine, language, operating system, etc. This has produced an underlying pattern of continuous development of programming methods.

In aerospace another factor leading to a stable situation at least in the context of programming, is the long timespan for a typical project. As many as ten years can elapse between initial design and entry into service. Since engineering computing relates directly or indirectly to current or future projects, it is quite common for programs to be in use for equally long spells.

Perhaps 80% of work consists of modification to existing programs. This includes routine changes to improve efficiency, and adaptations to increase versatility. The other 20% concern development of, and research into, new methods needed to solve existing or new problems.

With project-related computing, consistency can often be more important than accuracy. This may constrain the way an existing program is modified. The need for consistency can even preclude the replacement of an old method, whatever the advantages of a newer alternative.

3.2 Other Souces of Software

Most software is produced internally, but engineering needs are many and various, and several important programs have been obtained from external sources. Formal exchanges within consortia working on particular projects were referred to in Section 2 above. In addition, within the aerospace industry, there is some informal sharing of methods but on the gas-turbine side, perhaps because of the very small number of large companies and high degree of specialization, this is not very extensive.

Government agencies in the United Kingdom, the United States and elsewhere have supported the development of advanced computing methods. Partly through this, and partly through direct contact, the Company has acquired several useful items of engineering software from various universities. In the main these sources have been within the engineering departments of universities. Smaller, self contained programs may have been produced in the course of their normal research activity, but with larger software items it has become the practice for universities to form specialist groups to sell particular services.

Other sources include software houses specializing in engineering science and computer aided design.

4. NUMERICAL METHODS IN ENGINEERING

4.1 Growth in Use of Numerical Methods

The jet engine was introduced not long before computers became available to industry. Its subsequent development has, in part, been made possible by their use. The technical computing load grows by about 40% per annum, which corresponds to almost a thousand-fold increase in twenty years. Initially, this growth was due to the development of methods most fundamental to gasturbine engineering. Then it became necessary to keep pace with the demands of an advancing technology. For example, it would not have been possible to design a three-shaft engine using first generation computers. More recently, growth has been due mainly to the broadening of the front on which computing is applied to engineering, and the introduction of on-line and interactive techniques.

Numerical routines are used more intensively, in line with the observed growth rate. The number and variety of routines has grown, although not by so high a factor.

4.2 Provision of Numerical Subroutines

It is helpful in assessing provision to consider historical perspective. At Bristol, the first computer to be installed, twenty years ago, was an English Electric D.E.U.C.E. The manufacturer supplied a set of numerical subroutines that would now be seen as rather basic, but at the time was adequate for most of the general needs. The set was in machine orders, but portability was not then a consideration.

The machine which replaced D.E.U.C.E. was a KDF9. Once more the manufacturers provided a set of subroutines, again in machine orders, with a somewhat expanded coverage. By the late 1960's transportability as outlined in Section 2 above, was a requirement. Another disadvantage of a machine-dependent library was realized: it tends to become frozen several years before the useful life of the machine is over.

Numerical methods have also been implemented within the Company throughout the twenty year period. Sometimes this has been to augment the range of general purpose subroutines then in the Company's library. More commonly, numerical techniques thought to be particular to aeronautical engineering were implemented. Especially in the earlier years, the engineering rathe than the numerical literature was the main source of methods.

One example of an engineering bias in approach is in the method of characteristics. This method was in use for solving supersonic inviscid gas flows before computers were introduced. It is suitable for hyperbolic partial differential equations more generally, but implementations in the Company have been specifically for the equations of supersonic flow, as set out by Ferri (1955).

An engineering approach may be forced if no numerical method is available to match the needs. Such a case concerned the flow of a chemically reacting gas mixture through an engin exhaust nozzle. The ordinary differential equations which aris from the normal physical model of the problem form a stiff syst In the early 1960's, no numerical methods that could tackle stiffness effectively had been developed. The engine exhaust problem was identified and solved by Bray and Appleton (1961), who modified the physical model in such a way that only non-stiff equations appeared. For the modified model, standard methods of integration were effective, and gave sufficiently accurate solutions. Bray and Appleton's method was implemented successfully in KDF9 User code within the Company.

Other, incidental sources of numerical subroutines are groups which supply technical software (see Section 3.2). In this case, subroutines are embedded in the programs supplied, and the Company has not hitherto sought to influence the suppliers over their sources of numerical software.

The first item of numerical software to be purchased deliberately, and the only one not in FORTRAN, was the KDF9 Partial Differential Equations Scheme. Subsequently the constrained optimizer REQP was purchased from the Numerical Optimization Centre, Hatfield, and a curve fitting package from the National Physical Laboratory, and several subroutines for unconstrained optimization were received from the Atomic Energy Research Establishment, Harwell. These more recent acquisition reflected the growing availability of FORTRAN implementations of numerical algorithms. It was perhaps no coincidence that the three sources mentioned are institutions that have been consistently helpful to industrial mathematicians.

The newest source of numerical software is IMSL.

4.3 *Organization of Numerical Working inside the Company*

Before 1970, there was no group at Bristol dealing specifically with numerical methods. Most of the technical departments investigated their own numerical problems, seeking assistance as they saw fit from the individuals working in the Engineering Computing Centre. Sometimes, as instanced above this was the best, or only, way to proceed. At other times, methods that could have been implemented as general purpose numerical subroutines were instead written as integral parts of particular applications programs.

Failure to recognize generality led to some duplication. Numerical methods treated as special, were thereby rendered vulnerable to changing engineering and political situations. Their development might be curtailed, and their maintenance was not provided for.

By 1970, the subroutines supplied earlier for the KDF9 were capable of fulfilling a diminishing fraction of growing general needs. A small group, which now carries the title of the Computing Standards Group, was then set up and assigned the tasks of surveying needs, providing subroutines to fill the most pressing gaps in capability, and advising on future policy.

5. ACQUIRING A GENERAL PURPOSE NUMERICAL LIBRARY

5.1 *The Decision to Lease a Library*

Despite success with a number of implementations, the Computing Standards Group soon decided that a slow build up of the capability of the old KDF9 library was not a sound policy. Once a need had emerged, a subroutine was wanted quickly. Contact with numerical analysis groups outside the Company led to the Engineering Computing Centre purchasing the individual items of numerical software referred to in Section 4.2.

Useful though these items have been, the Computing Standards Group came to the conclusion that there were limitations in proceeding only in this manner. It took time to consider each separate item and more time to make a technical case for acquisition. The Group decided in 1972 that a general-purpose numerical library would be required.

Because of the growing use of transported programs, the Group advised that the library should be in FORTRAN. The large company can, if it chooses, deploy enough people to implement such a library. At least one aerospace company and several other technical and scientific establishments have undertaken this task successfully.

At this juncture the IMA announced their Conference on Software for Numerical Mathematics, see Evans (1974).

In April 1973, two members of the Group, and two representatives of engineering departments attended the Conference, and recommended that a general purpose library be leased for whatever computer would replace the KDF9 at Bristol. At Derby, the situation was not so desperate. The library then available from IBM was quite useful, and rather larger than the KDF9 library at Bristol. Nevertheless, the advantages in both Computing Centres having the same library were considerable (see Section 2), and subsequently a decision was taken that the Derby Engine Group also should lease IMSL.

5.2 Choice of Library

The next task of the Computing Standards Group was to choose a library. The basic requirements of Industry for numerical libraries are much as described in Ford (1972), and Ford and Hague (1974). In summary, the Group was looking for a professional implementation of a properly selected, comprehensive range of high quality validated algorithms, by an organization dedicated to the task, and to the continuing maintenance and enhancement of their product. The library would need to be internally self-consistent, well and uniformly documented for the user, and in every way suitable for being made an integral part of the engineering computer service.

With these criteria, and the added condition of availability on several machine ranges, the Group found itself with a shortlist of two libraries: NAG and IMSL.

One factor which affected the Group's assessment was that, in 1973, the NAG Library was available only to universitie
Nearly three years were to elapse before NAG Limited, the organization through which this Library was to become commercia

available, was formally set up. During the assessment, the Group came gradually to the view that the relative remoteness of the IMSL Office from Bristol was not a significant factor. In the Group's judgement, most needs would be met if the Company received an informative reply by return airmail to any technical query correctly transmitted. This has been borne out by subsequent experience: there has not so far been any occasion to contact IMSL by telephone. In the computing and software environment outlined in Sections 2 and 3, the Group already had the responsibility of providing a numerical methods advisory service. Leasing a library has made this task much easier.

The assessment was concluded at the end of 1974, and the decision taken that the Company should lease IMSL.

5.3 Introducing IMSL into the Computing System

There was a breathing space for the Computing Standards Group to explore basic features of IMSL Library before others at Bristol had access. This was arranged by initially leasing the IBM version, and mounting it at Derby. There was the added advantage that the Engineering Computing Department at Derby were able to experiment with the library, and it was during this period that they concluded that their Centre as well should lease IMSL.

Even during the period of familiarization the usefulness of the library was made apparent. A number of problems referred by users were resolved by running IMSL routines, which either gave good answers where none previously had been obtained, or indicated the cause of instability (e.g. near singular matrices).

6. IMSL AS PART OF THE COMPANY'S COMPUTING SERVICE

6.1 Initial Impact

The CYBER 74 computer was installed in the Engineering Computing Centre at Bristol in the Autumn of 1975. Because of the prior period of familiarization, it was possible to make IMSL available on CYBER before users switched from KDF9.

It was a novel experience for users to find some four hundred numerical subroutines available to them instead of the mere handful. Initially this caused not a little bewilderment. Several chapters dealt with topics that were quite foreign to many users. The policy of the Computing Standards Group was to make a general announcement of the availability of IMSL, and to circulate the eighty departmental representatives with the full list of subroutine titles. Some caution was exercised in releasing any of the three copies provided of the two-volume library manual.

It was felt that, at least in the first few months, users should come to the Group with their problems and be helped to use the manual. Subsequently four extra copies of the manuals were purchased for siting in user areas. When Edition 6 was released, the collected set of introductions to chapters was added to the basic information circulated to the departmental representatives.

As word spread of the availability of IMSL a pattern of usage began to emerge. Three separate categories were identified and are considered in Sections 6.2 - 6.4 below.

6.2 Numerical Methods in Long Established Programs

Long established, regularly used production programs embody a great deal of engineering and numerical experience. Opportunities to incorporate new routines are limited. Nevertheless, departments responsible for these programs do carry out research into techniques, and from time to time this results in new versions of the programs. Computing Standard Group decided that their rôle was to assist in numerical aspects of this research, rather than to promote a separate exercise of replacement of all existing numerical subroutines with IMSL equivalents.

To give an example, the program package which computes engine performance using thermodynamic equations is the most fundamentally important technical system in the Company. As such it is the subject of continuous study, and regular development. Numerically, it embodies the solution of a set of non-linear, algebraic equations.

The existing method, a compound Newton-Raphson algorithm, had been in use for ten years, and had been highly adapted to the particular equations and physical context. It was as robust as it could be made. For conventional engines, the routine worked, given adequately close starting guesses, that is, provided engine conditions were not altered too substantially between consecutive computed points. For less orthodox engine cycles, failures were not uncommon, because mechanisms for providing good starting guesses were not available. When investigating unorthodox cycles it could require several attempts with altered guesses provided by knowledgeable engineers to obtain a successful computation.

The performance engineers substituted IMSL routine ZSYSTM in the experimental version of the program and found the dependence on good starting guesses to be less severe. For conventional engines, this program need no longer restrict the changes in conditions between computed points, and no failures have been encountered with unorthodox cycles.

The first fruit of this research was a new more robust version of the production system. The research continues, because there are prospects of producing a version that would also be faster than the original.

Other well established program systems, such as some dealing with finite element applications, have been examined with similar results. One of the values of having a library has been the impetus given to research into methods in use.

6.3 New Programs

Section 3 above attributed 20% of all programming effort to work on new methods. Sometimes new methods are applied to problems of long-standing. Some ten years ago the finite element method was introduced into the Stress Department, with far reaching effects. There has been an upsurge of time series analysis applications in vibration engineering coinciding with the acquisition of IMSL. On other occasions the problems are newer than the methods, for example radiography applied to engine testing is leading to new applications of digital picture processing. With a library, new problems can be approached more positively. From the outset it is clearer what is likely to be practicable, and how much work will be entailed. Lack of experience in a numerical area is no longer the handicap it used to be.

The rôle of the Computing Standards Group here, is to ensure that all working on new applications are well aware of the availability of IMSL and have ready access to manuals. Users are encouraged to seek help from the Group whenever they encounter difficulties in method selection or interpretation of results.

6.4. Programs that are Suspect

All instances of possible numerical malfunction are referred to the Group. The user himself or the Computing Centre may have implemented the program to be scrutinized, or the source may be an external one. Unless the program is new, it will probably not contain IMSL subroutines. The Group wastes little time exploring the intricacies of questionable subroutines it finds in the programs. All too often the methods are old, the subroutines undocumented, and the original implementers no longer with the Company. Almost invariably, the best course of action is to substitute the appropriate IMSL subroutines, and then to examine the answers that they give. Then it only remains to establish, to the Group's and users' mutual satisfaction, that these answers are the valid ones. Where the implementers are still with the Company, such investigations are an effective means of weaning them away from old methods.

It is interesting that several cases of malfunction have come to light as a consequence of the move to CYBER 74 and the change in wordlength from 48 to 60 bits. Changes to answers, perhaps of the order of 1%, have caused the engineers to report their unease to the Group. Sometimes the investigation reveals hitherto unsuspected ill-conditioning, at other times numerical methods are shown to be poor, in which case the library provides a quick remedy.

If the suspect program is part of a large package obtained from an external source, it may be possible to spot the cause, but not to propose a remedy. Most external sources are other engineering groups, and it is not politic for the Engineering Computing Centre to propose that they too should take their numerical subroutines from a reputable library. A solution to this general problem has yet to be found.

6.5 Other Technical Aspects

The library was received in source, and compiled using CDC's FTN Compiler at Optimizer Level 2. The source is protected, and listings are held only by Computing Standards Group. If a user has a need to know something of the way the algorithm works, the Group will refer to the source if appropriate. So far this has happened on only a few occasions.

The complete library has been compiled, even though some chapters have not yet been accessed by any user. As there is no intention of deleting any chapter from the on-line file, no statistics have been gathered on the use of individual routines, although a mechanism is available for doing so. It is regarded as more important to have support at hand for new applications that can arise suddenly and could involve use of new chapters. Furthermore some chapters, notably those dealing with certain statistical methods, cover topics that may have been neglected, and which could be relevant to the Company. There should be no discouragement to individual engineers willing to learn how methods new to them could be applied to their problems.

7. CONCLUSIONS

1. A comprehensive numerical library is an essential feature of an engineering computing service operated internally by a large company.

2. It is beneficial to a company with international commitments if the library of its choice is available and known internationally, and has a large and broadly based set of customers, so that all chapters are in regular use and are subject to effective maintenance and enhancement.

3. A large company can usually mount the entire library on-line, and should do so.

4. A large company having its own numerical supporting services, is less likely to call for guidance from the Library Office, but it is important that this Office can respond adequately and promptly to both written and telephoned technical enquiries.

5. A library that is available across most mainframes is well suited to the needs of companies large enough to have problems of internal portability.

6. It would be of immense benefit to the Company if engineering based organisations which supply applications software took the trouble to avail themselves of good quality numerical libraries.

7. If a company has a good library its engineers spend less time implementing numerical techniques and more on engineering.

8. ACKNOWLEDGEMENTS

We are appreciative of the contribution of our late colleague, Richard Graham-Brown, who carried out all preliminary work on IMSL before the CYBER 74 computer was installed at Bristol, and who then ensured the smooth introduction of IMSL into the services operated by Engineering Computing Centre.

Many in E.C.C. have helped the Computing Standards Group, in particular B.J. Banes, A.H.O. Brown, D. Furze and D.J. Holmes In the Engineering Computing Department at Derby, IMSL has been the responsibility of D. Gale and R. Price, and the Group were indebted to them for their assistance in the preliminary phase, when the Library was only at Derby.

The Chief Computing Engineer at Bristol, D.C. Boston, with whose permission this paper is presented, first identified the need for a library, and A.C. Clarke, now with Honeywell Information Services worked hard, before and after the 1973 IMA Conference on Software for Numerical Mathematics, to establish the particular requirements of engineers at Bristol.

Outside the Company, the authors wish to acknowledge the help given by Dr. B. Ford, during the 1973 Conference and on subsequent occasions during 1973 and 1974. Although these discussions did not lead to the Company subscribing to NAG Ltd., Dr. Ford's advice has always been valued, and has influenced the Company's thinking on libraries and software generally.

Similarly, Dr. E.L. Battiste, President of IMSL has shown a thorough understanding of the needs of engineers. Since subscribing to IMSL, the Company has been pleased with the way this understanding is reflected in the high quality of service provided. Distance has caused no problems of communication.

9. REFERENCES

Bray, K.N.C. and Appleton, J.P. (1961) "Atomic recombination in nozzles: Methods of analysis for flows with complicated chemistr A.R.C. CP 636.

Evans, D.J. (1974) (Ed.) "Software for Numerical Mathematics", Academic Press.

Ferri, A. (1955) "General Theory of High Speed Aerodynamics", Volume VI of "High Speed Aerodynamics and Jet Propulsion",

(W.R. Sears, Ed.) Oxford University Press.

Ford, B. (1972) "Developing a numerical algorithms library", *IMA Bulletin* **8**, pp 332-336.

Ford, B. and Hague, S.J. (1974) "The organisation of numerical algorithms libraries", in "Software for Numerical Mathematics", Ed. D.J. Evans, pp 357-372, Academic Press.

I. 6

USING THE NAG LIBRARY IN INDUSTRIAL RESEARCH

J.P. Whelan

(Philips Research Laboratories, Redhill, England)*

1. INTRODUCTION

In this chapter we will present the experiences of using the NAG numerical software library at Philips Research Laboratories* (Redhill) over a period of two years. Our aim will be to highlight some of the benefits, and difficulties, encountered in using a leased software library at an industrial research laboratory.

1.1. Historical Background

In the spring of 1975 Philips Research Laboratories (PRL) obtained the numerical software library of the Numerical Algorithms Group Ltd (NAG) during a trial release to industrial research centres. Naturally there existed already a local software library for the 600 PRL staff using the mainframe computer, an ICL 1904S. By Autumn 1976 sufficient favourable experience had been gained to justify the adoption of NAG as the standard, or preferred, library and the majority of routines in the local library were removed from the user library manual.

Early in 1977 Philips chose NAG as its research standard library at all of its laboratories in Europe, even though this involved making implementations of the library on Philip's computers at Hamburg and Eindhoven.

1.2. The Leased Numerical Software Library

In an era when rapid advances are being made in numerical mathematics it would require a large and specialized manpower effort to produce an up to date software library of proven integrity. Any industrial concern must therefore consider the possibility of leasing an established software library. The most

---

* Previously Mullard Research Laboratories

important reasons for choosing NAG were firstly its high reputation for quality and robustness and secondly the availability
of ALGOL 60 and FORTRAN versions both of which are widely used
languages at PRL. An additional factor to consider is the impl
mentations of NAG which exist on a wide range of computers, and
PRL being part of a multi-national Company lends added weight
to this argument.

1.3 Scope of this Paper

We will first discuss the relevance of the portability of
the library under the heading of "Changing Computers". The
other topics dealt with are, in order of presentation, the user
documentation subset, a user advisory service, dealing with
difficult problems and, finally, monitoring library usage.

2. CHANGING COMPUTERS

In Spring 1971 the ICL 1904A, as it then was, replaced an
Elliot 503. Floating point arithmetic was to 11 significant
figures on the new computer, a gain of 2 figures. Many numeric
incompatibilities were probably avoided by this improved precis
However, the differing word lengths of 24 and 39 respectively,
meant that a new random number generator had to be provided,
because the one provided by ICL was inadequate. In other respe
the ICL library was obviously out of date and not sufficiently
comprehensive for the needs of a research laboratory. Indeed,
its inadequacy was one of the deciding factors in the formation
of what is now the Numerical Algorithms Group. However in the
absence of manpower resources for such a specialized project
PRL decided to take the best from the Elliot 503 and the ICL
libraries. For those users using FORTRAN, a language which had
not been available on the old computer, special difficulties
existed because PRL library programs written in ALGOL could
sometimes not be called from FORTRAN programs. There were two
main reasons: either a failure exit had been made to an ALGOL
label rather than an integer or logical flag or, in other case
a real variable had been used rather than a real function. The
latter occurred in a PRL quadrature routine declared by

```
real procedure RSIMPS (F, X, A, B, ACC, EPS);
value A, B, ACC; real F, X, A, B, ACC, EPS;
```

Since this routine was far superior to the best ICL quadrature
routine it would have been best to modify its coding so that F
was provided as a function. However existing programs using

the old version were still extant so an ICL routine was provided for FORTRAN users. In this instance, and others, the acquisition in 1972 of the Harwell library of FORTRAN subroutines was a great help. Although many important areas such as linear algebra contained system dependent routines, often because double precision inner product routines had been written in assembly language, nevertheless the Harwell library provided much software at very low cost and of a quality not easily obtainable elsewhere. As we shall see in a later section of the paper some routines in the Harwell library still offer facilities not yet available in NAG and it is no secret that many Harwell routines form the basis of NAG routines.

Given the background we have just described to the change from one mainframe computer to another it is obvious that had the NAG library been available to industry earlier it would almost certainly have been acquired by our laboratory.

If the scientist is to use the NAG library he must be provided with adequate documentation and this we discuss in the following section.

3. THE USER DOCUMENTATION SUBSET

There is a difficulty in trying to provide for the user a reasonably compact library documentation. This arises because the write-ups for the 300 or so routines comprising a single language version of the NAG library fill three standard size ring binders - a total of 2000 pages. In fact this problem is not really one confined to NAG. For example the PRL library manual issued to users contained write-ups for only a selected number of routines chosen from the total available software. To select such a documentation subset, as we shall call it, from the complete NAG library is a more difficult task because the NAG library is a purpose made product deliberately providing many routines for each problem area. For example in the quadrature chapter there are six routines for computing definite integrals. One is for a function defined numerically and was included in the subset. Of the other five, for analytically defined functions, four are adaptive routines. These are, in order of compiled size for ALGOL (size in 24 bit words in brackets).

(1) Romberg's method (4416)
(2) Clenshaw-Curtis method (5248)

(3) Oliver's doubly adaptive method (6144)
(4) Patterson's method (7360)

In FORTRAN the ordering is different - Patterson's methoc comes in second place. We wish to choose one routine for the subset and for this purpose we observe that the local requirement is for only modest relative accuracy, say 10^{-4}, and the smallest program so that turn around time is minimized. On these criteria the Clenshaw-Curtis method would provide a good balance between reliability and size. Unfortunately, the NAG version does not provide an error estimate so to be sure the answer _is_ reliable a further quadrature is necessary. The corresponding FORTRAN routines have sizes 3456 and 4800 respectively so this finally decided us to put the Romberg method in the subset. If the integration proves difficult the user is strongly advised to consult an expert on quadrature. Clearly the usage of this quadrature routine should be carefully monitored, a subject which we will discuss in a general way later in this paper.

The library manual contains not only the subset of routir but also a complete catalogue of all the library. Access to the complete three volume documentation is given by placing set in each room, ten in all, where there are computer terminals. While the subset can be revised in the light of experience the is need for reasonable stability so that the initial choices are likely to remain for at least a year or so. At this point therefore, we ought to mention a few other possibly contentiou choices, bearing in mind always that the user is made fully aware of the alternative routines which exist.

For ordinary differential equations of the initial value type Merson's form of the Runge-Kutta method is chosen rather than a routine which can cope with stiff systems. Again the reasons cited above, of compactness and only moderate accuracy needs, guide us in our choice. If stiffness is suspected then monitoring of the step length which is recommended as a matter of course, will usually decide the matter.

For systems of linear equations, whether overdetermined or not, those routines which give an iteratively refined solution are chosen for the subset, even though more compact routines without such refinement are available. This choice was made because experience has shown that, apart from inherent il conditioning, users often introduce ill-conditioning themselve by using inappropriate methods. In the users' defence, sometimes, as we shall see later, considerations of program size force the attempt to use such methods. Since the routines yielding refined solutions can often detect ill-conditioning they are far superior in practice.

Finally, we consider the area of data fitting by cubic splines. Here the Harwell library has useful facilities not available in the NAG routines. These include the possibility of obtaining derivatives or integrals for the fitted spline and the automatic positioning of knots. A possibly useful feature of the NAG routines, not available in the Harwell library, is that by taking coincident knots, cusps and other discontinuities in the spline can be easily built in if the data suggest such a shape.

4. A USER ADVISORY SERVICE

Philips Research Laboratories provides a mathematical, problem solving and software advisory service in the computer department. The need for this service has increased as users have tackled more difficult problems - and the existence of a comprehensive library such as NAG has drawn people into areas of numerical mathematics new to them. We have already cited the possibility of ill-conditioning occurring and it is appropriate here to discuss in more detail the reasons why some users who are aware of the dangers nevertheless try to solve ill-conditioned sets of linear equations. Consider the problem of finding the solution in the L_2 norm of an overdetermined set of linear equations with m rows and n columns ($m >> n$). The numerically stable method provided by NAG involves the pseudo inverse matrix and requires storage of order mn. However the normal equations require storage of order n^2 only. In one project the outcome was that the program using the former method was too large to be run in time sharing mode from a terminal. The user had to submit a batch job and suffered the necessary limitations of having his program run only in the evenings. In striking contrast the program using the second method - normal equations - was small enough to be run from the terminal. The user could deal with many sets of data in a working day and have the convenience of making immediate alterations to his program. To the user "ill-conditioning" was as much in his restricted mode of working as in the real numerical problem of solving normal equations.

However, users must of course be aware of the limitations of routines and when, as occasionally happens, this is not pointed out in a particular write-up an advisory service is essential. To take a trivially simple example, the NAG Gauss-Laguerre routine gives the following integral to only 0.3% relative accuracy.

$$\int_0^\infty e^{-x} \sqrt{x}\, dx.$$

The numerical analyst is not surprised, but the user is, and probably needs some convincing of the integrity of the library.

The advisory service we have discussed so far is local but it is important to have recourse to an outside source of advice and this NAG provides. It cannot be stressed too much how important such a back up is, for it enables the local advisory service to apply the most appropriate NAG routine in areas of numerical analysis outside their experience. Additionally, problems which have defied solution can be given to outside consultants with whom NAG has intimate contact. But this is part of the subject of our next topic, that of solving "difficult problems".

5. DIFFICULT PROBLEMS

To say that a problem is difficult is a subjective statement in that it depends on the expertise of the problem solver. However most numerical analysts would agree that double quadrature, two point boundary value problems, non-linear singular integral equations and, finally, non-linear partial differential equations may all constitute difficult problems in the following ways. Firstly reliable software of the standard which users of NAG have become accustomed to is not generally available. Secondly, numerical analysis theory is not in a sufficiently advanced state to provide robust methods, nor even methods which can be guaranteed to converge for a wide class of problems. At PRL we have found three possible avenues of escape in tackling such problems.

(1) NAG provides an advisory service itself but if their analysts are unable to solve the problem they pass it to a suitable expert who is often a contributor to their library in that particular field.

(2) The Numerical Analysis Division of the National Physical Laboratory (NPL) provides to industry a consultancy service which is extremely flexible and lately has been of material use to PRL.

(3) The Harwell subroutine library often has software, as we have already said, which is not available elsewhere and can provide solutions to difficult problems, if only for limited ranges of the physical parameters.

A few examples illustrating the above methods of attacking problems will now be given. Firstly, consider the fitting of a polynomial surface to a set of data so as to obtain the best fit in the L_2 norm. A straightforward analysis had been done by a user but there seemed no obvious minimum degree polynomial. The problem was somewhat complicated because for the first time in two years a NAG routine (for solving linear systems of equations) was found to be in error because of a fault in implementation. Independently of this the users' numerical method was not suitable and NAG suggested that NPL take on the task. By using products of Chebyshev polynomials the minimum degree fit could be found by inspection of the Chebyshev coefficients (Hayes, 1977, NPL Consultancy, Private Communication).

Secondly, two users wanted to do a double quadrature where their integrands were not well behaved. Both the NAG and Harwell routines failed for important ranges of the physical parameters involved. With advice from NPL one of these integrals was solved but the other remains unsolved.

Thirdly, a non-linear Volterra integral equation with a singular kernel proved difficult but after consultation with NPL the numerical problem was identified and solutions have now been obtained (Miller, 1977, NPL Consultancy, Private Communication).

Having given some examples of how difficult problems may be dealt with we can draw one important conclusion. Such problems often indicate which areas of the NAG library require additional or improved routines. Thus in our example on integral equations the NAG library contains one routine only - for Fredholm non-singular equations of second type. This is one area of the library which is undergoing expansion and feedback from users should be valuable to the library contributor. Also it is clear that considerable improvement is required in the double quadrature routine. However there is yet another way in which feedback from the users of the library to NAG may be made, and that is by automatically monitoring the usage of its contents. This will be the next and final topic in our paper.

6. MONITORING LIBRARY USAGE

In this section we turn our attention to automatically monitoring the usage of the routines in the library. The precis method will depend on the operating system of the computer. On the ICL 1904S sufficient information about users' activities is retained by the operating system on backing store. A program is run at monthly intervals to extract the names of users and their programs whenever they contain a call to the NAG library.

A given routine is counted only on its first successful consolidation into the named program. There are some pitfalls: firstly if the user changes the name of his program but makes only a trivial change to the coding, the NAG routines are counted again - although in reality this is not at all a new usage of the routines. However there is no reason to suppose that such bias will appear preferentially in one area of the library and so relative comparisons should be meaningful. Secondly some routines are called often as auxiliaries to other: This applies particularly in the decomposition of matrices into a product of factors. The specific use of such routines in their own right cannot really be monitored therefore.

The main points which emerged are as follows.

(1) Users are not inhibited by the choice of routines in the documentation subset since they have used others. For example the minimization routine in the subset does not require derivatives to be given explicitly. Yet users frequently use routines requiring derivatives and so gain considerable advantages of improved convergence and increased speed.

(2) The most frequently accessed routines are determined. At PRL these were in decreasing use: quadrature, ordinary differential equations, non-linear equations, simultaneous linear equations and special functions.

(3) Areas of the library rarely accessed are found.

In Table I we give the results for the six months from February 1977. One point must be made about the interpretation of these figures. Low usage of a chapter does not in itself mean that those routines are redundant. For example under the heading curve and surface fitting, routines have been used to fit splines in the L_2 norm. Quite often, as was the case here, such infrequently used routines are applied in research areas of importance and their presence in the library is vital.

Table I

Monitoring Analysis

Chapter Name	Usage as % of Total Programs containing NAG Routines
Quadrature	18.5
Special Functions	14.5
Zeros of one or more Transcendental Equations	14.1
Ordinary Differential Equations	13.2
Simultaneous Linear Equations	12.1
Random Number Generators	8.8
Minimization	6.0
Fast Fourier Transforms	4.5
Zeros of Polynomials	2.4
Curve and Surface Fitting	2.4
Sorting	2.4
Correlations and Regression "Analysis"	1.2

While results for one site may, in themselves, be of only local interest, feedback of monitoring information from all NAG sites could well give direction to future development of the library. We would argue that automatic monitoring is important and should be incorporated into the operating system whenever a site acquires any leased software library.

7. SUMMARY

The Numerical Algorithms Group library of scientific routines has been used successfully at Philips Research Laboratories for over two years. A local advisory service which deals with users' mathematical problems and with queries on how to make the best use of the library is required. Additionally a back-up service such as is provided by NAG is essential and it should bring the local service into contact with appropriate numerical analysis consultants when necessary. The rôle of the Numerical Analysis Division of the National Physical Laboratory has been discussed in this connection. The provision of a documentation subset and the need for automatic monitoring of library usage have also been discussed.

PART II

LINEAR AND NON-LINEAR ALGEBRA

II. 1

SINGULAR-VALUE DECOMPOSITION - BASIC ASPECTS

J.H. Wilkinson

(National Physical Laboratory, Teddington)

1. INTRODUCTION

Perhaps the most fundamental problem in classical linear algebra is the solution of the linear system

$$Ax = b \tag{1.1}$$

where A is an $m \times n$ matrix, x an n vector and b an m vector, each with elements over a field F. Here we shall be concerned only with the complex and real number fields. The existence and nature of the solution are discussed in terms of the concept of matrix rank. For convenience we give a brief exposition of the results in terms which seem best related to the requirements of this paper.

The concept of column rank, $cr(A)$, of an $m \times n$ matrix A is first introduced and typically is defined to be the maximum number of linearly independent columns of A, *i.e.*, the dimension of the subspace spanned by these columns. Clearly $cr(A) \leq \min(m,n)$. In a similar way the row rank, $rr(A)$, is defined in terms of the rows of A and again $rr(A) \leq \min(m,n)$. It is then shown that $rr(A) = cr(A) = r(A)$ where $r(A)$ is defined to be <u>the</u> rank of A. It is proved that $r(A)$ is the highest order of non-vanishing minors of A.

Now if P and Q are square non-singular matrices of orders m and n, respectively, we may consider the related linear system

$$PAQy = Pb. \tag{1.2}$$

Clearly if y is a solution of (1.2) then Qy is a solution of (1.1) while if x is a solution of (1.1) then $Q^{-1}x$ is a solution of (1.2). In this sense the problems are fully equivalent. We may in particular take P and Q to be permutation matrices. The use of P merely changes the ordering of the equations while the use of Q implies that the elements of y are those of x with a different ordering. From the definition of column rank it is easy to show that $cr(A)$ and hence $r(A)$ are unaffected by a pre-multiplication of A with any non-singular P while from the definition of row rank it follows that $rr(A)$ and hence $r(A)$ are unaffected by a post-multiplication of A with any non-singular Q.

The fundamental result in the classical theory is that the system (1.1) has a solution if and only if $r(A) = r(A|b)$ where $(A|b)$ is the augmented matrix of dimension $m \times (n+1)$. When this condition is satisfied the nature of the set of solutions is as follows.

If A is of rank r then there exists a non-vanishing minor of order r. By suitable permutations of rows of (1.1) and of the elements of x the corresponding sub-matrix can be brought into the top left-hand corner of A and we shall assume this is true initially. It can then be shown that each of equations $r+1$ to m is a linear combination of the first r equations; hence x is a solution of the complete system if and only if it is a solution of the first r equations. We may write these first r equations in the form

$$[M_r | N_{n-r}]x = b^{(r)} \tag{1.3}$$

where M_r is an $r \times r$ matrix (non-singular from our assumption), N_{n-r} is an $r \times (n-r)$ matrix and $b^{(r)}$ is the vector consisting of the first r elements of b. If we partition x into $x^{(r)}$ and $x^{(n-r)}$ in the obvious way then (1.3) states that

$$M_r x^{(r)} = b^{(r)} - N_{n-r} x^{(n-r)} . \tag{1.4}$$

Clearly the $n-r$ elements of $x^{(n-r)}$ may be chosen arbitrarily and then $x^{(r)}$ is given by the solution of the $r \times r$ system (1.4); this system has a non-singular matrix of coefficients, M_r, and hence each choice of $x^{(n-r)}$ gives a unique vector $x^{(r)}$.

In theory the determination of $r(A)$ appears to be quite a simple problem and indeed it is effected by several of the factorizations of A which are in common use in linear algebra. Suppose for example we have determined permutation matrices P and Q and a non-singular matrix X such that

$$PAQ = XT \tag{1.5}$$

where T is a trapezoidal matrix of the form

$$T = \begin{array}{l} s \{ \\ m-s \{ \end{array} \left[\begin{array}{c|c} T_{11} & T_{12} \\ \hline O & O \end{array} \right]. \quad \begin{array}{cc} s & n-s \end{array} \tag{1.6}$$

Hence T_{11} is a non-singular $s \times s$ upper-triangular matrix and T_{12} is an $s \times (n-s)$ matrix, not otherwise specialized. Then since

$$T = (X^{-1}P)A(Q) \tag{1.7}$$

$r(T) = r(A)$ and it is trivial to show that $r(T) = s$.

Now such a factorization of A is achieved, for example, by the process of Gaussian elimination with complete pivoting. In this process the matrix $A \equiv A^{(r)}$ is reduced progressively to the forms

$$A^{(k)} = \begin{array}{l} k \{ \\ m-k \{ \end{array} \left[\begin{array}{c|c} T_{11}^{(k)} & T_{12}^{(k)} \\ \hline O & T_{22}^{(k)} \end{array} \right] \quad k = 1,2, \text{ etc.} \quad \begin{array}{cc} k & (n-k) \end{array} \tag{1.8}$$

where $T_{11}^{(k)}$ is a non-singular upper-triangular matrix. At a typical stage one determines the element of $T_{22}^{(k)}$ of maximum modulus. If this is in row u and column v of $A^{(k)}$ then rows (k+1) and u and columns (k+1) and v are interchanged. Multiples of row (k+1) are then taken from each of rows (k+2) to n in order to produce zeros in column(k+1)in each case. The matrix is now in the form required for $A^{(k+1)}$. The row and column interchanges involve pre-multiplications and post-multiplications by elementary permutation matrices and the subsequent row subtractions are achieved by a pre-multiplication with an elementary non-singular matrix. The process terminates either when $T_{22}^{(k)}$ is a null-matrix or when (m-k) or (n-k) has become zero. In the two latter cases the final $A^{(k)}$, *i.e.* $A^{(r)}$, is of the forms

$$\left[T_{11} \middle| T_{12} \right] \text{ and } \left[\frac{T_{11}}{O} \right], \tag{1.9}$$

respectively, T_{11} having the maximum rank, min(m,n), consistent with the dimensions of A. In the most common case we have m=n and if r=n the final matrix is merely a non-singular upper-triangular matrix. In this case A is itself non-singular and therefore of rank n.

2. PRACTICAL DIFFICULTIES

As described in the previous section not only is the concept of rank quite a simple one but we have quite a straightforward algorithm for determining it. In practice difficulties arise for the following reasons.

(i) The elements of A may well not be given exactly and even if they are they may not be exactly representable on a computer. What then do we mean by r(A)? In the mathematical concept we have distinguished only between completely null matrices $T_{22}^{(k)}$ and those with some "non-zero" elements. If we reach a $T_{22}^{(k)}$ all of whose elements are "small" in some sense, it may be appropriate to regard $T_{22}^{(k)}$ as null and to take r(A)=k.

(ii) Even if A is exactly known and representable the execution of the algorithm we have described will usually involve rounding errors. If the elements of A are small integers it may be possible to perform a related reduction on the computer within the ring of integers, but we shall not concern ourselves here with that possibility and hence even if some rows of A are exactly linearly dependent this exact linear dependence will not persist in the reduced $A^{(k)}$ and we shall not reach a null $A^{(r)}$.

It might be felt that these misfortunes are no more serio than those inevitably encountered in practical computations and in so far as we are dealing with linear operations they will be somewhat simpler to overcome than is often the case. Unfortunately this is not true.

We are tempted to argue that if perturbations of order ε in elements of A would make some rows linearly dependent then during the course of reduction this would lead to a matrix $T_{22}^{(k)}$ with elements of the order of magnitude ε. The choice at each step of the maximum element in the $T_{22}^{(i)}$ as the "pivotal" element would seem to enhance this probability. A trivial example will serve to show the extent of the difficulty.

Consider the $n \times n$ upper-triangular matrix A with $a_{ii} = 1$, $a_{ij} = -1$ $(j > i)$. If we factorize this matrix by the algorithm we have described then no computation is required and A itself is the final matrix $A^{(n)}$. Clearly A is non-singular and its rank is n. However, even for quite modest n, A is very close to a matrix of rank n-1. In fact if one perturbs A by $-2^{-(n-2)}$ in its (n,1) element, the matrix is exactly singular and is, in fact, of rank n-1. No hint of the proximity to singularity is provided by the factorization of A itself; at each stage in the reduction every element of $T_{22}^{(k)}$ is ± 1. Note that rounding errors do not contribute to the "deception". It might be objected that at each stage during the reduction all elements competing for pivotal element are ± 1 and we have assumed that the (k,k) element which is +1 is taken. However, if one takes the diagonal elements to be $1+\varepsilon_i$ where ε_i is very small, we can remove any ambiguity from the choice and the matrix is still almost singular.

The plain fact is that the simplest of the factorizations give no reliable indication of the proximity of a matrix to rank deficiency even when rounding errors are not involved. The great virtue of the singular value decomposition (SVD) is that it does give a perfectly reliable indication and it can be performed in such a stable manner that rounding errors do not bring any significant complications.

3. THE SINGULAR VALUE DECOMPOSITION

The singular value decomposition is based on the following theorem.

Any $m \times n$ matrix A with complex elements can be factorized in the form

$$A = U\Sigma V^H, \tag{3.1}$$

where U and V are unitary matrices of order m and n, respectivel and Σ is an $m \times n$ "diagonal" matrix with non-negative "diagonal" elements σ_i, *i.e.* with $\Sigma_{ii} = \sigma_i$ and $\Sigma_{ij} = 0$ $(i \neq j)$.

The ordering of the elements can clearly be chosen arbitrarily since if P and Q are permutation matrices

$$A = U\Sigma V^H = (UP^T)(P\Sigma Q)(Q^T V^H) \tag{3.2}$$

and UP^T and $Q^T V^H$ are both still unitary. If P and Q are chosen so as to make the row interchanges in Σ the same as the column interchanges, $P\Sigma Q$ is still "diagonal" with the diagonal element reordered. (N.B. Since P is of order m and Q is of order n we cannot say that $Q = P^T$!) Since $r(\Sigma) = r(A)$ both are determined by the number of strictly positive σ_i. We shall assume from now on that they are ordered so that

$$\sigma_1 \geq \sigma_2 \geq \dots \geq \sigma_r > 0, \ \sigma_{r+1} = \dots = \sigma_{\min(m,n)} = 0. \tag{3.3}$$

The σ_i are called the <u>singular values</u> of A.

Notice that since Σ is an $m \times n$ matrix it cannot be strictly diagonal unless $m = n$. The two matrices

$$\Sigma = \begin{bmatrix} \sigma_1 & 0 & 0 & 0 & 0 \\ 0 & \sigma_2 & 0 & 0 & 0 \\ 0 & 0 & \sigma_3 & 0 & 0 \\ 0 & 0 & 0 & 0 & 0 \end{bmatrix} \text{ and } \Sigma = \begin{bmatrix} \sigma_1 & 0 & 0 & 0 \\ 0 & \sigma_2 & 0 & 0 \\ 0 & 0 & \sigma_3 & 0 \\ 0 & 0 & 0 & 0 \\ 0 & 0 & 0 & 0 \end{bmatrix} \tag{3.4}$$

illustrate the form of Σ for the cases $m = 4$, $n = 5$, $r = 3$ and $m = 5$, $n = 4$, $r = 3$ respectively. From (3.1) it is evident that

$$A = \sum_{i=1}^{r} \sigma_i u_i v_i^H \tag{3.5}$$

where u_i is the ith column of U and v_i is the ith column of V; hence A does not really involve the last m-r columns of U or the last n-r columns of V. From (3.5) we may write

$$A = U^{(r)}D(V^{(r)})^H \tag{3.6}$$

where $U^{(r)}$ is an $m \times r$ matrix with columns $u_1, \ldots, u_r$, $V^{(r)}$ is an $n \times r$ matrix with columns $v_1, \ldots, v_r$ and D is the $r \times r$ positive diagonal matrix $\mathrm{diag}(\sigma_i)$. The factorization (3.6) itself is sometimes referred to as the SVD of A.

When A is a positive semi-definite $n \times n$ matrix then A may be expressed in the form

$$A = X\,\mathrm{diag}(\lambda_i)X^H \tag{3.7}$$

where the λ_i are the non-negative <u>eigenvalues</u> of A. This factorization is, of course, also the SVD of A, and we have $V = U = X$ and $\lambda_i = \sigma_i$. More generally if A is normal we have the factorization (3.7) but now the λ_i are, in general, complex. However if

$$\sigma_i = |\lambda_i| \text{ and } \lambda_i = \sigma_i \exp(i\theta_i) \tag{3.8}$$

we have

$$A = X\,\mathrm{diag}(|\lambda_i|)\ \mathrm{diag}(\exp(i\theta_i))X^H \tag{3.9}$$

and since $\mathrm{diag}(\exp(i\theta_i))$ is itself unitary, (3.9) gives the SVD of A with $U = X$, $V = \mathrm{diag}(\exp(-i\theta_i))X$, $\Sigma = \mathrm{diag}(|\lambda_i|)$. In general though the SVD of A and its eigenvalues and eigenvectors have no simple relationship.

The proof of the existence of the SVD is quite simple, though one must take care to deal satisfactorily with zero singular values and to take full account of the fact that m and n may be different. We observe first that if $A = U\Sigma V^H$ then

$$A^HA = V\Sigma^HU^HU\Sigma V^H = V(\Sigma^H\Sigma)V^H. \tag{3.10}$$

The matrix $\Sigma^H\Sigma$ is an $n \times n$ diagonal matrix; its diagonal elements are σ_i^2 $(i = 1, \ldots, r)$ augmented (if necessary) by n-r zero elements. Hence if the SVD exists, the columns v_i of the matrix V must be an orthonormal set of eigenvectors of the semi-definite

Hermitian matrix A^HA and the corresponding eigenvalues must be σ_i^2.

This provides the motivation for the proof. Let

$$A^HA = V \operatorname{diag}(\lambda_i) V^H \tag{3.11}$$

be a spectral decomposition of A^HA where the λ_i have been ordered so that

$$\lambda_1 \geq \lambda_2 \geq \dots \geq \lambda_r > 0, \ \lambda_{r+1} = \dots = \lambda_n = 0. \tag{3.12}$$

(We say <u>a</u> spectral decomposition since if there are any multiple eigenvalues, including, in particular, multiple zero eigenvalues the unitary matrix V will not be unique). Anticipating our result we write

$$\sigma_i = +\lambda_i^{\frac{1}{2}} \ (i = 1, \dots, r), \ \sigma_{r+1} = \dots = \sigma_n = 0. \tag{3.13}$$

Now (3.11) implies that

$$V^HA^HAV = \operatorname{diag}(\lambda_i) = \operatorname{diag}(\sigma_i^2) = D^2 \tag{3.14}$$

where D is an $n \times n$ non-negative diagonal matrix. If we write

$$AV = P \tag{3.15}$$

then (3.14) states that $P^HP = D^2$; denoting the columns of P by p_i this implies that

$$\left.\begin{aligned} \|p_i\|^2 &= \sigma_i^2 \quad (i = 1, \dots, r) \\ p_i &= 0 \quad\ \ (i = r+1, \dots, n) \\ p_i^Tp_j &= 0 \quad\ \ (i \neq j) \end{aligned}\right\} \tag{3.16}$$

and these relations in turn imply that if we write

$$u_i = p_i/\sigma_i \quad (i = 1, \dots, r) \tag{3.17}$$

these r vectors are orthonormal. Equation (3.15) may therefore be written in the form

$$AV = [\sigma_1 u_1 | \sigma_2 u_2 | .. | \sigma_r u_r | 0 | 0 | .. | 0] \quad (3.18)$$

and since AV is an $m \times n$ matrix this means that there are n-r null columns. Now the p_i, and therefore the u_i, are m-vectors and since $r \leq m$ we can certainly choose m-r further u_i to make up a complete orthonormal set. Let this complete set of u_r define the unitary matrix U. We cannot in general express (3.18) in the form

$$AV = [\sigma_1 u_1 | \sigma_2 u_2 | ... | \sigma_r u_r | 0u_{r+1} | 0u_{r+2} | ... | 0u_m] \quad (3.19)$$

since AV is an $m \times n$ matrix, not an $m \times m$ matrix. However, since $r \leq \min(m,n)$, whatever the relative sizes of m and n the right-hand side of (3.18) is expressible as $U\Sigma$ since we have

$$U\Sigma = [u_1 | u_2 | ... | u_m] \left[\begin{array}{ccc|c} \sigma_1 & & & \\ & \cdot & & 0 \\ & & \cdot \; \sigma_r & \\ \hline & \underbrace{0}_{r} & & \underbrace{0}_{n-r} \end{array}\right] \begin{matrix} \} r \\ \\ \} m-r \end{matrix}$$

$$= [\sigma_1 u_1 | \sigma_2 u_2 | .. | \sigma_r u_r | 0 | .. | 0] \quad (3.20)$$

where, as in (3.18), there are n-r null columns. Hence we have

$$AV = U\Sigma. \quad (3.21)$$

Notice that in this last relation we could replace U by <u>any</u> $m \times m$ matrix having $u_1, ..., u_r$ as its first r columns since when we post-multiply by Σ the last m-r columns of U are multiplied only by the zero elements in the last m-r rows of Σ. Naturally it is convenient to choose the last m-r columns of U so as to make the complete U a unitary matrix. However, when we have built up U (if necessary) in this way, (3.21) implies that

$$AA^H = U\Sigma V^H V\Sigma^H U^H = U\Sigma\Sigma^H U^H \quad (3.22)$$

so that this U must give a complete set of orthonormal vectors of the semi-definite $m \times m$ Hermitian matrix AA^H. The matrix $\Sigma\Sigma^H$ is an $m \times m$ diagonal matrix, the non-zero diagonal elements being precisely the σ_i^2 $(i = 1, \ldots, r)$.

From now on we shall denote the $r \times r$ non-singular matrix $\text{diag}(\sigma_i)$ by D_r and we have

$$A = U\left[\begin{array}{c|c} D_r & O \\ \hline O & O \end{array}\right] V^H. \tag{3.23}$$

4. ADVANTAGES OF THE SVD

The SVD is a somewhat more complicated factorization than that given by Gaussian elimination or by the QR algorithm but it has a number of advantages, perhaps chief of which is its abilit to diagnose near rank degeneracy. We now discuss this point. For convenience in the remainder of this section we assume that $m \geq n$. Since $A^H = V\Sigma^T U^H$ (N.B. Σ is real) there is no essential loss of generality.

The sensitivity of the singular values σ_i of a matrix A to perturbations in its elements is readily derived if we observ that the eigenvalues of the Hermitian matrix H of order $m + n$ defined by

$$H = \left[\begin{array}{c|c} O & A \\ \hline A^H & O \end{array}\right] \tag{4.1}$$

are $\pm\,\sigma_i$ $(i = 1, \ldots, n)$ supplemented by $m-n$ zeros. Now the Hoffman-Wielandt theorem states that if X, Y, Z are Hermitian matrices with eigenvalues x_i, y_i, z_i, each set being arranged in non-increasing order and $X-Y = Z$ then

$$\Sigma(x_i - y_i)^2 \leq \Sigma z_i^2 = \|Z\|_F^2, \tag{4.2}$$

where $\|\cdot\|_F$ denotes the Frobenius norm.

Taking

$$X = \left[\begin{array}{c|c} O & A+E \\ \hline (A+E)^H & O \end{array}\right], \quad Y = \left[\begin{array}{c|c} O & A \\ \hline A^H & O \end{array}\right], \quad Z = \left[\begin{array}{c|c} O & E \\ \hline E^H & O \end{array}\right] \tag{4.3}$$

and denoting the singular value of A+E by $\tilde{\sigma}_i$ we have

$$2\Sigma(\tilde{\sigma}_i - \sigma_i)^2 \leq 2\|E\|_F^2. \qquad (4.4)$$

If then A+E is of rank r, *i.e.* $\tilde{\sigma}_i = 0$, $i = r+1, \ldots, n$ then from (4.4)

$$\sum_{r+1}^{n} \sigma_i^2 \leq \|E\|_F^2. \qquad (4.5)$$

In other words proximity to rank deficiency of A must be reflected in a corresponding "smallness" in the appropriate number of its singular values.

There remains the danger that in practice rounding errors may lead to a fatal weakening of this result. Fortunately the algorithm in most common use for the SVD, which was discovered by Golub (1970), is extremely stable and it can be shown that the computed σ_r are the exact singular values of some matrix A+G where

$$\|G\|_2 \leq \beta^{-t} \|A\|_2 \; f(n,m), \qquad (4.6)$$

β^{-t} being the computer precision and f(n,m) a very "modest" function of n and m. Hence if A is close to rank degeneracy this will be revealed reliably <u>even by the computed σ_i</u>. In fact when A arises from physical measurements then, unless the computer precision is exceptionally low, the equivalent perturbation in A resulting from rounding errors will usually be far smaller than the original perturbation in A arising from inaccuracies in the data.

5. THE GOLUB ALGORITHM

The Golub algorithm (1970) consists of two main stages. For the sake of generality we describe the algorithm in terms of a complex matrix A. If A is real all of the unitary matrices referred to below are real and therefore orthogonal. The first stage is direct and reduces A to bi-diagonal form, *i.e.* to a matrix B in which b_{ii} and $b_{i,i+1}$ are the only non-zero elements. The required zeros are produced successively in column 1, row 1, column 2, row 2 etc. If we denote the initial matrix by A_0, then at the end of stage s the current matrix A_s will have the required zeros in rows and columns 1 to s and will be typically of the form illustrated when m = 6, n = 5, s = 2 by

$$A_2 = \begin{bmatrix} X & X & O & O & O \\ O & X & X & O & O \\ O & O & X & X & X \\ O & O & X & X & X \\ O & O & X & X & X \\ O & O & X & X & X \end{bmatrix}. \qquad (5.1)$$

The matrix A_s is then pre-multiplied by a Householder matrix of the form $I-2uu^H$ designed to produce zeros in column s+1 in elements s+2, ..., m. This replaces rows s+1 to m by appropriate linear combinations of themselves and hence leaves earlier zeros undisturbed. The resulting matrix is then post-multiplied by a second Householder matrix which is designed to produce zeros in row s+1 in elements s+3, ..., n. When m > n the need for post-multiplications will cease at an earlier stage than that for pre-multiplication. When m < n the reverse will be true.

This is the form of the algorithm as described by Golub. However when m >> n it is more convenient first to reduce A to upper-triangular form by n pre-multiplications with Householder transformations. The reduction to bi-diagonal form is then completed by alternate post-multiplication and pre-multiplication with Householder matrices starting with a post-multiplication. Since rows n+1 to m remain null after the reduction to triangular form, this alternative is much more economical when m >> n; this is frequently the case.

The reduction of the bi-diagonal form to diagonal form is iterative in character, which is to be expected since we are effectively determining eigenvalues of A^HA. In general the bi-diagonal matrix will be complex but if we pre-multiply it and post-multiply it with appropriate diagonal matrices having diagonal elements of modulus unity, we can convert it to real form without affecting its singular values. We refer to the real bi-diagonal matrix from now on as J. Observe that J^TJ is a real symmetric tri-diagonal matrix, and its eigenvalues give the σ_i^2 for A. The eigenvalues of a symmetric tri-diagonal matrix may be found very efficiently by the QR algorithm with shifts, which preserves tri-diagonal form. However it is unsatisfactory to find the σ_i in this way. If $\sigma_i \lesssim \beta^{-\frac{t}{2}} \sigma_1$ then $\sigma_i^2 \lesssim \beta^{-t}\sigma_1^2$ and σ_1^2 will therefore be present only at noise level in J^TJ. It will therefore be impossible to determine any singular value smaller than $\beta^{-\frac{t}{2}} \sigma_1$ with any accuracy at all.

Golub devised an algorithm which enabled one to obtain all the advantages of the rapid convergence of the QR algorithm as applied to J^TJ while working with J itself. Let us rename J as J_0 and write

$$M_0 = J_0^T J_0 . \tag{5.2}$$

If the implied QR algorithm with shifts is applied to M_0 we obtain a sequence of tri-diagonal matrices M_s each of which is similar to M_0. Successive M_s satisfy a relation

$$M_s = Q_s^T M_{s-1} Q_s, \tag{5.3}$$

where Q_s is an orthogonal matrix. In Golub's algorithm a sequence of bi-diagonal matrices J_s is produced such that at each stage $J_s^T J_s$ (which is symmetric and tri-diagonal) is the same as the matrix M_s which would have been derived by applying the implicit QR algorithm with shifts to M_0.

Suppose we have achieved this up to J_{s-1}. In the next stage J_s is defined by a relation

$$J_s = U_s J_{s-1} V_s, \tag{5.4}$$

where both U_s and V_s are orthogonal matrices the derivation of which will be described below. Hence

$$J_s^T J_s = V_s^T J_{s-1}^T U_s^T U_s J_{s-1} V_s$$
$$= V_s^T J_{s-1}^T J_{s-1} V_s = V_s^T M_{s-1} V_s, \tag{5.5}$$

the last relation following from the inductive hypothesis. Now from (5.3)

$$Q_s^T M_{s-1} = M_s Q_s^T, \tag{5.6}$$

while from (5.5)

$$V_s^T M_{s-1} = (J_s^T J_s) V_s^T. \tag{5.7}$$

Now Francis (1961, 1962) has shown in his establishment of the explicit QR algorithm that if H is upper-Hessenberg and

$$Q^T A = HQ^T \tag{5.8}$$

where Q is orthogonal, then the whole of H and Q are determined when the first row of Q^T is prescribed. Since J_s is to be bi-diagonal, $J_s^T J_s$ is tri-diagonal and therefore certainly upper-Hessenberg. Hence if V_s^T has the same first row as the Q_s^T which is used in the implicit QR algorithm for M_{s-1} (*i.e.* by our inductive assumption $J_{s-1}^T J_{s-1}$) then, provided J_s is bi-diagonal, $J_s^T J_s$ is the same as M_s.

Now in the QR algorithm Q_s^T is determined as the product of rotations in planes (1,2), (2,3), ..., (n-1,n) and the first row of Q_s^T is determined entirely by the (1,2) rotation. In Golub's algorithm V_s^T is also derived as the product of rotations in the (1,2), (2,3), ..., (n-1,n) planes and if we make the (1,2) rotation the same as in the QR algorithm applied to $J_{s-1}^T J_{s-1}$ the whole of V_s will be the same as Q_s and $J_s^T J_s$ will be the same as M_s.

The Golub algorithm therefore proceeds as follows. The (1,2) plane rotation involved in the implicit QR algorithm for $J_{s-1}^T J_{s-1}$ is determined and J_{s-1} is post-multiplied with it. This induces a non-zero element in the (2,1) position. Thus typically

$$\begin{vmatrix} X & X & & & \\ & X & X & & \\ & & X & X & \\ & & & X & X \\ & & & & X \end{vmatrix} \text{ becomes } \begin{vmatrix} X & X & & & \\ + & X & X & & \\ & & X & X & \\ & & & X & X \\ & & & & X \end{vmatrix}, \tag{5.9}$$

where we use a + sign to denote an induced non-zero element. The resulting matrix is then pre-multiplied with a rotation in the (1,2) plane to remove this element,thereby inducing a non-zero element in the (1,3) position. Typically

$$\begin{vmatrix} X & X & & & \\ + & X & X & & \\ & & X & X & \\ & & & X & X \\ & & & & X \end{vmatrix} \text{ becomes } \begin{vmatrix} X & X & + & & \\ 0 & X & X & & \\ & & X & X & \\ & & & X & X \\ & & & & X \end{vmatrix}, \tag{5.10}$$

where the zero indicates the position of the annihilated element and the + sign denotes the induced non-zero element. We continue with a post-multiplication in plane (2,3) annihilating element (1,3) and inducing a non-zero in (3,2) following with a pre-multiplication annihilating this (3,2) element and inducing a non-zero element in (2,4). Continuing with successive post-multiplications and pre-multiplications the extraneous element is pushed first to one side of the bi-diagonal and then to the other until finally it disappears off the bottom of the matrix. The final bi-diagonal is J_s where

$$J_s = U_s J_{s-1} V_s . \tag{5.11}$$

Here U_s and V_s are the products of the pre-multiplications and the post-multiplications respectively. The first column of V_s (*i.e.* the first row of V_s^T) comes only from the first (1,2) rotation which by design is the (1,2) rotation used in the QR algorithm applied to $J_{s-1}^T J_{s-1}$. Now to determine this plane rotation we need only compute the first column of $J_{s-1}^T J_{s-1}$ (and this has only 2 non-zero elements) and the shift which is derived from the bottom 2×2 matrix of $J_{s-1}^T J_{s-1}$.

The super-diagonal elements in successive J_s diminish rapidly, the (n-1,n) particularly so, and when it is negligible the (n,n) element gives a singular value and we can continue with the remaining bi-diagonal matrix of order n-1. Thus as in the QR algorithms we have a natural method of deflation when a singular value has been determined. As in the QR algorithm if any other super-diagonal element of any J_s becomes negligible we may split J_s into the direct sum of two smaller bi-diagonal matrices.

However there is a further device of a similar nature for reducing the volume of work. Suppose a diagonal element of a J_s is zero. Then since the product of the σ_i is equal to the product of the diagonal elements of J_s one of the σ_i must be zero. In this case the following gives a direct process for deflating J_s by removing this zero singular value. The process is sufficiently illustrated by considering a J of order 6 in which J_{33} is zero so that J is of the form

$$\begin{bmatrix} X & X & & & & \\ & X & X & & & \\ & & O & X & & \\ & & & X & X & \\ & & & & X & X \\ & & & & & X \end{bmatrix}. \qquad (5.12)$$

This matrix may be pre-multiplied successively by rotations in the planes (3,4), (3,5), (3,6). The matrix takes successively the forms

$$\begin{bmatrix} x & x & & & & \\ & x & x & & & \\ & & 0 & \underline{0} & \underline{+} & \\ & & & \underline{x} & \underline{x} & \\ & & & & x & x \\ & & & & & x \end{bmatrix}, \begin{bmatrix} x & x & & & & \\ & x & x & & & \\ & & 0 & 0 & \underline{0} & \underline{+} \\ & & & x & x & \\ & & & & \underline{x} & \underline{x} \\ & & & & & x \end{bmatrix}, \begin{bmatrix} x & x & & & & \\ & x & x & & & \\ & & 0 & 0 & 0 & \underline{0} \\ & & & x & x & \\ & & & & x & x \\ & & & & & \underline{x} \end{bmatrix}. \tag{5.13}$$

The induced non-zero elements are denoted by a + sign and those elements which were changed in the last transformation are underlined. On completion the row containing the original non-zero diagonal has become null. The matrix then splits up into the direct sum of two smaller bi-diagonal matrices of which the upper one has a zero singular value clearly exhibited.

If at any stage a diagonal element is ε we may replace this by zero, thereby changing each of the singular values by no more than ε. The above process then reduces the matrix.

6. LINEAR SYSTEMS AND LEAST SQUARES

The solution of a linear system $Ax = b$ may be analyzed very simply in terms of the SVD of A. The system is equivalent to

$$U^H Ax = U^H b \quad \textit{i.e.} \quad U^H AV(V^H x) = U^H b \tag{6.1}$$

and hence to

$$\Sigma y = c \quad \text{where } y = V^H x \text{ and } c = U^H b. \tag{6.2}$$

Equation (6.2) gives

$$\sigma_i y_i = c_i \quad (i = 1, \ldots, r), \quad 0 = c_i \quad (i = r+1, \ldots, m). \tag{6.3}$$

Hence for compatibility we require $c_i = 0$, $i = r+1, \ldots, m$. Notice that $y_{r+1}, \ldots, y_n$ are arbitrary and may be taken to be zero for minimal $\|y\|_2$ and hence for minimal $\|x\|_2$.

In practice we are unlikely to obtain a sharp cut off between zero and non-zero σ_i and we shall usually have min(m,n) values of σ_i some of which may be regarded as zero having regard to uncertainties in the initial A and rounding errors made in the reduction. Since $c = U^H b$ and U is unitary $\|c\|_2 = \|b\|_2$ and hence we are in a good position to determine whether an element c_i can be regarded as negligible. (When $c = U^H b$ for some U but we have no control on $\|U^H\|$ then this decision is much less simple.) There is no doubt that the SVD is an admirable tool for analysis of a general $m \times n$ system with inexact data.

Turning now to the least squares problem

$$Ax = b, \tag{6.4}$$

we require the vector x for which $\|r\|_2$ is a minimum, where

$$r = b - Ax. \tag{6.5}$$

When this x is not unique we require the member of the acceptable set such that $\|x\|$ is a minimum. This is referred to as <u>the minimal least squares solution</u>. Now we have

$$\|b-Ax\|_2 = \|U^H(b-Ax)\|_2 = \|c-\Sigma(V^H x)\|_2 = \|c-\Sigma y\|_2 \tag{6.6}$$

where

$$c = U^H b \text{ and } y = V^H x. \tag{6.7}$$

Since $\|y\| = \|x\|$ we require the y of minimum norm which minimizes $\|c-\Sigma y\|_2$ and noting that

$$\|c-\Sigma y\|_2^2 = \sum_{i=1}^{r} (c_i - \sigma_i y_i)^2 + \sum_{r+1}^{m} c_i^2 \tag{6.8}$$

we see that the minimal solution is

$$y_i = c_i/\sigma_i \ (i = 1, \ldots, r), \quad y_i = 0 \ (i = r+1, \ldots, n), \tag{6.9}$$

the minimal $\|r\|$ being $\left(\sum_{i=r+1}^{m} c_i^2\right)^{\frac{1}{2}}$. Again in practice there will not usually be a clear cut distinction between zero and non-zero

σ_i. Unless the c_i corresponding to small σ_i are small the system is highly unsatisfactory. If one decides to regard such a σ_i as non-zero the corresponding y_i will be very large; if on the other hand one regards it as zero (and therefore contributing to rank deficiency) the corresponding c_i is merely part of the residual. The decision will therefore make a vast difference to the solution. This is not in any sense a weakness of the SVD. On the contrary it presents the difficulties inherent in such a system with startling clarity!

7. THE PSEUDO-INVERSE

If we express the SVD in the form

$$A = U\left[\begin{array}{c|c} D_r & O \\ \hline O & O \end{array}\right]V^H \tag{7.1}$$

we see that minimal least square solution is given by

$$(7.2) \qquad y = \left[\begin{array}{c|c} D_r^{-1} & O \\ \hline O & O \end{array}\right] c \text{ or } x = \left\{V \left[\begin{array}{c|c} D_r^{-1} & O \\ \hline O & O \end{array}\right] U^H\right\} b. \qquad (7.3)$$

The matrix X in brackets in (7.3) is called the <u>pseudo-inverse</u> of A. It will readily be verified that

$$\left.\begin{array}{ll} \text{(i)} \quad AXA = A & , \text{(ii)} \quad XAX = X \\ \text{(iii)} \quad AX = (AX)^H & , \text{(iv)} \quad XA = (XA)^H \end{array}\right\}. \tag{7.4}$$

When A is a non-singular $n \times n$ matrix, A^{-1} satisfies these four relations so that the pseudo-inverse is a natural generalization of the ordinary inverse.

The pseudo-inverse is often <u>defined</u> as the matrix X satisfying these relations. This seems to me to be an extremely artificial definition and I prefer to introduce it as the matrix X such that Xb is the solution of the minimal least squares problem.

It is by no means obvious that there is an X which satisfies the four equations (7.4). However, one can deduce this from the existence of the SVD of A. Clearly X must be an $n \times m$ matrix and any such matrix is expressible in the form

$$X = VYU^H \tag{7.5}$$

where Y is an $n \times m$ matrix. (Obviously $Y = V^H XU$ is just such a matrix!) Let Y be partitioned in the form

$$Y = \underset{\quad r \quad m-r}{\left[\begin{array}{c|c} Y_1 & Y_2 \\ \hline Y_3 & Y_4 \end{array}\right]} \begin{array}{l} \}\ r \\ \}\ n-r \end{array} \tag{7.6}$$

and now substitute X in each of the four equations. The first gives

$$U \left[\begin{array}{c|c} D_r & O \\ \hline O & O \end{array}\right] V^H V \left[\begin{array}{c|c} Y_1 & Y_2 \\ \hline Y_3 & Y_4 \end{array}\right] U^H U \left[\begin{array}{c|c} D_r & O \\ \hline O & O \end{array}\right] V^H = U \left[\begin{array}{c|c} D_r & O \\ \hline O & O \end{array}\right] V^H \tag{7.7}$$

or

$$\left[\begin{array}{c|c} D_r Y_1 D_r & O \\ \hline O & O \end{array}\right] = \left[\begin{array}{c|c} D_r & O \\ \hline O & O \end{array}\right], \tag{7.8}$$

showing that $Y_1 = D_r^{-1}$. The second gives

$$V \left[\begin{array}{c|c} D_r^{-1} & Y_2 \\ \hline Y_3 & Y_4 \end{array}\right] U^H U \left[\begin{array}{c|c} D_r & O \\ \hline O & O \end{array}\right] V^H V \left[\begin{array}{c|c} D_r^{-1} & Y_2 \\ \hline Y_3 & Y_4 \end{array}\right] U^H = V \left[\begin{array}{c|c} D_r^{-1} & Y_2 \\ \hline Y_3 & Y_4 \end{array}\right] U^H \tag{7.9}$$

or

$$\left[\begin{array}{c|c} D_r^{-1} & Y_2 \\ \hline Y_3 & Y_3 D_r Y_2 \end{array}\right] = \left[\begin{array}{c|c} D_r^{-1} & Y_2 \\ \hline Y_3 & Y_4 \end{array}\right], \tag{7.10}$$

showing that $Y_3 D_r Y_2 = Y_4$. The third gives

$$AX = U \left[\begin{array}{c|c} I & D_r Y_2 \\ \hline O & O \end{array}\right] U^H = (AX)^H = U \left[\begin{array}{c|c} I & O \\ \hline (D_r Y_2)^H & O \end{array}\right] U^H, \tag{7.11}$$

and hence $D_r Y_2 = O$, *i.e.* $Y_2 = O$.

Finally the fourth gives

$$XA = V \left[\begin{array}{c|c} I & O \\ Y_3 D_r & O \end{array}\right] V^H = (XA)^H = V \left[\begin{array}{c|c} I & (Y_3 D_r)^H \\ \hline O & O \end{array}\right] V^H \qquad (7.12)$$

and hence $Y_3 D_r = O$ giving $Y_3 = O$. The relation $Y_4 = Y_3 D_r Y_2$ derived earlier therefore gives $Y_4 = O$ and

$$X = V \left[\begin{array}{c|c} D_r^{-1} & O \\ \hline O & O \end{array}\right] U^H. \qquad (7.13)$$

8. ANALYSIS IN TERMS OF THE SVD

In recent years the SVD has been used increasingly in the analysis of matrix algorithms. We may express the SVD in either of the forms

$$AV = U\Sigma \text{ or } U^H A = \Sigma V^H. \qquad (8.1)$$

If the columns of U and V are u_i and v_i, respectively, the first of these gives

$$Av_i = \sigma_i u_i \quad (i = 1, \ldots, n) \qquad (8.2)$$

and the second gives

$$u_i^H A = \sigma_i v_i^H \quad (i = 1, \ldots, n), \quad u_i^H A = O \quad (i = n+1, \ldots, m) \quad (8.3)$$

where we assume for convenience that $m \geqslant n$. (The modifications when $m < n$ are obvious.)

Now the systems u_i and v_i have the advantage of being orthonormal. Any unit m-vector b may be expressed in terms of u_i in the form

$$b = \sum_{i=1}^{m} \alpha_i u_i \text{ where } \sum |\alpha_i|^2 = 1 \qquad (8.4)$$

while any unit n-vector d may be expressed in the form

$$d = \sum_{i=1}^{n} \beta_i v_i \text{ where } \sum |\beta_i|^2 = 1. \tag{8.5}$$

If we attempt to solve

$$Ax = b \tag{8.6}$$

we see that there can be no solution if the expansion of b in terms of the u_i contains any components of $u_{n+1}, \ldots, u_m$ since Ax lies in the subspace spanned by $u_1, \ldots, u_n$ for any n-vector x. If

$$b = \sum_{i=1}^{n} \alpha_i u_i \tag{8.7}$$

then

$$x = \sum_{i=1}^{n} (\alpha_i/\sigma_i) v_i \tag{8.8}$$

which is valid provided no σ_i is zero. If any σ_i is zero then there will be a solution if and only if all corresponding α_i are zero and the corresponding components of v_i in the solution will then be arbitrary.

When solving

$$x^H A = d^H \tag{8.9}$$

the situation is simpler and we have

$$x^H = \sum_{i=1}^{n} (\bar{\beta}_i/\sigma_i) u_i^H \tag{8.10}$$

provided none of the σ_i is zero. Notice that for all n-vectors d the corresponding solution contains components of $u_1, \ldots, u_n$ only. If a σ_i is zero there will be no solution unless the corresponding β_i is zero and the solution may then contain an arbitrary component of u_i.

Returning again to equations (8.2) and (8.3) we see that when $m \geqslant n$ the right-hand null space of A is spanned by those v_i, if any, which correspond to zero values of σ_i. The left-hand null space is spanned by those u_i, if any, corresponding to zero values of σ_i, augmented by $u_{n+1}, \ldots, u_m$.

When A is an $n \times n$ matrix, expansion in terms of the u_i and v_i is particularly valuable and for many purposes has replaced expansion in terms of the left-hand and right-hand systems of eigenvectors. The latter expansions suffer not only from the fact that if A is defective the systems of eigenvectors are incomplete, but even when A is not defective several of the eigenvectors may be almost linearly dependent. If

$$Ax_i = \lambda_i x_i, \quad \|x_i\|_2 = 1, \quad i = 1, \ldots, n, \tag{8.11}$$

any unit vector b may be expressed in the form

$$b = \sum_{i=1}^{n} \alpha_i x_i, \tag{8.12}$$

but if the vectors 1, ..., r are almost linearly dependent we have a relation of the form

$$\sum_{i=1}^{r} \beta_i x_i = \varepsilon u \quad (\|u\|_2 = 1) \tag{8.13}$$

where ε is very small. Hence for the unit u we have

$$u = \sum_{i=1}^{r} (\beta_i/\varepsilon) x_i \tag{8.14}$$

and its expansion in terms of the x_i has very large coefficients. This makes analysis in terms of the x_i extremely deceptive.

9. INVERSE ITERATION

Perhaps the greatest triumph for the use of the vectors u_i and v_i is in the analysis of inverse iteration. In this context it was natural to use the systems of eigenvectors since it is expansion in this form which provided the original motivation for inverse iteration, as follows.

Let z_0 be an arbitrary unit vector and μ an approximation to the eigenvalue λ_k. In inverse iteration the sequence of iterates is defined by

$$(A-\mu I)w_{s+1} = z_s, \; z_{s+1} = w_{s+1}/\|w_{s+1}\|_2. \qquad (9.1)$$

If $z_0 = \sum_1^n \alpha_i x_i$ where x_i is the right-hand eigenvector corresponding to λ_i then it is easy to see that

$$z_s = K_s \sum_1^n (\alpha_i/(\lambda_i-\mu)^s) x_i \qquad (9.2)$$

where K_s is a scale factor. Hence if $|\lambda_k-\mu| << |\lambda_i-\mu|$ $(i \neq k)$ the vector z_s is rapidly dominated by the component of x_k. One might expect steady and rapid progress to x_k and that the residuals corresponding to successive z_s would diminish progressively.

When λ_k is an ill-conditioned eigenvalue and the approximation μ is "very good" (in the sense that it is exact for some $A+E$ where $\|E\|$ is of noise level relative to $\|A\|$) then these expectations are not realized. For almost all z_0 the first iteration leads to a z_1 with a negligible residual, but in general all subsequent z_s give much larger residuals. This phenomenon has nothing to do with rounding errors.

A complete understanding of inverse iteration evolved only slowly, an elucidation of the above mentioned phenomenon being the last step in the process. The explanation in terms of the SVD of $A-\mu I$ is particularly pleasing. We note first that

$$(A-\mu I)w_1 = z_0, \; z_1 = w_1/\|w_1\|_2, \qquad (9.3)$$

and hence

$$r_1 = (A-\mu I)z_1 = z_0/\|w_1\|_2, \; \|r_1\|_2 = 1/\|w_1\|_2. \qquad (9.4)$$

The larger the vector $\|w_1\|_2$ the smaller $\|r_1\|_2$ will be. Similarly at each subsequent iteration

$$\|r_s\|_2 = \|(A-\mu I)z_s\|_2 = 1/\|w_s\|_2. \tag{9.5}$$

We therefore have to explain why $\|w_1\|_2$ may be much larger than subsequent $\|w_s\|_2$. Now we have

$$(A-\mu I)v_i = \sigma_i u_i \quad (i = 1, \ldots, n) \tag{9.6}$$

and hence if we write

$$z_0 = \sum \alpha_i u_i, \; \sum |\alpha_i|^2 = 1 \tag{9.7}$$

$$w_1 = \sum (\alpha_i/\sigma_i) v_i. \tag{9.8}$$

By assumption $A+E-\mu I$ is exactly singular and hence the smallest singular value σ_n will be at noise level relative to $\|A\|$. Unless α_n is very small $\|w_1\|_2 \doteq |\alpha_n|/\sigma_n$ and will be very large. For almost all z_0 therefore, $\|w_1\|_2$ will be very large and $\|r_1\|_2$ will be correspondingly small.

Now z_1 arises from normalizing w_1 and since w_1 is, in general, dominated by $(\alpha_n/\sigma_n)v_n$ we shall have

$$z_1 = \beta v_n + \sum_1^{n-1} \varepsilon_i v_i \tag{9.9}$$

where β is of order unity and all the ε_i are negligible. Unfortunately we have z_1 expressed in terms of the v_i. In order to solve $(A-\mu I)w_2 = z_1$ we require it to be expressed in terms of the u_i. If we write

$$v_i = \sum_j p_{ij} u_j, \; \sum_j |p_{ij}|^2 = 1, \; (i = 1, \ldots, n) \tag{9.10}$$

then

$$z_1 = \beta[\sum_j p_{nj}u_j] + \sum_{i=1}^{n-1} \varepsilon_i \ (\sum_j p_{ij}u_j), \qquad (9.11)$$

and z_1 is a unit vector in which the coefficient of u_n is

$$\beta p_{nn} + \sum_{i=1}^{n-1} \varepsilon_i p_{in}. \qquad (9.12)$$

Since the ε_i are all small, if z_1 is to have a respectable coefficient of u_n we require that βp_{nn} is not small, *i.e.* that p_{nn} is not small.

Unfortunately if μ corresponds to an ill-conditioned eigenvalue p_{nn} will be small. For we have

$$(A-\sigma_n u_n v_n^H)v_n = \mu v_n \qquad (9.13)$$

$$u_n^H(A-\sigma_n u_n v_n^H) = \mu u_n^H \qquad (9.14)$$

and hence v_n and u_n are the right-hand and left-hand <u>eigenvectors</u> of $A-\sigma_n u_n v_n^H$ corresponding to its exact eigenvalue μ. Now our assumptions are that σ_n is negligible and μ is an approximation to an ill-conditioned eigenvalue of A. Here μ must also be an ill-conditioned eigenvalue of the neighbouring matrix $A-\sigma_n u_n v_n^H$. It is well known that the left-hand and right-hand eigenvectors corresponding to an ill-conditioned eigenvalue are almost orthogonal and hence p_{nn} is small!

We have the ironical situation that the z_1 produced by the first iteration is virtually certain to be extremely deficient in its component of u_n, the very component which must be of reasonable size if w_2 is to be large and $\|r_2\|$ to be small.

10. REFERENCES

Francis, J.G.F. (1961, 1962) "The QR transformation, Parts I and II", *Computer J.*,**4**, 265-271, 332-345.

Golub, G.H. and Kahan, W. (1965) "Calculating the singular values and pseudo inverse of a matrix", *J. SIAM. Numer. Anal.*, Ser. B, **2**, 205-224.

Golub, G.H. and Reinsch, C. (1970) "Singular value decomposition and least squares solutions", *Num. Math.*, **14**, 403-420.

Lawson, C.L. and Hanson, R.J. (1974) "Solving Least Squares Problems", Prentice-Hall, New Jersey.

II. 2

SINGULAR VALUE DECOMPOSITION IN MULTIVARIATE ANALYSIS

C.F. Banfield

(Rothamsted Experimental Station, Harpenden)

1. INTRODUCTION

The singular value decomposition of a real matrix A having m rows and n columns ($m \geq n$) can be defined as

$$A = USV'$$

where $U'U = V'V = VV' = I_n$. S is a diagonal matrix holding the n positive singular values s_i, i=1, ..., n which are sometimes referred to as the zeroes of A. The orthonormal matrix U (m×n) holds the latent vectors associated with the n largest latent roots of the product AA' and similarly V (n×n) holds the latent vectors of $A'A$. The singular values are the positive square roots of the latent roots of $A'A$.

Because of these properties and the fact that the decomposition is a most reliable method for determining the effective rank of a matrix (see Wilkinson, Chapter II.1), the singular value decomposition is a suitable basis for many multivariate analysis techniques where latent roots and vectors or underlying dimensionality are the key feature. This is particularly true of ordination methods where representations of samples or individuals as configurations of points in a small dimensional space are required. Efficient algorithms for obtaining the latent roots and vectors of symmetric matrices are well known but the singular value decomposition can operate on rectangular matrices which need not be square or symmetric. This makes possible the analysis of asymmetric matrices and the rectangular submatrices of symmetric matrices as is necessary in multivariate methods such as correspondence analysis and metric multidimensional unfolding. For those with computers of limited storage or lacking double precision arithmetic the singular value decomposition can prove the most, if not the only, efficient method of obtaining a multiple regression or a principal component analysis if large numbers of variates are involved. The decomposition is also useful for computing generalized or pseudo-inverses, which often occur in practical multivariate analysis, and can play an important rôle in solving other least-squares problems (see for example Golub (1969)).

2. MULTIVARIATE TECHNIQUES USING SINGULAR VALUE DECOMPOSITION

2.1 Procrustes Rotational Fitting

Schönemann and Carroll (1970), and independently Gower (1971), showed that to match two configurations of n points located in a p-dimensional space the singular value decomposition could be used to obtain the orthogonal rotation matrix that gives the optimum least-squares fit. The sum of the squared distances between the pairs of corresponding points of the two configurations is minimized by fixing one configuration and rotating the other about their common centroid. Explicitly Gower (1971) stated that if X and Y are matrices having n rows and p columns and holding the coordinates of the two configurations each centred about the origin, then the decomposition

$$X'Y = USV'$$

is used to obtain the orthogonal rotation matrix $H=VU'$ which minimizes the residual sum of squares, Trace $[(X-YH)'(X-YH)]$.

Such a fitting process is usefully employed when comparing the results of two different multivariate analyses of the same set of variates each having produced a configuration of n individuals in p dimensions. The residual sum of squares, which Gower calls M^2, quantifies the consistency of the two analyses. Similarly the technique can be used to compare the results of the same analysis of two different sets of variates measured on the same n individuals, as might happen when two assessments are made of the same objects (see for example, Banfield and Harries (1975)).

Gower (1975) has since generalized the fitting technique to rotate several configurations simultaneously to fit a single "consensus" configuration that has been produced by averaging the configurations.

2.2 Correspondence Analysis

Correspondence analysis is a method by which both the individuals and the discrete variates observed for those individuals can be simultaneously ordinated and often displayed on a single map. It was originally developed by Hirschfeld (1935) and Fisher (1940) and then rediscovered by Benzécri (1969) and his colleague Escofier-Cordier (1969) who named the technique "*L'Analyse Factorielle des Correspondances*". Hill (1973, 1974) more recently has retitled the method "reciprocal averaging", which seems more appropriate as the coordinates obtained for the individuals are averages of the coordinates obtained for the variates and reciprocally the variate-coordinates are the averages of the individual-coordinates.

Hill (1973) specified this procedure as

$$x = R^{-1}Ay$$

$$y = C^{-1}A'x$$

where A is the n×v matrix of v discrete variates observed for n individuals, R^{-1} and C^{-1} are diagonal matrices holding the reciprocals of the n row and v column sums, respectively, of A, and x and y are column vectors eventually holding the coordinates of the individuals and variates, respectively, in any one dimension. Considering x only, the next value, say x_2, in the iterative procedure can be expressed as

$$x_2 = R^{-1}AC^{-1}A'x_1$$

where x_1 is the previous value of x. Hill (1973) shows that the whole set of v x-vectors that are the solution to the reciprocal averaging problem can be obtained by the singular value decomposition of

$$R^{-\frac{1}{2}}AC^{-\frac{1}{2}} = USV'$$

in which case X, the n×v matrix of coordinates for the individuals having the v x-vectors as its columns, is given by X=RU. Similarly it is shown that Y, the v×v matrix of coordinates for the variates having the v y-vectors as its columns, is given by Y=CV. The first singular value in S is necessarily equal to one and the corresponding first columns of both X and Y are constant.

The coordinates in X and Y are scaled such that they can be displayed on the same map by plotting the pairs of columns of X and Y corresponding to the largest singular values held in S, excluding the first. The close proximity of particular variates to particular individuals on the map can indicate the correspondence of that variate to that individual.

2.3 Analysis of Asymmetric Matrices

Given a square asymmetric matrix A, it can be partitioned into a symmetric and a skew-symmetric part by

$$A = \tfrac{1}{2}(A+A') + \tfrac{1}{2}(A-A').$$

If the rows and columns of A classify the same entities then it is well known that the values of $\tfrac{1}{2}(A+A')$ expressing the symmetric associations between the entities can be represented, perhaps only approximately, by distances on a map using multidimensional scaling (see for example, Kruskal (1964) and Gower (1966)). However, Gower (1977) has recently shown that the skew-symmetric associations may also be represented approximately on a map. Such a technique can prove useful when analysing matrices of confusion coefficients arising in sensory assessment, the response to stimulus i when following stimulus j often being different from that of j following i, or when the attraction of entity i to entity j is different from that of j to i, as can happen in biological, ecological and geological situations.

Gower (1977) notes that the singular value decomposition of the skew-symmetric part is

$$\tfrac{1}{2}(A-A') = USV' = USJU'$$

where S contains pairs of equal singular values $s_1, s_1, s_2, s_2,$ with a final 0 if n, the order of A, is odd, and J is the elementary block-diagonal skew-symmetric orthogonal matrix made up of 2×2 diagonal blocks of the form $\begin{pmatrix} 0 & 1 \\ -1 & 0 \end{pmatrix}$, with a final unit diagonal if n is odd. The Eckart-Young theorem (1936) then shows that the columns of U corresponding to the k largest pairs of singular values will give the best rank-2k approximation of $\tfrac{1}{2}(A-A')$. Further if the first pair of singular values are larg[e] compared to the rest then the first pair of columns of U give the coordinates of the n entities on a map such that the areas of triangles made by pairs of entities with the origin will give a good approximation of the skew-symmetric value between that pair of entities.

2.4 *Metric Multidimensional Unfolding*

Given the Euclidean distances between two distinct sets of p and q points, metric unfolding attempts to locate the points in a joint space of small dimension whilst giving good approximation to the inter-point distances. This should be compared with metric scaling (see for example, Gower (1966)) which is applicable when all the inter-point distances for the combined (p+q)-set are given. In its non-metric form, unfolding originated in the work of Coombs (1958) and Bennett (1956), the applications being to psychological data where only the rank order of the distances need be preserved. Later Ross and Cliff (1964) and Coombs (1964) considered the metric form.

It was shown that by factorizing the doubly-centred squared distance matrix a crude set of coordinates A ($p \times k$) and B ($q \times k$) could be obtained for the p-set and q-set of points, respectively, in a k-dimensional space. If D^2 is the $p \times q$ matrix of squared distances doubly centred and $C = -\frac{1}{2}D^2$, then $C = USV'$ is the singular value decomposition of C giving coordinates $A=U_k$ and $B=V_kS_k$, where the suffix indicates only the first k columns are considered. However, these coordinates do not have a common origin and are not unique in that

$$C = A_0 T^{-1} T B_0'$$

where $A_0 = AT$, $B_0 = BT^{-1}$, and T is any non-singular transformation, is an equally feasible factorization. Schönemann (1970) suggests a form of T such that A_0 and B_0 are the coordinates of the two sets of points in a joint space with common origin. T is however only determined up to an orthogonal rotation.

2.5 *Multivariate Analysis of the Behaviour of Mixtures*

Tso (1977) considers a linear regression model with constraints when analysing the responses to t tests performed on q mixtures formed from blending r substances in different proportions. If the $q \times t$ matrix Y holds the responses to the tests and $q \times r$ matrix X holds the proportions of the substances comprising each mixture then

$$Y = 1c' + XM + E$$

is the regression model, where c is a vector of test constants, 1 is a vector of ones and E is a $q \times t$ matrix of errors assumed to be uncorrelated and distributed $N(0,1)$. The vector c can be removed from the model by appropriate subtraction of column mean from both Y and X.

The $r \times t$ matrix M of coefficients in the above model is considered to be of less than full rank because linear relationships between the responses are a consequence of the tests mutual dependence on a small number, s say, of underlying "factors" which are some linear combination of the component substances. The rank of M is consequently $s<\min(r,t)$. Tso (1977) as a consequence shows that the matrix of unconstrained least-squares estimated regression coefficients, $\hat{M}$ must be further modified by obtaining the singular value decomposition of $\hat{Y}$, the predicted value of Y,

$$\hat{Y} = USV'$$

and then using the first s columns of V, corresponding to the s largest singular values, to obtain the modified coefficients $\overset{*}{M} = \hat{M}V_sV_s'$. The corresponding predicted values of Y are then

$$\overset{*}{Y} = X\overset{*}{M} = X\hat{M}V_sV_s' = \hat{Y}V_sV_s'.$$

2.6 Canonical Correlation Analysis

Given a set of p variates and a set of q variates $(p \geq q)$, each measured on n individuals, canonical correlation analysis finds the q linear combinations of the p-set that have maximum correlations with q linear combinations of the q-set. The technique was originally given by Hotelling (1936), and much later Golub (1969) described an alternative method using singular value decomposition which also obtains the canonical correlatio and the loadings for the linear combinations.

If X $(n \times p)$ and Y $(n \times q)$ are matrices having the p and q-se as their columns, respectively, then the solution to the equations

$$\left.\begin{aligned} X'YQ &= X'XPC \\ \text{and } Y'XP &= Y'YQC \end{aligned}\right\} \qquad (2.6.1)$$

will give the canonical correlations in the diagonal matrix

C ($q \times q$), the loadings for the p-set in matrix P ($p \times q$) and for the q-set in Q ($q \times q$).

Letting $X'X = D'D$ and $Y'Y = E'E$, where D and E are the respective "square-root" matrices, and normalizing so that

$$\left.\begin{aligned} P'(X'X)P &= I_q \\ \text{and } Q'(Y'Y)Q &= I_q \end{aligned}\right\} \qquad (2.6.2)$$

we obtain from equations (2.6.1) and (2.6.2) that

$$P'X'YQ = C$$

$$\text{therefore } P'DD^{-1}X'YE^{-1}EQ = C.$$

Now if $U = D'P$ and $V = EQ$, which from equations (2.6.2) are both orthogonal, then

$$D^{-1}X'YE^{-1} = UCV'.$$

So the singular value decomposition of $D^{-1}X'YE^{-1}$ gives the canonical correlations, and the loadings $P = D^{-1'}U$ and $Q = E^{-1}V$.

2.7 Multiplicative Models Fitted to Two-way Tables

It is often required to fit a multiplicative model to a two-way table, Y having m rows and n columns ($m > n$). Such a model

$$y_{ij} = \mu + \alpha_i + \beta_j + \gamma_{ij} + \varepsilon_{ij}$$

($i = 1, \ldots m$, $j = 1, \ldots, n$) has interaction parameters, γ_{ij} which are expressed as the sum of several successive multiplicative orthogonal contrasts. Fisher and Mackenzie (1923) first considered fitting such a model and Gollob (1968) shows how the interaction parameters can be estimated using singular value decomposition.

The least-squares estimates of μ, α_i, β_i and γ_i are given by

$$\left.\begin{aligned} \hat{\mu} &= y_{..} \\ \hat{\alpha}_i &= y_{i.} - y_{..} \\ \hat{\beta}_j &= y_{.j} - y_{..} \\ \text{and} \quad \hat{\gamma}_{ij} &= y_{ij} - y_{i.} - y_{.j} + y_{..} \end{aligned}\right\} \tag{2.7.1}$$

Define Z as the doubly centred m×n matrix formed from the two-way table such that

$$z_{ij} = \hat{\gamma}_{ij} = y_{ij} - y_{i.} - y_{.j} + y_{..} \tag{2.7.2}$$

then the singular value decomposition of Z yields

$$Z = USV' = CD'$$

$$\text{where } C = US^{\frac{1}{2}} \text{ and } D = VS^{\frac{1}{2}}.$$

Now expanding the matrix Z and combining with equations (2.7.1) and (2.7.2) we obtain

$$y_{ij} = \hat{\mu} + \hat{\alpha}_i + \hat{\beta}_j + \sum_{k=1}^{n-1} c_{ik} d_{jk}$$

which is a model where the interaction parameter has been completely factored. Mandel (1971) and Corsten and van Eijnsberger (1972, 1973) show how the statistical significance of these factors may be assessed.

3. TECHNIQUES ON SMALL MACHINES USING SINGULAR VALUE DECOMPOSITION

3.1 Multiple Regression

Given a dependent variate y and v independent variates $x_1, x_2, \ldots, x_v$ each comprising n observations, we can express least-squares multiple regression as estimating a vector b of length v such that the residual sum of squares z is minimized, where

$$z = (y-Xb)'(y-Xb)$$

and X is the matrix having the x_i, i=1, ..., v as its columns.

The conventional approach to estimating b_{min} requires the formation of the sum-of-squares and products matrix, $X'X$, the solution of normal equations and the computation of the residual sum of squares, all of which on small computers lacking double precision arithmetic can introduce unacceptable inaccuracies. Nash and Lefkovitch (1976a) suggest an alternative method employing singular value decomposition in which sums of squares and products are unnecessary.

They show that b_{min} can be estimated by

$$b_{min} = (X'X)^{-1}X'y = V(S^{+})^{2}V'X'y = X^{+}y$$

where X^{+} is the Moore-Penrose pseudo-inverse of X such that $X^{+} = VS^{+}U'$ where $X = USV'$ is the singular value decomposition of X, and

$$s_i^{+} = \begin{cases} 1/s_i, & s_i > 0 \\ 0 & s_i = 0. \end{cases}$$

A suitable compact algorithm to obtain the decomposition on a small machine is also described by Nash and Lefkovitch (1976a) which first reduces X to triangular form by a sequence of Given's plane rotations then row-orthogonalizes the triangular form. The multiple regression employing this algorithm can be sequentially updated. It should be noted however, that the algorithm does not use the more preferable square-root free forms of the Given's plane rotations (see for example, Gentleman(1 (1973).)

3.2 Principal Components Analysis

Given a data matrix X of v variates measured on n individuals the underlying algorithm of principal components analysis is the computation of the latent roots and vectors of the v×v sums-of-squares and products matrix, $X'X$. However, as mentioned above, the computation of sums of squares and products is inadvisable on a small computer that has only single precision accuracy so an alternative method using singular value decomposition may be used.

The decomposition of X gives

$$X = USV' \quad \text{with } U'U = V'V = I$$

and we know that the columns of V are the latent vectors of $X'X$ because $X'X = VSU'USV' = VS^2V'$, so they are the required principal component loadings, the first column of V correspondi[ng] to the first principal component and so on. Again the algorith[m] described by Nash and Lefkovitch (1976a) may be used to obtain the decomposition on a small machine and so compute the principal components.

4. STATISTICAL SOFTWARE FOR SINGULAR VALUE DECOMPOSITION

Several algorithms have been published for the computation of a singular value decomposition. Golub and Reinsch (1970) give an ALGOL procedure, SVD which reduces the given $m \times n$ ($m \geq n$) matrix A to bi-diagonal form using Householder transformations and then uses a QR-algorithm to obtain the singular values of the bi-diagonal form. It does not however, order the singular values according to their relative magnitudes and if $m < n$ A must first be transposed. Nash and Lefkovitch (1976b) give documented listings of the algorithm mentioned in section 3.1 in both ANSI Standard FORTRAN and a subset of BASIC for use on the Data General NOVA and Hewlett-Packard 9830A machines. These are extensions of an earlier algorithm by Nash (1975) whi[ch] employs post-multiplication by plane rotations to orthogonalize the columns of the given matrix. Earlier algorithms by Forsyth[e] and Henrici (1960), Hestenes (1958) and Kogbetliantz (1955) als[o] use plane rotations.

The GENSTAT program (Nelder *et al.* (1977)), which has its own user language although is written in ANSI Standard FORTRAN, adopts the algorithm of Golub and Reinsch (1970) mentioned abov[e] and not only provides singular value decomposition as an individual matrix operation but uses the method to provide many of the multivariate techniques described earlier. Procrustes rotational fitting is provided as a single "directive" and the analysis of asymmetric matrices, correspondence analysis, canonical correlation analysis and metric unfolding are available as macros of the language. GENSTAT is available on many machine ranges including ICL 470/472/475, IBM 360/370, UNIVAC 1108, CDC 7600 and ICL 1900.

5. ACKNOWLEDGEMENT

I wish to thank Mr. John Gower for his very helpful suggestions as to the contents of this paper.

6. REFERENCES

Banfield, C.F. and Harries, J.M. (1975) "A technique for comparing judges' performance in sensory tests", *J. Fd. Technol.*, **10**, 1-10.

Bennett, J.F. (1956) "Determination of the number of independent parameters of a score matrix from the examination of rank orders", *Psychometrika*, **21**, 383-393.

Benzécri, J.P. (1969) "Statistical analysis as a tool to make patterns emerge from data", in "Methodologies of Pattern Recognition", Ed. S. Watanabe, pp. 35-60, Academic Press, New York.

Coombs, C.H. (1958) "Inconsistencies of preferences in psychological measurement", *J. Experi. Psychol.*, **55**, 1-7.

Coombs, C.H. (1964), "A Theory of Data", New York, Wiley.

Corsten, L.C.A. and van Eijnsbergen, A.C. (1972) "Multiplicative effects in two-way analysis of variance", *Statistica Neerlandica*, **26**, 61-68.

Corsten, L.C.A. and van Eijnsbergen, A.C. (1973) Addendum to "Multiplicative effects in two-way analysis of variance", *Statistica Neerlandica*, **27**, 51.

Eckart, C. and Young, G. (1936) "The approximation of one matrix by another of lower rank", *Psychometrika*, **1**, 211-218.

Escofier- Cordier, B. (1969) "L'analyse factorielle des correspondances", *Cah. Bur. Univ. Rech. opér. Univ. Paris*, **13**.

Fisher, R.A. (1940) "The precision of disciminant functions", *Ann. Eugen. Lond.*, **10**, 422-429.

Fisher, R.A. and Mackenzie, W.A. (1923) "Studies in crop variation, II, The manurial response of different potato varieties", *J. agric. Sci.*, **13**, 311-320.

Forsythe, G.E. and Henrici, P. (1960) "The cyclic Jacobi method for computing the principal values of a complex matrix", *Proc. Amer. Math. Soc.*, **94**, 1-23.

Gentleman, W.M. (1973) "Least squares computations by Givens transformations without square roots", *J. Inst. Maths. Applics.*, **12**, 329-336.

Gollob, H.F. (1968) "A statistical model which combines features of factor analytic and analysis of variance techniques", *Psychometrika*, **33**, 73-115.

Golub, G.H. (1969) "Matrix decompositions and statistical calculations", in "Statistical Computing" Eds. R.C. Milton and J.A. Nelder, pp. 365-397, Academic Press, New York.

Golub, G.H. and Reinsch, C. (1970) "Singular value decompositio and least squares solutions", *Numer. Math.*, **14**, 403-420.

Gower, J.C. (1966) "Some distance properties of latent root and vector methods in multivariate analysis", *Biometrika*, **53**, 325-338.

Gower, J.C. (1971) "Statistical methods of comparing different multivariate analyses of the same data", in "Mathematics in the Archaeological and Historical Sciences", Eds. F.R. Hodson, D.G. Kendall, P. Tautu, pp. 138-149, Edinburgh University Press.

Gower, J.C. (1975) "Generalized Procrustes Analysis", *Psychometrika*, **40**, 33-51.

Gower, J.C. (1977) "The analysis of asymmetry and orthogonality in "Recent Developments in Statistics", Ed. J. Barra, North Holland, Amersterdam.

Hestenes, M.R. (1958) "Inversion of matrices by biorthogonalization and related results", *J. Soc. Indust. Appl. Math.*, **6**, 51-90.

Hill, M.O. (1973) "Reciprocal Averaging: An eigenvector method of ordination", *J. Ecol.*, **61**, 237-249.

Hill, M.O. (1974) "Correspondence analysis: a neglected multivariate method", *Appl. Statisti.*, **23**, 340-354.

Hirschfeld, H.O. (1935) "A connection between correlation and contingency", *Proc. Camb. Phil. Soc.*, **31**, 520-524.

Hotelling, H. (1936) "Relations between two sets of variates", *Biometrika*, **28**, 321-377.

Kogbetliantz, E.G. (1955) "Solution of linear equations by diagonalization of coefficients matrix", *Quart. Appl. Math.*, **13**, 123-132.

Kruskal, J.B. (1964) "Multidimensional scaling by optimising goodness-of-fit to a non-metric hypothesis", *Psychometrika*, **29**, 1-27.

Mandel, J. (1971) "A new analysis of variance model for non-additive data", *Technometrics*, **13**, 1-18.

Nash, J.C. (1975) "A one-sided transformation method for the singular value decomposition and algebraic eigenproblem", *Computer J.*, **18**, 74-76.

Nash, J.C. and Lefkovitch, L.P. (1976a) "Principal Components and regression by singular value decomposition on a small computer", *Appl Statist.*, **25**, 210-216.

Nash, J.C. and Lefkovitch, L.P. (1976b) "Programs for sequentially updated principal components and regression by singular value decomposition", Mary Nash Information Services, 188 Dagmar, Vanier, Ontario, Canada, KIL 5T2.

Nelder, J.A. and members of the Rothamsted Statistics Department(1977) "GENSTAT Manual", Rothamsted Experimental Station, Harpenden, UK.

Ross, J. and Cliff, N. (1964) "A generalization of the interpoint distance model", *Psychometrika*, **29**, 167-176.

Schönemann, P.H. (1970) "On metric multidimensional unfolding", *Psychometrika*, **35**, 349-366.

Schönemann, P.H. and Carroll, R.M. (1970) "Fitting one matrix to another under choice of a central dilation and a rigid motion", *Psychometrika*, **35**, 245-255.

Tso, M.K.S. (1977) "Multivariate analysis of the behaviour of mixtures", Unpublished report Shell Research Limited, Chester, UK.

II. 3

SOFTWARE FOR SPARSE MATRICES

J.K. Reid

(Computer Science and Systems Division, AERE Harwell, Oxfordshire)

1. INTRODUCTION

Sparse problems arise in a wide variety of application areas, including management science, power systems analysis, surveying, circuit theory and structural analysis, for example. In fact any problem that involves a very large number of variables must have some simplifying feature if it is to be tractable for computer solution and this feature is almost always sparsity. Known relationships can usually be expressed so that each involves only a small number of variables.

To exploit this sparsity it is necessary to use code that avoids storing the zeros and operating with them or at least mostly does so. The code must also avoid an unreasonable increase in the number of non-zeros when some transformation, such as a step of Gaussian elimination, is applied to the problem. These aims can sometimes be achieved quite simply, as for example in the successive over-relaxation method for iterative solution of linear equations. At other times, for example when using Gaussian elimination to solve a general sparse set of linear equations, full advantage of sparsity can only be taken with the aid of sophisticated data structures and quite complicated code. In such a case the availability of reliable code that is easy to use and well documented is essential. Our aim here is to review those algorithms that are suitable for implementation in library software now or in the near future. There is a place for library software even in apparently simple cases such as successive over-relaxation (SOR). Here good algorithms exist (Carré 1961; Reid 1966) for choosing the relaxation parameter, but my belief is that they are not widely used because they are not present in libraries.

The user interface is a key problem in the design of successful and widely used software. If the code being called is complicated, it is important that the user is isolated from this complication and does not need himself to become a top expert in the field for which the software was written. He needs his time to become a top expert in chemical engineering or whatever and not in sparsity. A good interface is equally important in an apparently simple case such as choosing the SOR relaxation parameter, for there is otherwise every likelihood of the user writing his own ad hoc procedure. For these reasons we devote special attention to the user interface in this paper.

2. DIRECT SOLUTION OF LINEAR SETS OF EQUATIONS

2.1 Fixed bandwidths

It is very straight-forward to exploit the sparsity of band matrices of fixed band-width and codes to do so have been available since the earliest days of computing. If the diagonal elements can be used in order as pivots (for example when the matrix is symmetric and positive-definite or when it is diagonally dominant) then no fresh non-zeros (fill-ins) are created outside the band structure. Otherwise partial pivoting by rows or columns may be used to ensure a stable factorization and the factors will have about 50% more non-zeros than those resulting from pivoting down the diagonal. Most libraries contain codes for the general band matrix and the symmetric positive-definite band matrix.

2.2 Variable bandwidths (profiles)

Pivoting straight down the diagonal also preserves the "profile", that is it avoids any fill-ins in front of the first non-zero in each row and in front of the first non-zero in each column. Profiling subroutines are very little more complicated than fixed-band ones (see Jennings (1977),155-156) and they often give worthwhile gains since in practice there is usually significant variation in band-width. The computational gain can be achieved within a fixed-band storage pattern by testing for zeros, but really a packed storage scheme should be used. This packed storage is slightly more cumbersome for the user. The Harwell code (MA15, Reid 1972) attempts to alleviate this by requiring the rows to be passed across one-by-one through subroutine calls, but it is probably no easier to write such a subroutine than to master the row-by-row packed storage scheme. Corr and Jennings (1976) have published such code.

2.3 Out of main storage

Band routines can operate very successfully even when main storage is not available for holding the whole factorized form. If there is room for the active part of the matrix then no more transfers out of main store should be needed than those that write the factors. It seems to me to be important that library software exploits this feature. Matrix elements must be accessed in such a way that a virtual memory system will not "thrash" when there is really plenty of real memory for the active part. I am not clear whether virtual memory systems will eventually become so common-place and successful that such code will suffice for all problems. For the present I am convinced that there is a need for code including explicit input-output statements. For the symmetric positive-definite case the Harwell subroutine MA15 has all these features.

2.4 Ordering for bandwidth

Sometimes it is easy to order the variables manually for small band-width, but often it is preferable to have an automatic procedure do this. Good algorithms, for example that of Gibbs, Poole and Stockmeyer (1976), are now available but I do not know of any readily available software. It is to be hoped this will be remedied soon.

2.5 General sparse techniques

A quite separate approach is to treat the sparsity in a general way by using a data structure that holds only the non-zeros and permits insertions and deletions as fill-ins and eliminations occur. For unsymmetric cases a pivotal strategy that takes account of both stability and likely fill-in is needed. A criterion on the relative size of the pivot compared to the largest element in its row or column has proved adequate for stability and the Markowitz criterion of minimizing the product of number of non-zeros in the pivot row with number of non-zeros in the pivot column (excluding the pivot in each case) has proved a very successful sparsity criterion. A subroutine with these strategies and using a doubly linked list for internal storage has been in use for some years at Harwell as subroutine MA18 and it is present in the Numerical Algorithms Group (NAG) library as subroutines F03AJF, F03AKF and F04APF. Duff (1977) has recently provided a replacement, MA28, which is both faster and more convenient to use (see section 7). The speed is gained by using an ordered list storage scheme and various other less fundamental changes. A similar storage scheme is used by the IBM SL-Math routines and that at the Technical University of Denmark (Zlatev and Thomsen 1976).

2.6 Permutation to block triangular form

If the matrix of a set of linear equations can be permuted to block triangular form then there is every advantage in doing so since then a block forward (or backward) substitution can be used and only the diagonal blocks need be factorized. The block triangular form is essentially unique and can be obtained very economically in two stages. The first stage involves using unsymmetric permutations to place non-zeros on the diagonal. Modern implementations by Duff (1978) and Gustavson (1976) of the algorithm of Hall (1956) usually involve a number of operations that is a small multiple of the number of non-zeros τ, although examples exist of computational complexity $n\tau$. The algorithm of Hopcroft and Karp (1973) has computational complexity $n^{\frac{1}{2}}\tau$ but Duff (1978) finds that even with very careful implementation this algorithm executes significantly slower on all his test examples except those constructed to demonstate the $O(n\tau)$ complexity of the Hall algorithm. The second stage involves symmetric permutations to the block triangular form and the algorithm of Tarjan (1972) is extremely satisfactory, requiring only $O(n+\tau)$ operations. It is described by Duff and Reid (1977), who give Fortran code to implement it. The storage needed for both stages is only $O(n)$ locations beyond that required for the non-zero pattern of the matrix. They are included as an optional first stage in the Harwell package MA28.

2.7 Symmetric general sparse matrices

Perhaps surprisingly, techniques are not so well established for the direct solution of symmetric general sparse sets of equations. A number of codes exist for choosing diagonal pivots by the criterion of minimal degree, that is at each pivotal step the diagonal element with least non-zeros in its row is chosen for pivot. This is the restriction of the Markowitz strategy to choosing diagonal pivots. Examples of such codes are Harwell's MA17 (see Reid 1972) and that in IBM's SL Math. These routines, however, are like their un-symmetric counterparts in that choosing all the pivots and following all the fill-ins is very expensive compared with repeating the factorization when these are known. However George and McIntyre (1976) have shown that this pivotal choice can be made much more cheaply, in fact with only O(n logn) operations and O(n) extra storage. George (1977) has suggested a way of automating the method of nested dissection, also without great expense in storage or computation. Since this gives favourable operation counts in problems arising from partial differential equations, it seems likely to be a powerful contender in the general case.

Stable factorizations of non-definite symmetric matrices can be obtained by including some 2x2 pivots (see Duff, Munksgaard, Nielsen and Reid 1977) and Munksgaard (Technical University of Denmark) has written code to implement this. It is competitive with Harwell's MA17, but I anticipate future codes for the positive definite case will be much faster.

3. ITERATIVE SOLUTION OF LINEAR SETS OF EQUATIONS

3.1 Introduction and discussion of SOR

Iterative methods are capable of solving much larger sets of equations than can direct methods and they have been in use for far longer. However little has appeared in any of the libraries. I believe that the reasons for this are the relative simplicity of the iterations themselves and the need for cunning on the data handling side of each iteration. To hand control over to a library subroutine at first seems neither necessary nor practical. But even successive over-relaxation (SOR), one of the simplest successful algorithms, requires estimation of the relaxation parameter ω. It is quite easy to invent ad hoc procedures for doing this, but they are unlikely to be as good as established techniques (for example Carré 1961, Reid 1966) where the matrix is symmetric and possesses Young's (1954) "Property A". The increasing availability of virtual storage makes storage of the matrix in packed form for hand-over to a library subroutine practical for a wider range of problem sizes. Such a subroutine could check that Property A holds and could exploit the SOR ability to perform several iterations for each transfer to actual memory provided the matrix is banded (if the semi-bandwidth is s then $x_i^{(k+1)}$ can be calculated as soon as $x_{i+s}^{(k)}$ is available since it does not require $x_j^{(k)}$, $j>i+s$). The case for such a library subroutine seems strong.

3.2 Stone's semi-implicit iteration

Another method for which library software seems very desirable is the semi-implicit method of Stone (1968) for solving sets of finite-difference equations. It is often very successful and is not restricted to problems whose matrix is symmetric. It is rather more complicated than SOR, and it is therefore more obviously a candidate for library software. As with SOR, there seems no difficulty provided we can allocate storage, perhaps in virtual memory, for the matrix. In both cases there is merit in arranging for a call to perform only a few iterations rather than continue to convergence for then a non-linear problem can be treated by the user altering the matrix between calls.

3.3 Conjugate gradients

Another reason for little standard software appearing is that iterative methods often exploit particular features of the problem under solution. The methods of both the last sub-sections, for instance, make assumptions about the structure of the matrix. On the other hand, the method of conjugate gradients (Hestenes and Stiefel 1952), for systems with symmetric positive-definite matrices, does not have such a restriction. It is particularly effective if most of the eigenvalues of the matrix of the system lie in a small number of groups. To be precise, it is necessary that it be possible to find a low order polynomial P_m, normalised so that $P_m(0)=1$ which has small values at all the matrix eigenvalues λ_i. Meijerink and Van der Vorst (1976) have proposed the simple device of transforming the system

$$A\underline{x} = \underline{b} \tag{3.3.1}$$

to

$$(L^{-1}AL^{-T})\underline{y} = \underline{c} \tag{3.3.2}$$

where LL^T is an approximate factorization of A, $\underline{y}=L^T\underline{x}$ and $\underline{c}=L^{-1}\underline{b}$. If L has the same sparsity pattern as the lower triangular part of A then the work per conjugate gradient iteration on (3.3.2) is only about double that on (3.3.1) and the experience of both Meijerink and Van der Vorst (1976) and of Kershaw (1977) is that most eigenvalues of $L^{-1}AL^{-T}$ are near unity and consequently the number of iterations is quite modest. This algorithm therefore has the potential of being the foundation of a very successful piece of software.

Kershaw (1977) also reports good results when applying the method to the system of normal equations

$$A^T A \underline{x} = A^T \underline{b} \qquad (3.3.3)$$

in which form we do not need A to be symmetric, and can even perform least squares solutions of overdetermined systems.

4. LEAST SQUARES PROBLEMS

It is much easier to lose sparsity in a linear least squares problem than when solving a set of linear equations. This is illustrated, for example, by comparing the effect of an elementary Gaussian elimination operation on two sparse row vectors with the effect of an elementary orthogonal matrix. In the first case one row is unchanged and the other takes the sparsity pattern of their union, whereas in the second both take the pattern of the union. Therefore the overall gain from exploiting sparsity in least squares problems is not so dramatic, but nevertheless it can be worthwhile.

Band matrices can be treated effectively either by using Givens orthogonal reductions or by using carefully chosen and slightly unconventional Householder reductions (see Reid 1967). Such an algorithm could be made available as a library subroutine, but I do not know of one. Such gains are available for a wider class of matrices by using sequences of Givens orthogonal reductions, as explained for example by Gentleman (1976). Their great merit is flexibility for it is easy to add or remove equations and storage is required only for the reduced upper triangular system. A library subroutine could be written to accept a sequence of equations with varying weights (negative if removal is required) and produce the solution whenever required. Its drawback would be that the arithmetic performed would depend both on column ordering and the order of equation presentation. Often, however, good orderings are obvious and in such cases the storage gains of this approach are very significant.

A number of techniques for the general sparse case where the matrix can be stored and automatic ordering is required were compared by Duff and Reid (1976), whose overall conclusions on the whole favoured the algorithm of Peters and Wilkinson (1970), though this did not perform best on all their test cases. The algorithm involves using sparsity techniques, as in Gaussian elimination,to produce the factorization

$$A = PLUQ \tag{4.1}$$

where P is an mxm permutation matrix, L is mxn lower trapezoidal, U is nxn upper triangular and Q is an nxn permutation matrix. The least squares set

$$A\underline{x} = \underline{b} \tag{4.2}$$

is then equivalent to

$$L(UQ\underline{x}) = P^T\underline{b} \; . \tag{4.3}$$

If we write $UQ\underline{x}$ as

$$\underline{y} = UQ\underline{x} \tag{4.4}$$

then the normal equations corresponding to (4.3) are

$$L^TL\underline{y} = L^TP^T\underline{b} \; . \tag{4.5}$$

These are solved by explicitly forming and factorizing the matrix L^TL and the solution is found by backsubstitution in (4.4). The method has satisfactory sparsity and stability properties and deserves its place in a library.

5. NON-LINEAR SETS OF EQUATIONS

Sparse non-linear sets of equations are often solved by Newton's method so that available software for sparse linear sets of equations can be used. Convergence is obtained by ensuring that the initial estimate is close to the solution. Sometimes such an estimate is readily available, as for example when solving a sequence of problems in which a parameter is being varied slowly. Otherwise it may be necessary to construct a sequence of such problems artificially. The resulting algorithm is not robust and may be very expensive.

There is a great need to follow the success that has been obtained with software for sets of non-linear equations whose Jacobian matrix is full, with something for the sparse case. It is my belief that it would not be enormously difficult to produce robust software that executes with the speed of Newton's method once the iterates are sufficiently close, perhaps by following the ideas of Powell (1970) in his "dog-leg" algorithm.

For such a library subroutine to be widely used it would be necessary for it to be written "inside-out", to use the term of Mallin-Jones in the next paper of this volume (this is also called "reverse communication"). A conventional library subroutine for solving non-linear sets of equations requires a a user-written subroutine (often called CALFUN and we will refer to it by this name) to evaluate non-linear expressions that are required to have the value zero. Once the library subroutine has been called it continues execution (calling CALFUN) until a sufficiently accurate solution is obtained. This is awkward for the user, because the argument list for CALFUN has to be fixed a priori with the design of the library subroutine and may not suit his need, and it leaves him no control to terminate early because of some reason that could never have been anticipated by the writer of the library software. Also it does not permit full advantage to be taken of structure (and existing software) for solving the linearized set of equations. With "inside-out" communication, the user calls the library subroutine repeatedly and on each return a flag indicates what the subroutine wants next from the user's code, which is probably in line. It may be to place the current function values in an argument array F, it may be to form and solve the current linearized system (Newton step) or it may be to continue with the next problem because it regards this one as solved. This interface is convenient to the user because all the information he needs is likely to be at hand without the need for passing information in COMMON with the restrictions that this implies. More importantly, it permits any linear equation solver to be used for the Newton step. This may be a band matrix technique, in or out of main store, or a general sparse technique or anything else specially suited to the problem in hand. The one disadvantage of this approach is that code within the library subroutine is not so well structured.

For the non-linear least squares problem I have written a version of the Levenberg-Marquardt algorithm (Reid 1972). It was written before I appreciated fully the advantages of "inside-out" communication and does not handle the linearized least-squares problem as well as it might, but it is available as NS03 in the Harwell library. It has been used for cases with no more equations than unknowns and has proved itself robust here, but the cost in storage and computing time of going unncessarily through the least squares problem is quite high.

6. EIGENVALUE PROBLEMS

It is usual for only a group of eigenvalues (usually the smaller ones) and their associated eigenvectors to be wanted for sparse matrices. For the symmetric problem and provided there is storage available to perform a triangular factorization of an associated matrix with the same sparsity pattern, there are three competing algorithms, all related to inverse iteration. They are simultaneous iteration (Clint and Jennings 1970 and Rutishauser 1970), block Lanczos (Cullum and Donath 1974 and Underwood 1975) and the sectioning method of Jensen (1972). It is not clear yet whether any of these three has significant advantages over the others. For the present I think it is desirable for any library to have an implementation of one of them. Variants of simultaneous iteration are also available for the unsymmetric problem (see Stewart 1976 and Jennings and Stewart 1975).

Where the matrix is band these methods will be able to take advantage of this feature when solving associated sets of equations, but an alternative is available in the symmetric case. Schwarz (1968) shows that a sequence of plane rotations may be used to reduce a band matrix to tridiagonal form without destroying the structure, and he gives an Algol 60 procedure to do this. All the eigenvalues of the resulting tridiagonal matrix may be obtained cheaply by the QR algorithm, so it is practical to find all eigenvalues of a band matrix.

Another possibility for band matrices, including the generalized problem

$$A\underline{x} = \lambda B\underline{x} \tag{6.1}$$

is the use of the Sturm sequence property, discussed by Peters and Wilkinson (1969). Two levels of library subroutine could be provided. The inner one would find how many eigenvalues lie above a given value p and the outer one would call this repeatedly to find an individual eigenvalue or all eigenvalues in a given range.

Algorithms for the really huge eigenvalue problem have not yet been developed to the stage of reliable software. It is my belief that the algorithm of Lanczos (1950), in its original form, could be used for the symmetric problem. The difficulty lies in separating double copies of a single root from genuine multiple roots and ensuring that all wanted eigensolutions are found. Careful a posteriori analysis of the computed results is necessary.

7. THE USER INTERFACE

The designer of any library software must pay careful attention to the user interface if his code is to be widely used (and liked). This is particularly so with sparse matrix code, where internal data structures often need to be complicated and it is not reasonable to expect the user to master them. My own experience with code for solving general sparse sets of linear equations (Curtis and Reid 1971 and Reid 1972) bears this out. Subroutine MA18 was designed originally for incorporation into a package for solving stiff sets of ordinary differential equations, where the sparse matrices were machine generated, coming naturally by columns. It was easy to generate them by columns in natural order initially and in pivotal order once this was known, and this is an economical form for final storage. Therefore this was required by MA18. It is not convenient, however, in other applications and the different ordering required for the original entry and the factorization of another matrix having the same pattern was particularly resented by users and unfortunately I used the same interface for the symmetric version, MA17. Therefore for Harwell's new code (Duff 1977) we require for both entries that the non-zeros be placed in any order in a real array with associated row and column numbers placed in the corresponding positions of two integer arrays. This form of input demands an initial sort, thereby slightly increasing the execution time. To avoid this in cases where the data can easily be generated in order, we also provide a direct entry.

There still remains a problem over argument lists being longer than is really desirable. An interesting solution to this problem has been obtained by George and Liu (1977). All information about the problem in hand is held in scalars in COMMON and as vectors in a user array called (say) S, whose length is itself held in COMMON. The user makes a sequence of calls of the following subroutines:

1) SUBROUTINE INIT(S)
   This initializes certain variables, loads installation-dependent constants and sets default values, all in COMMON.

2) SUBROUTINE INIJ(I,J,S)
   This specifies that there is a non-zero in row I and column J.

3) SUBROUTINE IJEND(S)
This specifies the end of input of pairs (I,J).

4) SUBROUTINE ORDERi(S)
This finds an ordering for elimination. Different versions are needed for each ordering algorithm.

5) SUBROUTINE INAIJi(I,J,VALUE,S)
This adds a numerical value into the storage position allocated for a_{ij}. This permits assembly to take place in finite-element applications and "non-zeros" a_{ij} with the numerical value zero need not be input. The versions correspond to the versions ORDERi.

6) SUBROUTINE INRHSi(I,VALUE,S)
This adds in a numerical value for a right-hand side coefficient.similarly.

7) SUBROUTINE SOLVEi(S)
This places the solution at the beginning of S.

8) SUBROUTINE SAVE(K,S)
This saves the current state of the package on unit K.

9) SUBROUTINE RESTRT(K,S)
This restores the package from unit K.

Finally I would like to mention the use of reverse communication. Its advantages were explained in section 5 in connection with solving non-linear sets of equations, but its usefulness is not restricted to this one area. The greatly enhanced flexibility that it gives is an advantage in many areas. It deserves wider use.

8. ACKNOWLEDGEMENT

I would like to thank I.S. Duff for reading a draft of this paper and suggesting several improvements.

9. REFERENCES

Carré, B.A. (1961). "The Determination of the Optimum Accelerating Factor for Successive Over-relaxation. Comput. J. 4, 73-78.

Clint, M. and Jennings, A. (1970). "The Evaluation of Eigenvalues and Eigenvectors of Real Symmetric Matrices by Simultaneous Iteration". Comput. J. 13, 76-80.

Corr, R.B. and Jennings, A. (1976). "A Simultaneous Iteration Algorithm for Symmetric Eigenvalue Problems". Int. J. Num. Math. Engrg. 10, 647-663.

Cullum, J. and Donath, W.E. (1974). "A Block Lanczos algorithm for Computing the q Algebraically Largest Eigenvalues and a Corresponding Eigenspace of Large, Sparse, Real Symmetric Matrices". Proc. IEEE Conf. on Decision and Control, Phoenix, Arizona.

Curtis, A.R. and Reid, J.K. (1971). "Fortran Subroutines for the Solution of Sparse Sets of Linear Equations". Harwell report AERE-R.6844. HMSO, London.

Duff, I.S. (1977). "MA28 - a Set of FORTRAN Subroutines for Sparse Unsymmetric Linear Equations". Harwell report AERE-R.8730. HMSO, London.

Duff, I.S. (1978). "On Algorithms for Obtaining a Maximum Transversal". To appear.

Duff, I.S., Munksgaard, N. and Reid, J.K. (1977). "Direct Solution of Sets of Linear Equations whose Matrix is Sparse, Symmetric and Indefinite". Submitted to J. Inst. Maths. Applics.

Duff, I.S. and Reid, J.K. (1976). "A Comparison of some Methods for the Solution of Sparse Overdetermined Systems of Linear Equations". J. Inst. Maths. Applics., 17, 267-280.

Duff, I.S. and Reid, J.K. (1977). "An Implementation of Tarjan's Algorithm for the Block Triangularization of a Matrix". To appear in ACM TOMS.

Gentleman, W.M. (1976). "Row Elimination for Solving Sparse Linear Systems and Least Squares Problems". In "Numerical Analysis, Dundee 1975" (G.A. Watson, ed.). Springer-Verlag, Berlin, Heidelberg and New York.

George, J.A. (1977). "Solution of Linear Systems of Equations: Direct Methods for Finite Element Problems". In "Sparse Matrix Techniques" (V.A. Barker, ed.) Springer-Verlag, Berlin, Heidelberg and New York.

George, J.A. and Liu, J.W.H. (1977). "The Design and Implementation of a Sparse Matrix Package". University of Waterloo report CS-77-21.

George, J.A. and McIntyre, D.R. (1976). "On the Application of the Minimum Degree Algorithm to Finite Element Systems". University of Waterloo report CS-76-16.

Gibbs, N.E., Poole, W.G. Jnr. and Stockmeyer, P.K. (1976). "An Algorithm for Reducing Bandwidth and Profile of a Sparse Matrix". SIAM J. Num. Anal. 13, 236-250.

Gustavson, F.G. (1976). "Finding the Block Lower Triangular Form of a Matrix". In "Spase Matrix Computations" (J.R. Bunch and D.J. Rose, eds.), Academic Press, London and New York.

Hall, M. (1956). "An Algorithm for Distinct Representatives". Amer. Math. Monthly, 63, 716-717.

Hestenes, M.R. and Stiefel, E. (1952). "Methods of Conjugate Gradients for Solving Linear Systems". J. Res. Nat. Bur. Standards 49, 409-436.

Hopcroft, J.E. and Karp, R.M. (1973). "An $n^{5/2}$ Algorithm for Maximum Matchings in Bipartite Graphs". SIAM J. Comput. 2, 225-231.

Jennings, A. (1977). "Matrix Computation for Engineers and Scientists". John Wiley, London and New York.

Jennings, A. and Stewart, W.J. (1975). "Simultaneous Iteration for Partial Eigensolution of Real Matrices". J. Inst. Maths. Applics., 15, 351-361.

Jensen, P.S. (1972). "The Solution of Large Symmetric Eigenproblems by Sectioning". SIAM J. Numer. Anal. 9, 534-545.

Kershaw, D.S. (1977). "The Incomplete Cholesky-Conjugate Gradient Method for the Iterative Solution of Systems of Linear Equations". Submitted to J. Comp. Phys.

Lanczos, C. (1950). "An Iteration Method for the Solution of the Eigenvalue Problem of Linear Differential and Integral Operators". J. Res. Nat. Bur. Standards 45, 255-281.

Meijerink, J.A. and Van der Vorst, H.A. (1976). "An Iterative Solution Method for Linear Systems of which the Coefficient Matrix is Symmetric M-matrix". Math. Comp. 31, 148-162.

Peters, G. and Wilkinson, J.H. (1969). "Eigenvalues of $Ax=\lambda Bx$ with Band Symmetric A and B". Comput. J. 12, 398-404.

Peters, G. and Wilkinson, J.H. (1970). "The Least Squares Problem and Pseudo-inverses". Comput. J. 13, 309-316.

Powell, M.J.D. (1970). "A Hybrid Method for Non-linear Equations" and "A Fortran Subroutine for Solving Systems of Non-linear Algebraic Equations". In "Numerical Methods for Non-linear Algebraic Equations" (P. Rabinowitz, ed.), Gordon and Breach, London, New York and Paris.

Reid, J.K. (1966). "A Method for Finding the Optimum Successive Over-relaxation Parameter". Comput. J. 9, 200-204.

Reid, J.K. (1967). "A Note on the Least Squares Solution of a Band System of Linear Equations by Householder reductions". Comput. J. 10, 188-189.

Reid, J.K. (1972). "Two Fortran Subroutines for Direct Solution of Linear Equations whose Matrix is Sparse, Symmetric and Positive-definite". Harwell report AERE-R.7119. HMSO, London.

Rutishauser, H. (1970). "Simultaneous Iteration Method for Symmetric Matrices". Numer. Math. 16, 205-223.

Schwarz, H.R. (1968). "Tridiagonalization of a Symmetric Band Matrix". Numer. Math. 12, 231-241.

Stewart, G.W. (1976). "Simultaneous Iteration for Computing Invariant Subspaces of Non-Hermitian Matrices". Numer. Math. 25, 123-136.

Stone, H.L. (1968). "Iterative Solution of Implicit Approximations of Multi-dimensional Partial Differential Equations". SIAM J. Numer. Anal. 5, 530-558.

Tarjan, R. (1972). "Depth First Search and Linear Graph Algorithms". SIAM J. Comput. 1, 146-160.

Underwood, R. (1975). "An Iterative Block Lanczos Method for the Solution of Large Sparse Symmetric Eigenproblems". Ph.D. Thesis. Stanford Comp. Sci. Report STAN-CS-75-496.

Young, D.M. (1954). "Iterative Methods for Solving Partial Difference Equations of Elliptic Type". Trans. Amer. Math. Soc. 76, 92-111.

Zlatev, Z. and Thomsen, P.G. (1976). "ST - A Fortran IV Subroutine for the Solution of Large Systems of Linear Algebraic Equations with Real Coefficients by Use of Sparse Technique". Report 76-05. Institute of Numerical Analysis, Technical University of Denmark.

II. 4

NON-LINEAR ALGEBRAIC EQUATIONS IN PROCESS ENGINEERING CALCULATIONS

A.K. Mallin-Jones

(Imperial Chemical Industries Limited)

1. INTRODUCTION

This paper begins with an explanation of the term "Process Engineering". Then it describes the importance of steady state heat and mass balance calculations. These calculations usually require the solution of non-linear equations. The structure of our programs has implications for the way in which we code equation solving algorithms. We draw some comparisons between the way we code algorithms and the conventional way of coding them. Then we briefly list the algorithms we use. Finally we mention some of the alternative strategies for calculating a steady state heat and mass balance.

2. DESIGN OF CHEMICAL PLANT

We use the term "Process Engineering" to denote the application of chemical engineering principles to the design of chemical plant. This activity fits between the chemical and market research on the one hand, and the eventual mechanical design, construction and operation of the plant. Fig. 1 illustrates the relationship of these activities.

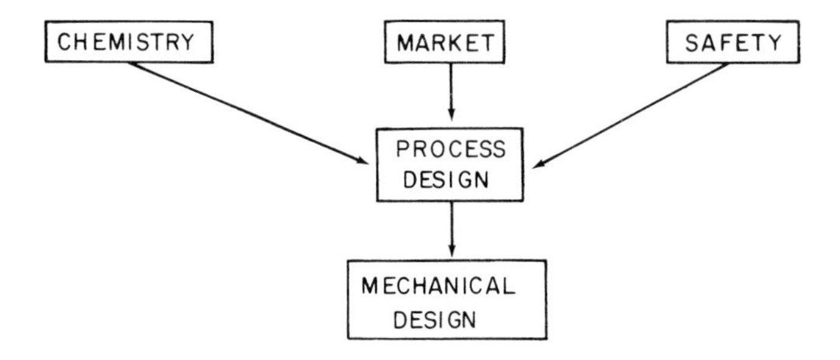

Fig. 1. Place of process design in creation of new chemical plant

So knowledge of the chemical and physical properties of materials tells us which chemical will produce a desired effect and what routes exist to reach that chemical from raw materials. Knowledge of the market tells us what demand there might be for the effect and what the raw materials and energy will cost. Safety standards constrain the possible routes to the chemical.

At the other end of design we have to consider the mechanical design of vessels, structures and foundations, pipe-routeing etc., leading eventually to the construction and operation of the plant.

Concentrating now on process design, we can roughly identify the tasks done by a chemical engineer. These are shown in Fig. 2.

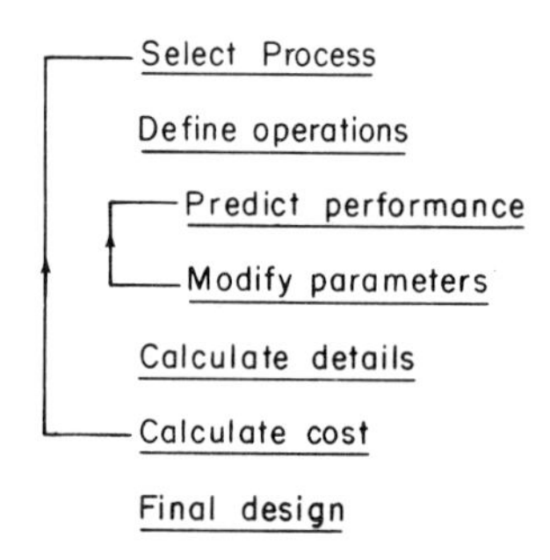

Fig. 2. Stages in process design

Thus the engineer selects one of the paths to the desired chemical. He defines the stages on this path in terms of distinct operations such as heat exchange, pumping, distillation. These are known as unit operations and will correspond to physically identifiable items in the eventual plant. The next stage is typically a mixture of predicting the performance of a collection of hardware and calculating the size of an item necessar to achieve some performance. Performance here generally means the steady state heat and material flows in the whole plant. The engineer then refines the description of all the items unti he has enough detail to calculate the cost of the plant. At any stage in the whole process he may loop back to a previous stage to evaluate an alternative - Figure 2 shows likely loops.

Clearly simulation of the plant is a very important task in process design. About half the computing effort goes in calculating heat and material balances over models of complete plants. This is done by steady state flowsheeting programs. In these programs the dominant numerical problem is the solution of algebraic equations. So the remainder of this paper is devoted to an examination of the steady state flowsheeting problem.

3. REPRESENTATION OF FLOWSHEETS

A chemical engineer will naturally represent his plant as a weighted directed graph. The nodes correspond to unit operations, and the arcs correspond to material or information flows from one unit to another. The weight of an arc is the number of values needed to represent the flow. For example one arc might represent 20 chemical component flowrates plus temperature and pressure. Another arc might represent a scalar control signal. A very simple flowsheet is shown in Fig. 3.

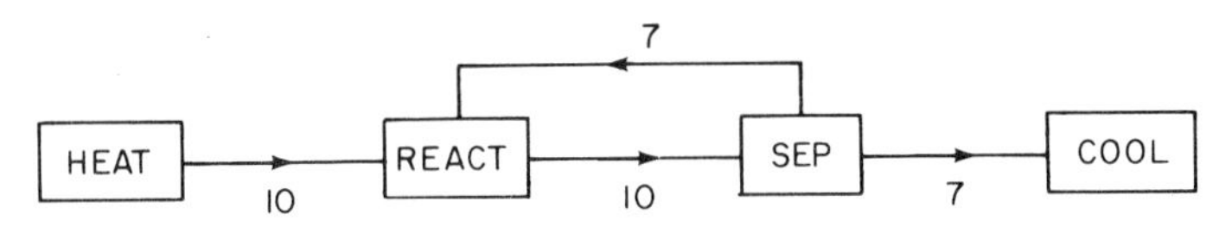

Fig. 3. A simple flowsheet

Typical flowsheets have between 10 and 50 units, rarer examples have hundreds. A closed loop in the graph shows the presence of equations. This is because the value of any stream in a loop depends upon the preceding streams, and this can be traced round until the chosen stream is seen to be a function of itself.

A loop results in the equations $u = f(u)$ where u is the stream vector, and f is the effect of the unit operations in the loop. Instead of using the loose term loop I should really say that equations arise from strong components in the graph; where a strong component is defined as a subgraph in which every node can be reached by a directed path from every other node in the subgraph.

Many flowsheeting systems will analyse networks as described by Johns (1970). Briefly this consists of two steps: partitioning and sorting. Partitioning is the identification of strong components in the graph. The result of partitioning the graph is an ordered list of unit operations in which the contents of strong components are grouped together. Except within strong components, all streams flow in the direction from beginning to end of the list. Figure 4 shows the effect of partitioning.

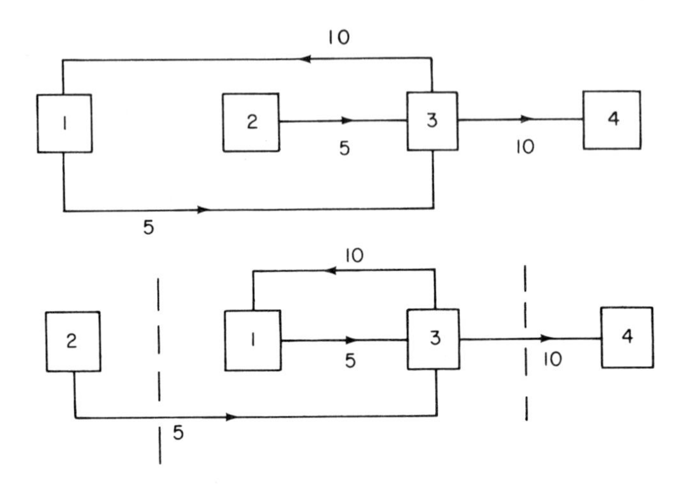

Fig. 4. Partitioning

This example is trivial, but a flowsheet of 50 units could be quite tedious to partition by hand. Partitioning can be described in terms of the operations on the incidence matrix corresponding to the graph. It is block triangularization of the graph, and a recent algorithm is described by Duff and Reid (1976).

The second step in analysing a network is to order the units within each strong component so as to minimize the number of backward flowing variables. If there were no such recycle variables we could simulate the network by calculating each unit operation once. To calculate the values of streams in a strong component we have to guess values for the recycle variables, calculate the unit operations, and hence get new values for the recycles. Adjusting the guessed values of the recycle variables until the resulting calculated variables match them amounts to solving the equations formed by the strong component. If we minimize the number of recycle variables then we minimize the storage required by the equation solver. Those equation solvers which store matrices use space of the order of the square of the number of recycle variables. For these cases the effect of minimizing the number of variables is quite dramatic. Another reason for this minimization is that if we choose a larger set of recycle variables then they cannot be independent. Lack of independence can upset some equation solving algorithms.

In Fig. 4 the minimal recycle set is given by swapping units 1 and 3. This results in 5 recycle variables. In the literature recycles are also known as cut or torn variables. There are many algorithms for finding a minimal cut set. We use programs developed by Muller and Johns, most recently published by Muller (1974). Other algorithms are described by Lapidus and Pho (1973), Upadhye and Grens (1972), Christensen and Rudd (1969).

4. PROGRAM STRUCTURE

Our flowsheeting programs operate in four phases as shown in Fig. 5.

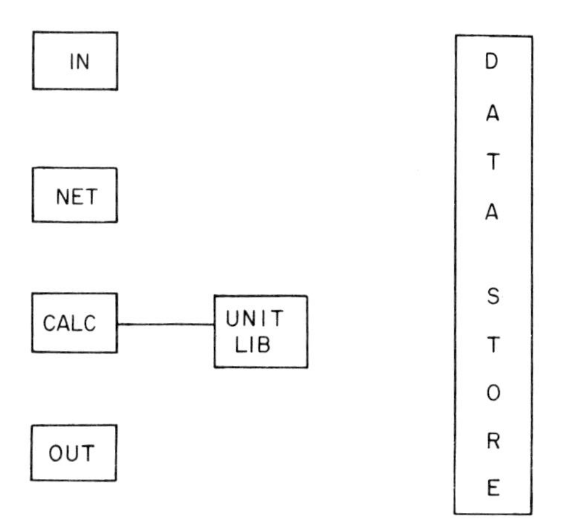

Fig. 5. Program structure

First we read a description of the flowsheet in some convenient code. Then we analyse the network as described above. Then we simulate the network and finally print the results. We simulate the network by calling a unit subroutine appropriate for each unit operation. The unit subroutine calculates its outputs from its inputs and parameters. This is the traditional method of simulating a network - I shall briefly mention other methods later. So the calculation phase is essentially very simple - it consists of calling unit subroutines in the order determined by NET. Each subroutine is given access only to its own input and output streams in the data store. At this stage the data store is a large array in core and the subroutines access various parts of this array via their argument lists. As well as partitioning and sorting the lists of units, NET divides up the calculation array into the areas required by the various streams, and generates a list of commands for the calculation phase. These commands make the calculation phase give areas containing the appropriate streams to the corresponding arguments of each unit subroutine in turn,

5. EFFECT OF PROGRAM STRUCTURE ON EQUATION SOLVERS

One function evaluation for an equation solver corresponds to the calculation phase working its way through the calls of unit subroutines modelling units in a strong component. It is not easy to write this task as a call of a standard subroutine from a normal equation solver. It would be easier if we could regard the equation solver as a special kind of unit operation. At the end of each partition we could put an equation solving operation which will be modelled by an equation solving subroutine. This concept fits neatly into the structure we already have of unit operations being modelled by unit subroutines. To make the concept work we take the following three steps:

(1) The user defines a SOLVE unit and chooses an equation solving subroutine to model it. He does not connect it to the network. Rather, after the automatic steps of partitioning and sorting, the NET phase inserts the equation solving unit at the end of each strong component into the recycle streams. So the network is amended as shown in Fig. 6, where the equation solving unit is denoted by △E.

(2) The calculation phase tests a flag set by the equation solving subroutine and repeats the sequence of units in the partition if another function evaluation is required.

(3) The equation solving subroutine is written "inside out". In order to ask for another function evaluation it returns with a flag set. This contrasts with a conventional equation solver which calls a subroutine for the next function evaluation.

6. EXPERIENCE OF USING "INSIDE OUT" EQUATION SOLVERS

We now have several programs which use "inside out" equation solving subroutines. Our experience of writing these programs and equation solvers is as follows:

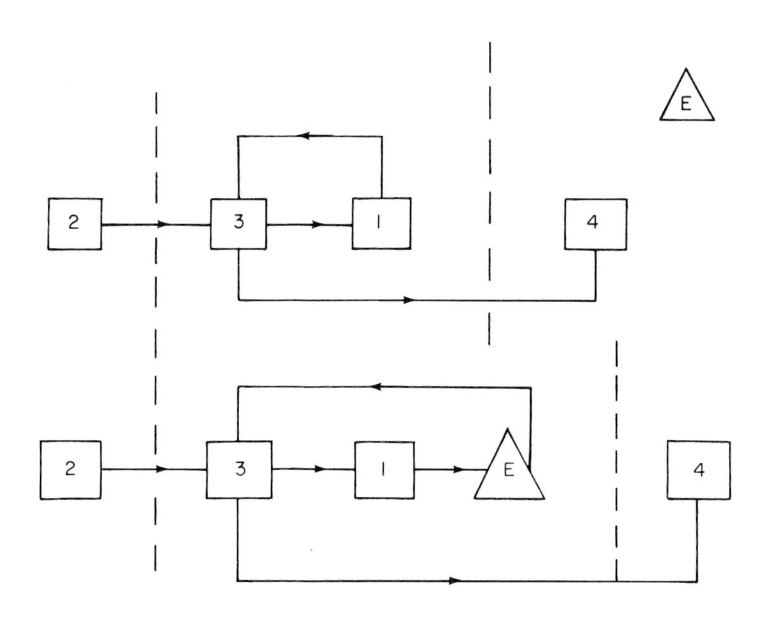

Fig. 6. Insertion of solve units into recycles

(1) In simple cases it is slightly more difficult to use an "inside out" routine in place of a conventional one. The FORTRAN needed is shown in Fig. 7.

For a large program this slight complexity is outweighed by the fact that it is no longer necessary to bend the program structure so that the function is calculated through a standard subroutine interface. When writing general purpose programs one often transmits variables by argument list rather than

```
C     INITIALIZE COUNT AND X
        IT = 1
        X(1) = ...
C     CALCULATE FUNCTION
    1  F(I) = ...  (X(J))
C      PREDICT NEXT POINT
       CALL INSOLV (X,F,N,CONV,IT,...)
       IF (AGAIN) GO TO 1
       IF (.NOT. OK) GO TO 999
C      USE RESULTS
       ...
```

Fig. 7. Using an "Inside out" equation solver

COMMON so as to use adjustable dimensions. It is also easier to trace the flow of data when using argument lists. If we were using conventional equation solvers we should have to keep amending them so as to cater for the particular argument list used for the current function subroutine.

(2) An "inside out" routine can be used to implement a conventional one but not vice versa. It is trivial to convert Fig. 7 to a sketch of an implementation of a conventional subroutine.

(3) We normally solve all recycles in one partition simultaneously. In network terms this means that we concatenate all recycle streams together into one large vector function value and one vector predicted point. However it is sometimes advantageous to nest the solution of one set of equations inside another set. For example suppose one unit subroutine takes much more CPU time than the others. We could probably reduce the number of times which it is called by reducing the dimensionality of the equations associated with it. We can do this by solving some equations in this partition at each step of the iterations involving the slow unit. Fig. 8 shows the sequence of subroutine calls needed in the conventional case, contrasted with the network illustrating the "inside out" case. Subroutine OUTER or unit O is a slow computation. Subroutine INNER or unit I represents faster parts of the calculation which generate equations (recycles) not involving OUTER (O).

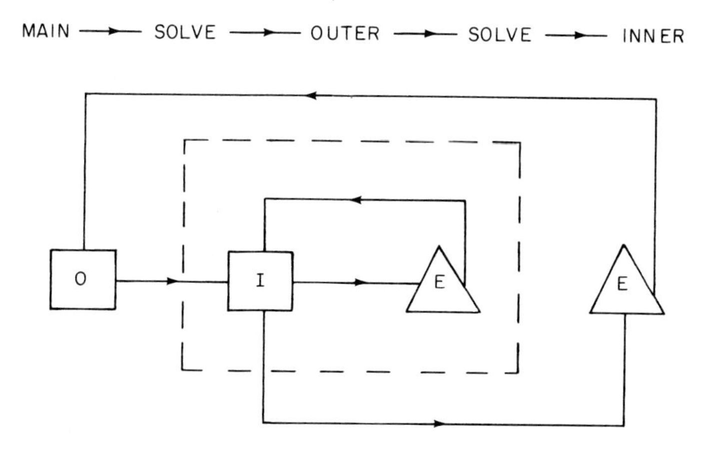

Fig. 8. Nesting equations

The conventional sequence of subroutine calls is unfortunately recursive and is therefore illegal FORTRAN. In the "inside out" network we have two distinct equation solving operations which happen to be modelled by the same equation solving subroutine. Calls of this routine are sequential and the two operations will have separate work spaces. There are other reasons for using nesting apart from reducing CPU time. If the equation solving subroutine needs to store matrices then we shall halve the total storage space if we divide the recycles into two equal groups. More importantly it may be that the inner set of equations represent some physically meaningful entity to the user. He may value a series of solutions for the inner equations even if the outer loop fails to converge. In our latest flowsheeting program the user may nest any group of units. The network analysis is modified so that it first treats the whole network with nests taken as pseudo units. Then the analysis is repeated for each nest. Nested nests are possible up to any level.

(4) When using the "inside out" method the iterative process remains under obvious control of the program. It is possible to stop iterating because of a time limit or because of some interactive command from the user. No special features have to be built into either the equation solving subroutine or into the function evaluation subroutine. Having stopped iterating the program can save the current state of the problem by writing all work spaces to files. The iterations can be resumed as if no interruption occurred, since all information

required by the equation solver must be in its work space. For the conventional method, such facilities have to be provided by adding options to the equation solving subroutine and function subroutine.

(5) A conventional equation solving subroutine is easier to implement. When programming an "inside out" subroutine one accumulates odd flags, counters, saved values and past points which must all go in the argument list so as to be remembered for the next iteration. To avoid unmanageably long argument lists these values tend to become elements of long general arrays, and so the FORTRAN becomes obscure. A "Dynamic Equivalence", "Map" or "Mode" statement in FORTRAN would help.

(6) Some algorithms, e.g. Davidenko (see Broyden (1969)) work by posing a series of subsidiary problems for another equation solver. This can be implemented in a modular way, i.e. we can write one Davidenko routine which can be used with alternative equation solvers. In the "inside out" case it can be used as shown in Figure 9.

```
INITIALIZE
DO
       CALCULATE FUNCTION
       . . .
       CALL DAVIDENKO  (. . .)
       CALL INSOLVE  (. . .)
UNTIL.NOT. AGAIN
IF  OK  THEN
       RESULTS
ELSE
       TROUBLE
ENDIF
```

Fig. 9. Posing a series of subsidiary problems

The Davidenko routine alters the calculated function values so as to pose the current subsidiary problem. It is necessary to provide communication between the routines in the form of an integer flag so that correct action is taken when a subsidiary problem is solved. However we have used such a structure successfully.

(7) My last point is that it is possible to make some economies of space by using an "inside out" structure. It is possible to share temporary work space between the function subroutine and the equation solving subroutine. It is also possible to overlay the function subroutine against the equation solver.

7. ALGORITHMS

One feature we put in all our algorithms is the possibility of failure to calculate the function. This can also be used to cover the case of failing to converge a nested loop. We are really trying to solve equations subject to constraints. For example, the variables must remain physically meaningful - and if they do not some calculations become impossible. However it is difficult to translate these constraints on all stream variables into constraints on the recycle variables. Instead we use a simple failure flag which signals that the function cannot be calculated at the current point. It is then the responsibility of the equation solver to backtrack towards the last successful points.

In some problems the equations are linear, and in many others one linearization will approximate the function closely over a wide range. For many problems the eigenvalues of the matrix in the linear approximation will all be less than one in modulus, so that the recycles are naturally convergent. We can solve a worth-while subset of the problems by taking advantage of such points. Our recommended methods are Quasi Newton algorithms which directly update an estimate of the inverse Jacobian matrix. Such methods are described by Fletcher and Powell (1963). Most flowsheets will converge by these methods. Those which do not usually have conflicting values for design variables. Others which fail initially often converge when given good enough starting values. A more serious problem at the moment is that some users are unable to use Quasi Newton methods because of restrictions in space. Some computer centres will not allow fast turn round on large jobs. In order to cope with these problems we have implemented a series of algorithms which store only a few vectors rather than matrices. For convergent equations we have Successive Overrelaxation and Chebyshev Semi-iteration (Varga, 1962). For some flowsheets we find that Aitken 3 or 5 point extrapolation formulae converge rapidly (Fox, 1964). The Single Dominant Eigenvalue method of Crowe and Nishio (1975) often converges fast enough to be useful.

8. ALTERNATIVE WAYS OF SIMULATING FLOWSHEETS

The method I have described of simulating flowsheets by calling unit subroutines to calculate outputs from inputs is generally called the modular method. Clearly it will be difficult to use on problems in which the output from the plant is specified and it is necessary to calculate the inputs. There are two other methods which we are aware of. One we know as the equation method. Here the program has access to individual equations in the sets defining the unit operations. Unit operations are described by sets of coded equations, not by unit subroutines. The program can choose which variable to solve each equation for. Examples are SPEED-UP by Leigh, Jackson and Sargent (1974), and the Generalized Interrelated Flow Simulation Program from the Japanese Union of Scientists and Engineers - described by Yajima (1970). Our limited experience of such systems is that at present the overheads of decoding stored equations outweigh the benefits of detailed knowledge of the equations. However Allen and Westerberg (1976) make some interesting observations on indexing coded equations. The method might be very powerful if combined with a program for automatic algebraic manipulation. The other method we call two tier because it operates on an inner linear level and an outer possibly non-linear level. Each unit operation is represented by a linear model and also by a subroutine which calculates coefficients for the model. An iteration consists of calculating a set of coefficients for the whole plant, solving the resulting sparse set of linear equations, and then recalculating the coefficients from the new stream values. An example is Quasilin by Hutchison and Gorczynski at Cambridge University, Chemical Engineering Department. An earlier example of Quasilin is described by Hutchison and Shewchuk (1974). As far as industrial use is concerned the main problem seems to be finding a simple way of writing the unit subroutines. Engineers should concentrate on making the model reflect reality adequately, rather than spend time modifying a natural way of writing the model. Again an algebra manipulation program might help.

9. CONCLUSION

The main purpose of this paper is to put a case for "inside out" interative routines. Most of the points raised apply to optimizers as well as equation solvers. There might also be some advantages in using the same program structure for integrating ordinary differential equations.

10. REFERENCES

Allen, G. L. and Westerberg, A. W. (1976), "Solution procedures for indexed equation sets: Structural considerations", *A.I.Ch.E. J.*, **22**(3), 549.

Broyden, C.G. (1969), "A new method for solving non-linear simultaneous equations", *Comp. J.*, **12**, 94.

Christensen, J.A. and Rudd, D.F. (1969), "Structuring design computations", *A.I.Ch.E. J.*, **15**(1), 94.

Crowe, C.M. and Nishio, M. (1975), "Convergence promotion in the simulation of chemical processes - The general dominant eigenvalue method", *A.I.Ch.E. J.*, **21**(3), 528.

Duff, I.S. and Reid, J.K. (1976), "An implementation of Tarjan's algorithm for the block triangularisation of a matrix", AERE Harwell Report CSS.29.

Fletcher, R. and Powell, M.J.D. (1963), "A rapidly convergent descent method for minimization", *Comp. J.*, **6**, 163.

Fox, L. (1964), "An Introduction to Numerical Linear Algebra", Clarendon Press, Oxford.

Hutchison, H.P. and Shewchuk, C.F. (1974), "A computational method for multiple distillation towers", *Trans. Instn. Chem. Engrs.*, **52**, 325.

Muller, F.R. (1974), "Optimal Computational Sequence for Use with Flowsheeting Program", Diplomarbeit Report, Eidgenossiche Technische Hochschule, Zurich, Technisch-Chemisches Laboratorium.

Johns, W.R. (1970), 97th event of the European Federation of Chemical Engineering "The use of computers in the studies preceding the design of chemical plants" at Florence, Italy.

Lapidus, L. and Pho, T.K. (1973), "Topics in Computer-Aided Design: Part 1. An optimum tearing algorithm for recycle systems", *A.I.Ch.E. J.*, **19**(6), 1170.

Leigh, M.J., Jackson, C.D.D. and Sargent, R.W.H. (1974), "SPEED-UP, a computer based system for the design of Chemical Processes" Int. Conf. and Exhib. on Computers on Engineering and Building Design.

Upadhye, R.S. and Grens, E.A. II (1972), "An efficient algorithm for optimum decomposition of recycle systems", *A.I.Ch.E. J.*, **18**(3), 533.

Varga, R.S. (1962), "Matrix Iterative Analysis", Prentice Hall.

Yajima, K. (1970), "JUSE-GIFS USERS MANUAL", Nippon Kaguka Gjutsu Kenshhujo Computation Centre (Institute of Japanese Union of Scientists and Engineers).

PART III

DATA-FITTING

III. 1

DATA-FITTING ALGORITHMS AVAILABLE, IN PREPARATION, AND IN PROSPECT, FOR THE NAG LIBRARY

J. G. Hayes

(National Physical Laboratory, Teddington)

1. INTRODUCTION

With a brief to discuss data fitting in the context of this Conference, and having some time ago accepted the rôle of Contributor for the EO2 chapter of the NAG library (a major source of the software of the Conference's title), my course seemed clear: to devote this paper to a summary of the routines in that chapter, both those currently available and those planned for the next one or two Marks of the library. I should explain that the NAG (Numerical Algorithms Group) library is subdivided on the basis of a subject-area classification into a number of sections, called chapters. The EO2 chapter covers the subject area "Curve and Surface Fitting". The chapter Contributor is the person responsible for the development of the contents of a particular chapter. The library is re-issued, with additional and perhaps some replacement routines, at approximately yearly intervals, the different issues being given consecutive Mark numbers. Marks 5 and 6 are the two I shall mention later, the former having appeared last year, the latter due shortly.

I should point out that there are a number of routines relevant to data fitting contained in some of the other chapters of the library. These chapters are EO4 - Minimizing or Maximizing a Function (which includes non-linear least squares), FO4 - Simultaneous Linear Equations (which includes routines for general linear least squares) and GO2 - Correlation and Regression Analysis. I shall be discussing, however, solely my own chapter, EO2. This is concerned, for the time being at any rate, primarily with polynomials and splines.

My responsibility for the chapter started with Mark 5, though in fact the decision to replace the previously existing routines (with the exception of one routine in Section 6 below) with a set we were developing at NPL had already been taken by my predecessor as Contributor. With, therefore, an essentiall fresh start, my initial objective was, and is, to get into the library as rapidly as possible a good set of basic routines for least-squares fitting with polynomial and spline functions of one and two variables. However, the plan has perforce to be fluid, since other demands on our efforts at NPL can cause priorities to be altered and other routines to become available. Thus for example the parametric spline routines in Section 3.3 below are almost complete though there is as yet no polynomial surface routine in the library. Moreover, routines from outside NPL can only be selected as and when they become available. It is because of these uncertainties that I have not attempted to give an order of availability of the prospective routines discussed, but have simply indicated either that they are in preparation or have not yet been started. In any case, our over-riding objective is to provide those routines which are of most practical value and so we welcome these external demands and any other evidence of practical need. On the other hand, the programme outlined below is itself based on our existing, fairly extensive, experience of such needs, and so hopefully will not be unduly disturbed.

One other detail should be mentioned. In the NAG scheme for naming routines, a name consists of six characters. The first three of these are the chapter name, EO2 in our case. Next come two letters which identify the individual routine, and lastly is another letter, A or F, specifying respectively Algol 60 or Fortran IV. In quoting routine names I shall omit this final letter: all routines exist in both versions.

In organising the paper, it proved more convenient to group the routines in sections related to subject matter rather than in the order suggested by the title of the paper. Thus the next four sections deal in turn with polynomial curves, spline curves, spline surfaces and polynomial surfaces, and are all concerned with least-squares methods. The concluding section is concerned with the use of the ℓ_1 and ℓ_∞ norms, mainly for the general linear case. The level of treatment of the numerical methods concerned is intentionally variable, more details being given for the newer material. The remaining methods have simply been outlined: further details can be obtained from the references cited, and in most cases from the survey by Hayes (1974).

Finally, I would like to acknowledge the fact that all the work described in the paper (except that in Section 6, from outside NPL) is in collaboration with my two colleagues G.T. Anthony and M.G. Cox.

2. POLYNOMIAL CURVES

2.1. Mark 5 Routines

Routine EO2AD computes least-squares polynomial approximations to an arbitrary set of data points (x_r, y_r), $r = 1,2, \ldots, m$, allowing a different weight for each point, when required. It gives least-squares polynomials of all degrees up to a user-specified degree k. Guidance on which of these polynomials to choose as the fit, as well as on the initial choice of k and of the weights, is given in the Chapter Introduction (part of the user documentation of the NAG Library).

The numerical method employed by the routine is the modification of Forsythe's technique due to Clenshaw (1960), that is to say, it is based on the generation of a set of polynomials mutually orthogonal with respect to summation over the data values of x, all these polynomials being represented in their Chebyshev-series form. The fitted polynomials are similarly represented, so that the fitted polynomial of degree i is in the form

$$a_{i0}T_0(X) + a_{i1}T_1(X) + \ldots + a_{ii}T_i(X). \qquad (2.1.1)$$

Here $T_j(X)$ is the Chebyshev polynomial of the first kind of degree j, and its argument is the independent variable x normalized so that its data values span the range -1 to + 1, that is

$$X = \frac{2x - x_{max} - x_{min}}{x_{max} - x_{min}}, \qquad (2.1.2)$$

x_{max} and x_{min} being respectively the largest and smallest of the x_r. (Note that the first coefficient in (2.1.1) is usually written $\frac{1}{2}a_{i0}$, for algorithmic convenience. For descriptive simplicity, here and elsewhere, I shall omit the factor $\frac{1}{2}$.)

The main output of the routine is the array of values of a_{ij}, $i=0,1, \ldots, k$, and $j = 0,1,\ldots,i$. For users unfamiliar with Chebyshev polynomials the a_{ij} for fixed i can be regarded simply as parameters which have to be supplied to the routine EO2AE to obtain the value of the fitted polynomial of degree i at any chosen value of x.

The other polynomial fitting routine which entered the library at Mark 5 is EO2AF. It is intended for use when the data are given in the form of a graph, or in any other situation in which the user can provide y-values at any desired set of x-values. The process carried out by the routine (see Clenshaw (1962)) was originally designed for the approximation of mathematical functions, in which case very accurate y-values can usually be provided, but it is equally useful for much less accurate data, such as experimental data.

Again taking X to denote the independent variable normalized according to (2.1.2), with x_{max} and x_{min} now simply defining the range over which the approximation is required, the user has to supply the routine with values y_r corresponding to values of X given by

$$X_r = \cos(\pi r/n), \qquad r = 0,1, \ldots, n, \qquad (2.1.3)$$

for some suitably chosen value of n. The routine then computes the polynomial of degree n which passes exactly through all the points provided. Again the polynomial is in Chebyshev-series form, and the output comprises the coefficients a_j in

$$a_0T_0(X) + a_1T_1(X) + \ldots + a_nT_n(X). \qquad (2.1.4)$$

For smoothing purposes, the significant feature of the polynomial (2.1.4) provided by the routine is that its truncation at any degree, i say, yields the least-squares fit of degree i to the data points supplied (each point given unit weight except the end ones with half weight). The use of the cosine points (2.1.3) not only gives this feature but also effectively removes the possibility of unwanted fluctuations in the polynomial finally chosen, as can happen with other data sets. The cosine points are therefore recommended whenever the abscissa values can be freely chosen.

The Chapter Introduction gives advice on the choice of n and of the degree at which to truncate the polynomial (2.1.4). The numerical value of the selected polynomial at any required X in its range can again be obtained using routine EO2AE.

2.2. *Constraints*

Currently being written is an enhancement to the polynomial fitting routine EO2AD which enables function and derivative constraints to be imposed on the polynomial. For example, the polynomial may be required to pass exactly through a given point or to join smoothly at the end of its range on to a given straight line. Denoting by $f^{(r)}(x)$ the rth derivative of the fitting polynomial, the constraints catered for are of the type: $f^{(r)}(x)$ to have a specified value at $x = t$ for each $r = 0,1,\ldots,p$, with p and t given ($p = 0,1$ or 2, and t at one end of the range of interest, are the most common requirements). This means, for example, that if the second derivative at a point is to be specified, then the function value and first derivative must also be specified. The constraints can be applied at any number of points, with a different p at each one, if required.

To apply such constraints, we consider the fitting polynomial in the form

$$f(x) = \mu(x) + \eta(x) \qquad (2.2.1)$$

where $\mu(x)$ is a polynomial (usually the one of lowest degree) which satisfies the constraints, and
$\eta(x)$ is the most general polynomial of the given degree k for which $\eta^{(r)}(x) = 0$ whenever $f^{(r)}(x)$ is specified.

Thus, if $f^{(r)}(x)$ is given at $x = t_i$ for $r = 0,1,\ldots,p_i$, where $i = 0,1,\ldots,$ then $\eta(x)$ takes the form

$$\eta(x) = \nu(x)\ g(x),$$

where

$$\nu(x) = \prod_i (x-t_i)^{p_i+1}$$

and $g(x)$ is a general polynomial of degree $k - \sum_i (p_i+1)$.

For example, if it is required that $f(0)=1$ and $f(1)=f'(1)=0$, we may take

$$\mu(x) = (1-x)^2$$

and $$\nu(x) = x(1-x)^2.$$

Having determined $\mu(x)$ and $\nu(x)$, we derive modified data Y_r defined by

$$Y_r = y_r - \mu(x_r)$$

and to these values we fit a polynomial having the form $\nu(x)g(x)$. The method of achieving this is described in Clenshaw and Hayes (1965). The facility was not included in EO2AD to avoid complicating that routine for the user with the more common unconstrained problem. The complete constrained polynomial $f(x)$ is again provided as a Chebyshev series.

2.3. Derivatives and Integrals

As already stated, routine EO2AE is available for computing values of a fitted polynomial given its Chebyshev coefficients. Often, derivatives and integrals of the polynomial are also required and it is our intention to provide this capability also. This will involve two simple routines which, from the coefficients of the given polynomial, derive the coefficients of the polynomial which is respectively the first derivative and the indefinite integral of the one given. Repeated use of these routines will provide Chebyshev coefficients for higher derivatives or repeated integrals. Routine EO2AE will then provide values of the derivatives or definite integrals corresponding to any value of X.

2.4. Simplified Routine

As indicated in Section 2.1, in the fitting routine EO2AD the choice of the initial degree, and of the degree finally chosen for the fit, is left to the user, as is necessary in a routine aiming to cover all data sets. However, we are considering a simplified version which will itself choose these degrees on the basis of some simple criteria, and also assume unit weights. Intended for the first-time user, it should be satisfactory in many cases and, when not, could hopefully lead such a user to fuller use of EO2AD.

3. CUBIC SPLINE CURVES

We may recall that a cubic spline curve is made up of seg-
ents each of which is a cubic polynomial arc. Adjacent arcs
join together with continuity, in general, up to and including
second derivatives. The positions (x-values) of the joins are
called knots. If, in using a cubic spline for fitting purposes,
t is required to have continuity lower than second derivative
at some point, this can be achieved by the artifice of placing
ore than one knot at the point.

3.1. *Mark 5 Routines*

Routine EO2BA computes a weighted least-squares approxim-
tion to an arbitrary set of data points (x_r, y_r), $r=1,2,\ldots,m$
y a cubic spline with knots specified by the user. Some
uidance on the choice of knots is given in the Chapter
ntroduction.

Just as polynomials can be expressed, in a variety of
ays, as linear combinations of sets of standard polynomials
uch as powers of x or Chebyshev polynomials, so cubic splines
an be expressed in terms of standard cubic splines. The stan-
ard splines preferred for most computational purposes are the
ormalized cubic B-splines (de Boor (1972)). These are uni-
odal, non-negative, and are non-zero only over four adjacent
egments of the spline they are used to represent (less if
nots coincide). Thus the cubic spline is represented in the
outine in the form

$$s(x) = c_1 N_1(x) + c_2 N_2(x) + \ldots + c_q N_q(x), \qquad (3.1.1)$$

ere the $N_i(x)$ denote the B-splines, and the main output is
e set of coefficients c_i. For users unfamiliar with B-splines, the c_i
an be regarded simply as parameters which have to be supplied
o the routine EO2BB in order to obtain the value of the fitted
line at any required value of x.

In order to define the full set of B-splines required for
e above representation, it is necessary to choose four extra,
tificial, knots at or outside each end of the data range.
utine EO2BA chooses each set of four to coincide at the end
int, a choice which provides a number of advantages. The
mputation proceeds by setting up the "observation equations"
th the aid of the recurrence relation for B-splines due to

Cox (1972) and de Boor (1972), and then obtaining the least-squares solution of these equations by a variant of Givens rotations (Gentleman (1973)), taking advantage of the special structure of the equations stemming from the use of B-splines.

3.2. Constraints

An enhancement of the spline fitting routine EO2BA is in preparation which allows function and derivative constraints to be imposed on the spline at the ends of its range. The constraints are of the same type as for polynomials (Section 2.2), namely all derivatives up to a chosen derivative (not greater than third) specified. The process follows the same route, considering the fitting function in the form (c.f. equation (2.2.1)),

$$s(x) = \mu(x) + \eta(x), \tag{3.2.1}$$

where $\mu(x)$ is a spline which satisfies the constraints,

and $\eta(x)$ is the most general spline (with the given knots) which has a zero value (or derivative) whenever $s(x)$ is required to have a specified value (or derivative).

These two functions can be derived very conveniently in the spline case if the artificial knots required to define the B-splines are taken to coincide at the end-points of the range. This is because of the particular pattern of zeros then exhibited by the B-splines and their derivatives at these points. For example, at the left-hand end-point $x=a$, say, we have

	$N_1(x)$	$N_2(x)$	$N_3(x)$	$N_4(x)$	$N_5(x)$	$N_6(x)$	...
Value at $x = a$	X	O	O	O	O	O	...
1st derivative	X	X	O	O	O	O	...
2nd derivative	X	X	X	O	O	O	...
3rd derivative	X	X	X	X	O	O	...

where the crosses denote non-zero values. We see from the first row of the table that $N_1(x)$ is the only B-spline to have a non-zero value at $x = a$, and thus specification of the value of $s(x)$ in (3.1.1) at this point directly determines the coefficient of $N_1(x)$. In particular, a spline which has zero value at $x = a$ must have $c_1 = 0$. In other words, if we simply require $s(a) = \alpha$, say, we must take

$$\eta(x) = c_2N_2(x) + c_3N_3(x) + \ldots + c_qN_q(x),$$

and we may set

$$\mu(x) = \frac{\alpha}{N_1(a)} N_1(x),$$

where $N_1(a)$ happens to be unity.

Similarly, we see from the first three rows of the table that only the first three B-splines can provide non-zero values for the spline $s(x)$ and its first two derivatives at $x = a$. Thus specification of these values directly determines the values of c_1, c_2 and c_3. In other words, if we require $s(a) = \alpha$, $s'(a) = \beta$ and $s''(a) = \gamma$, say, we must take

$$\eta(x) = c_4N_4(x)+c_5N_5(x) + \ldots + c_qN_q(x),$$

and we may set

$$\mu(x) = c_1N_1(x) + c_2N_2(x) + c_3N_3(x),$$

in which the coefficients are the solution of the equations

$$\begin{aligned} c_1N_1(a) &= \alpha \\ c_1N_1'(a) + c_2N_2'(a) &= \beta \\ c_1N_1''(a) + c_2N_2''(a) + c_3N_3''(a) &= \gamma, \end{aligned}$$

which involves only a forward substitution.

Thus, with end constraints on the spline, the determination of its coefficients splits into two parts. One set of coefficients is determined directly from the end conditions, and the remaining coefficients, contained in $\eta(x)$, are obtained by fitting $\eta(x)$ to modified data values Y_r, given by

$$Y_r = y_r - \mu(x_r).$$

The fitting process is the same as for the unconstrained case, except for the omission of the appropriate number of initial columns (and/or final columns if constraints at the right-hand end point are required) of the observation matrix, and the modification of the ordinate values.

3.3. *Parametric Spline Curves*

Another enhancement to the spline fitting routine EO2BA is being carried out to facilitate the fitting of data related to general curves (i.e. curves which cannot be expressed as single-valued functions of the independent variable), like the data in Fig. 1.

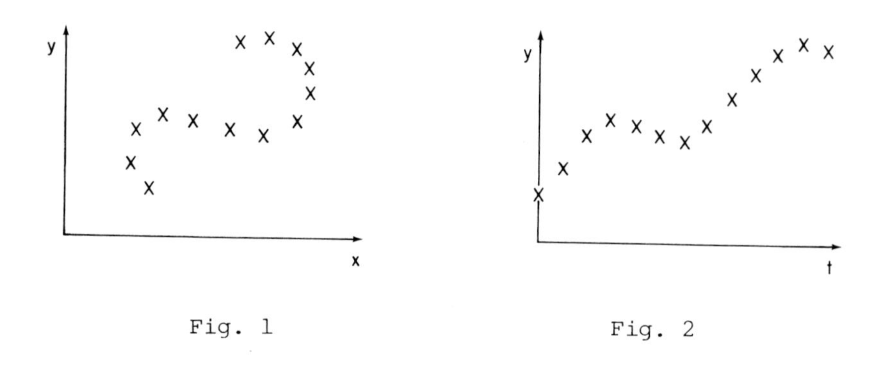

Fig. 1 Fig. 2

General curves in a plane can be represented in parametric form:

$$\begin{aligned} x &= f_1(t) \\ y &= f_2(t), \end{aligned} \tag{3.3.1}$$

where f_1 and f_2 are single-valued functions of some appropriate parameter t. This could, for instance, be arc-length, the distance measured along the curve from its end. Then as the parameter value moves over its range, the point (x,y), given by (3.3.1), describes a curve in the (x,y) plane.

For approximating data from such a curve we may use the parametric spline

$$\begin{aligned} x &= s_1(t) \\ y &= s_2(t), \end{aligned} \tag{3.3.2}$$

where s_1 and s_2 are cubic splines. Given a set of points (x_i, y_i), such as those in Fig. 1, with the suffix i indicating their order along the underlying curve, suppose first of all that we wish to find a parametric spline passing exactly through them. To each point we must first allocate a parameter value: we might take $t_i = i$. Then we can consider the points (t_i, y_i), illustrated in Fig. 2, and fit a spline $s_2(t)$ passing exactly through them. Similarly we can derive a spline $s_1(t)$ passing exactly through the points (t_i, x_i). Then, as t moves over its range, the point $(s_1(t), s_2(t))$ will describe a curve in the (x,y) plane passing through all the points (x_i, y_i). This is true whatever values we assigned to the t_i, but for the curve to pass through the points in the right order, the t_i values must be monotonic. In practice, however, to get a satisfactory curve it has been found advisable to use parameter values which are a reasonable approximation to arc length. Arc length is not available *ab initio* and so it is common practice to use cumulative chord-length instead, that is to say, distance along the polygon formed by joining adjacent points by straight lines. Curves in 3 dimensions can be treated in the same way by adding $z = s_3(t)$ to the equations (3.3.2).

When the data points have errors which we wish to smooth out, the position is not quite so straightforward. Nevertheless, if the errors are small compared with the distances between the data points, so that the latter still give reasonable approximations to arc length, the method of Earnshaw and Yuille (1971) works quite well. In this, the parameterization proceeds as described above for interpolation, but the point sets (t_i, x_i) and (t_i, y_i), and (t_i, z_i) if present, are fitted using least squares.

This fitting can be carried out using EO2BA with each set of points in turn. However, when the same knot-set is chosen for each variable, which is usual, the observation equations have the same left-hand sides in each case, so that EO2BA would simply be repeating most of the computation. Consequently, we are preparing a routine which will simultaneously fit a number of sets of data with the same abscissa values and knots. This, together with a small routine for setting up the parameter values, and an evaluation routine similarly dealing with several curves simultaneously, will form the basis of the new facility for parametric fitting. The fitting routine will allow end constraints, as described in Section 3.2.

3.4. Derivatives and Integrals

The evaluation routine mentioned at the end of the previous section will provide not only the value of one or more cubic splines at a given x-value but also the corresponding first three derivatives of each spline (fourth and higher derivatives being zero, of course).

A further routine will compute the definite integral of a spline over its complete range.

3.5. Simplified Routines

The spline fitting routine EO2BA leaves the choice of interior knots to the user. We find that, after a little experience, the user requires only one or two trials to arrive at a good choice of knots in most instances. With the first-time user in mind, however, we are experimenting with a driver for the above routine which will itself place the knots according to a simple strategy. This should yield a satisfactory fit in many cases, but at any rate provide the user with an initial set which he can improve upon. The criterion stems from the observations that

(a) knots need to be more closely spaced in regions where the underlying curve exhibits rapid changes in shape than where it changes slowly,
(b) in practice, data points are often supplied in a similar pattern, being more closely spaced around a sharp peak, for instance.

The strategy is then to arrange the knots to have a roughly equal number of data points between each adjacent pair of knots. The user, by specifying several values for the number of knots, can readily obtain some trial approximations.

The routine EO2BA is also effective for spline interpolation, i.e. passing a cubic spline exactly through the data points (x_r, y_r), $r = 1,2,\dots,m$. For this purpose, weights are irrelevant, the number of interior knots is known $(m-4)$, and a good choice of these knots for many problems is the set of data values x_r, $r=3,4,\dots,m-2$. A driver routine will be provided for the EO1 (Interpolation) chapter with a correspondingly reduced input parameter list.

4. SPLINE SURFACES

We may recall that a bicubic spline surface is defined on a rectangle in the (x,y) plane, where x and y denote the two independent variables. The rectangle is divided into panels (Fig. 3), and within each panel the spline is a bicubic

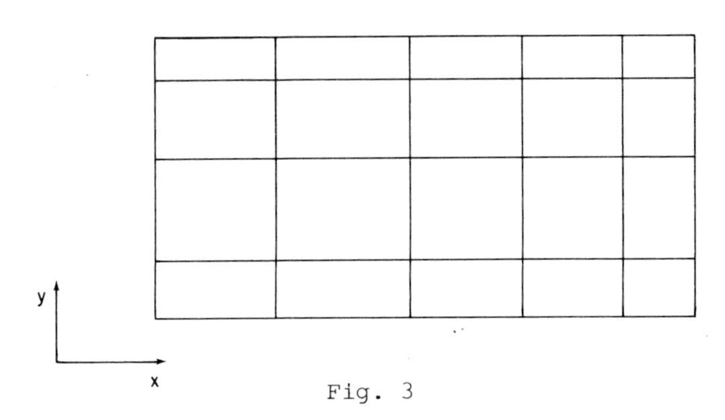

Fig. 3

polynomial, having the form

$$\sum_{i=0}^{3} \sum_{j=0}^{3} a_{ij} x^i y^j .$$

These polynomials join together with continuity up to second derivative across the panel boundaries. Such a spline, $s(x,y)$ say, can be represented in terms of cubic B-splines in each of the variables x and y separately:

$$s(x,y) = \sum_{i=1}^{p} \sum_{j=1}^{q} c_{ij} N_i(x) N_j^*(y), \tag{4.1}$$

where the knots on which the normalized B-splines $N_i(x)$ are defined are the x-values of the panel boundaries parallel to the y axis, plus the extra artificial knots at the ends of the x-range, and the normalized B-splines $N_j^*(y)$ are defined correspondingly with respect to the y-values of the panel boundaries parallel to the x axis.

4.1. Mark 6 Routines

Routine EO2DA fits bicubic splines by least squares to arbitrary data (x_r, y_r, z_r) with weights w_r, $r=1,2,\ldots,m$. The user chooses the two interior knot-sets within the respective data ranges, and then the computation proceeds as decribed in Hayes and Halliday (1974), with two subsequent improvements. The first is the use of normalized B-splines instead of the original B-splines, giving a better-balanced "observation matrix" (the 16 elements in each row summing to unity). The second is the use of Givens rotations instead of Householder transformations, substantially reducing the storage requirement (only one row of the observation matrix being stored at one time). The latter is particularly important in the surface case, where storage is often at a premium. Thus the basic pattern of the computation is similar to that for the curve case, except that, because it often arises in surface problems, the routine must be capable of dealing with the rank-deficient situation. In this event, the routine computes the minimal length solution, i.e. that least-squares solution which gives a minimum value to the sum of squares of the c_{ij}.

Routine EO2DB is provided to evaluate the fitted spline (4.1) at any given x and y values, once the c_{ij} have been obtained.

4.2. Data on Rectangular Mesh

A bicubic spline fitting routine is planned for the special case when the data points lie at the nodes of a rectangular mesh in the (x,y) plane, and are of equal weight. Again the knot-set for each of the two variables will be chosen by the user. We may denote the lines of the mesh by $x = x_r$ and $y = y_s$, for $r = 1,2, \ldots, m$ and $s = 1,2, \ldots, n$, and we are given a z value at each of the intersection points (x_r, y_s). This problem can be solved by two stages of curve fitting, one in the y direction, and one in the x direction. In the first stage, for each value of r in turn, a cubic spline in y, say $s_r(y)$, is fitted to the z-values on the line $x = x_r$, so that we derive the coefficients d_{jr} in

$$s_r(y) = \sum_{j=1}^{q} d_{jr} N_j^*(y), \qquad r=1,2, \ldots, m.$$

In the second stage, for each value of j in turn, a cubic spline in x, say $d_j(x)$, is fitted to the coefficients d_{jr}, $r = 1,2, \ldots, m$, yielding the coefficients c_{ij} in

$$d_j(x) = \sum_{i=1}^{p} c_{ij} N_i(x), \qquad j=1,2,\ldots, q.$$

The fitted surface is then given by

$$s(x,y) = \sum_{i=1}^{p} \sum_{j=1}^{q} c_{ij} N_i(x) N_j^*(y),$$

and is the same solution as would be obtained by fitting the data in a single stage using EO2DA (Section 4.1).

We note that all the curve fits carried out in the same direction have the same knots and the same values of the independent variable. Thus the basic subroutine required is that described in Section 3.3 for facilitating parametric fitting: the one which deals with several sets of dependent variable values simultaneously.

The routine outlined above is suited also to interpolating the same kind of data, i.e. data given at the nodes of a rectangular mesh. Therefore, as with univariate interpolation (see Section 3.5, last paragraph), a driver routine with a reduced list of input parameters is proposed for the EO1 (Interpolation) chapter. Again the driver will set the interior knots for the variable x to the values x_r, $r=3,4,\ldots,m-2$, and those for the variable y to the values y_s, $s=3,4,\ldots,n-2$.

5. POLYNOMIAL SURFACES

Currently there is no polynomial surface-fitting routine in the library.

5.1. Data on Lines

In preparation is a routine based on the method of Clenshaw and Hayes (1965) for the case when, viewed in the (x,y) plane (Fig. 4), the data points lie on a set of lines $y=y_s$ $(s=1,2, \ldots,n)$ parallel to the x axis. Such data arise, for example, in the common type of experiment in which the relationship between

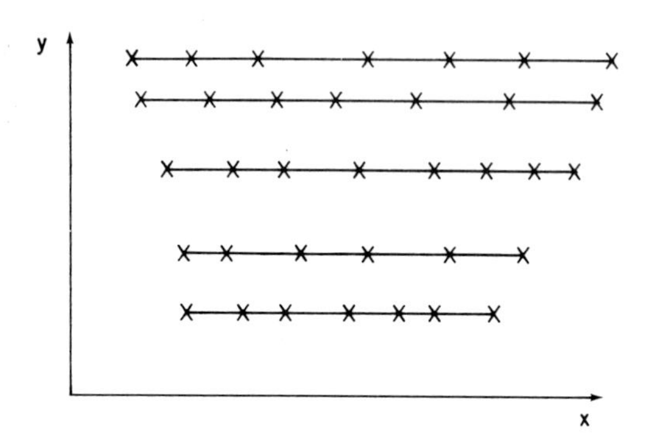

Fig. 4

the dependent and one independent variable investigated at each of a number of values of the other variable. In this method a polynomial $f_s(X)$ in Chebyshev-series form is fitted to the data on the line $y=y_s$, for each value of s in turn, and so we obtain the coefficients α_{is} in

$$f_s(X) = \sum_{i=0}^{k} \alpha_{is} T_i(X), \qquad s=1,2, \ldots, n,$$

where X is the variable x normalized so that its data values span the range -1 to +1. Then, for each value of i in turn, a polynomial $\alpha_i(Y)$ is fitted to the coefficients α_{is}, $s=1,2,\ldots,$ yielding

$$\alpha_i(Y) = \sum_{j=0}^{\ell} a_{ij} T_j(Y), \qquad i=0,1, \ldots, k,$$

where Y is the normalized y. The fitted surface is then

$$f(X,Y) = \sum_{i=0}^{k} \sum_{j=0}^{\ell} a_{ij} T_i(X) T_j(Y).$$

Unsatisfactory results can be obtained if the data region is not rectangular, that is, if the first (and similarly the last) x-value on every line is not roughly the same. Thus provision will be made to map the data region onto a rectangle before starting the fitting process.

A facility for applying function and derivative constraints will be added later.

5.2. Arbitrary Data

For the case of data scattered arbitrarily in the (x,y) plane, we are writing three routines with a view to comparing their performances and selecting the best one or two. They are based respectively on the following three methods.

(1) That of D.G. Hayes (1969). It is a generalization to two independent variables of the method of Clenshaw (1960), generating orthogonal bivariate polynomials which are represented in Chebyshev-series form. The computation is arranged, however, in a way which avoids the need to evaluate the orthogonal polynomials at the data points.
(2) A straightforward application of a linear least-squares algorithm (using Givens rotations on the observation matrix) to the form

$$\sum_{i,j} a_{ij} T_i(X) T_j(Y).$$

(3) A linear least-squares approach to the same form via the normal equations, but making use of the product rule for Chebyshev polynomials, thus substantially reducing the number of multiplications required (Hayes (1974) Section 3.1).

The final choice is likely to be a matter of balancing storage, computing time and stability considerations: for example, the last method will probably prove to be the fastest while being the least stable.

6. ℓ_1 AND ℓ_∞ NORMS

The only routine under this heading presently in the library is EO2AC. It fits a polynomial to data points (x_r, y_r), r=1,2, ...,m, using the ℓ_∞ criterion instead of the least-squares criterion. The polynomial is represented in power-series form

$$P(x) = b_0 + b_1 x + b_2 x^2 + \ldots + b_k x^k,$$

(so it is advisable to normalize the range to (-1,+1) before entering the routine), and the exchange algorithm (Stiefel (1959)) is used to determine the b_i to minimize the ℓ_∞ norm, namely

$$\max_i |P(x_i) - y_i| .$$

Currently being processed are two ℓ_1 routines for the general linear case, i.e. the approximating function L(x) has the form

$$L(x) = \sum_{i=1}^{k} a_i \phi_i(x)$$

where the ϕ_i are user-specified functions of a vector x of an arbitrary number of variables, and the coefficients a_i are determined so as to minimize the ℓ_1 norm f(a), where

$$f(a) = \sum_{r=1}^{m} |y_r - L(x_r)| .$$

Here, a is the vector of coefficients and the x_r and y_r are respectively data values of (the vector) x and of the dependent variable y.

One of these routines is by Barrodale and Roberts (1973), being based on a linear-programming formulation of classical type, in which all the unknowns must be cast in non-negative form. The simplex method for solving the linear programme is modified, however, in a way which allows several neighbouring simplex vertices to be passed through in a single iteration. It results in a very efficient routine.

The other routine is by R. H. Bartels and A. R. Conn of the John Hopkins University and the University of Waterloo respectively. It additionally allows linear inequality constraints,

$$C\,a \geq d, \quad \text{say},$$

to be imposed on the solution via an exact penalty function

$$g(a) = \gamma\, f(a) - \sum_{i=1}^{\ell} \min(0, c_i^T a - d_i),$$

where γ is a penalty parameter, c_i^T is the ith row of the matrix C, and d_i is the ith element of the vector d. The routine utilizes the stable updating procedures of an orthogonal factorization as developed in recent years in the optimization field (Gill, Golub, Murray and Saunders (1974)). It proceeds much like the simplex method, but does not move from vertex to vertex, nor require non-negativity.

The Barrodale-Roberts algorithm, though, has a much shorter code and has proved remarkably stable in practice.

Following these routines will be two for the general linear case using the ℓ_∞ norm, one from Barrodale and Phillips (1975) and one from Conn.

7. REFERENCES

Barrodale, I. and Phillips, C. (1975) "Algorithm 495, Solution of an Overdetermined System of Linear Equations in the Chebyshev Norm". *Assoc. comput. Mach. Trans. math. Software*, **1**, 264-270.

Barrodale, I. and Roberts, F.D.K. (1973) "An Improved Algorithm for Discrete ℓ_1 Linear Approximation". *SIAM J. numer. Anal.* **10**, 839-848.

Clenshaw, C.W. (1960) "Curve Fitting with a Digital Computer". *Comput. J.* **2**, 170-173.

Clenshaw, C.W. (1962) "Chebyshev Series for Mathematical Functions. NPL Mathematical Tables, Vol. 5". HMSO, London.

Clenshaw, C.W. and Hayes, J.G. (1965) "Curve and Surface Fitting". *J. Inst. Maths. Applics.* **1**, 164-183.

Cox, M.G. (1972) "The Numerical Evaluation of B-splines". *J. Inst. Maths. Applics.* **10**, 134-149.

Cox, M.G. (1977) "The Incorporation of Boundary Conditions in Spline Approximation Problems". National Physical Laboratory Report NAC 80.

De Boor, C. (1972) "On Calculating with B-splines". *J. Approx. Theory* **6**, 50-62.

Earnshaw, J.L. and Yuille, I.M. (1971) "A Method of Fitting Parametric Equations for Curves and Surfaces to Sets of Points Defining Them Approximately". *Comput. Aided Des.* **3**, 19-22.

Gentleman, W.M. (1973) "Least Squares Computations by Givens Transformations without Square Roots". *J. Inst. Maths. Applics.* **12**, 329-336.

Gill, P.E., Golub, G.H., Murray, W. and Saunders, M.A. (1974) "Methods for Modifying Matrix Factorizations". *Maths. Comput.* **28**, 505-535.

Hayes, D.G. (1969) "A Method of Storing the Orthogonal Polynomials Used for Curve and Surface Fitting". *Comput. J.* **12**, 148-150.

Hayes, J.G. (1974) "Numerical Methods for Curve and Surface Fitting". *Bull. Inst. Maths. Applics.* **10**, 144-152.

Hayes, J.G. and Halliday, J. (1974) "The Least-squares Fitting of Cubic Spline Surfaces to General Data Sets". *J. Inst. Maths Applics.* **14**, 89-103.

Stiefel, E.L.(1959) "Numerical Methods of Tchebycheff Approximation". In "On Numerical Approximation" ed. R. E. Langer, 217-232. University of Wisconsin Press, Madison.

III. 2

CURVE FITTING USING THE ROTHAMSTED MAXIMUM LIKELIHOOD PROGRAM

G.J.S. Ross

(Statistics Department, Rothamsted Experimental Station, Harpenden)

1. INTRODUCTION

The Rothamsted Maximum Likelihood Program (MLP) has been developed to provide the non-specialist user with a means of fitting a wide range of non-linear models to data. While the models arose originally in the context of biological and agricultural research, many of them can be applied to other fields such as medicine, economics and engineering, for the appropriate statistical framework is often the same.

The program fits either standard models or user-defined models. Standard models are requested by name and are given individual treatment by the program, whereas user-defined models are specified as a sequence of calculations giving access to the general optimization and output facilities but with no guarantee of satisfactory treatment. The standard models are the simplest non-linear (and linear) models that have been found useful in practical data analysis, but new models are being added as the program is developed.

In this paper I shall describe the curve-fitting facilities, for comparison with the polynomial and spline-fitting routines described by Hayes (197) in this volume.

1.1 The place of curve-fitting within MLP

MLP is organized as a tree-structure, as follows:

1. Root section: Program control and user-language, data manipulation, optimization, regression and graphics.
2. Model classes: Curve fitting, discrete distributions, continuous distributions, probit analysis, linear regression, genetic models, user-defined models, etc.
3. Curve fitting: Exponentials, growth curves, rational curves, polynomials etc.

Curves not included in the curve-fitting section must be fitted as user-defined models. At present splines and Fourier curves belong to this category, although the use of logical functions makes it a relatively simple matter to fit splines with unknown node positions.

Maximum likelihood estimation is a general method which applies whatever the probability distribution of the errors in the model. If the distribution is Normal the method is equivalent to least squares, and the standard curves in MLP are in fact fitted by the method of least squares. Other distributions such as the Binomial (for proportions) and the Poisson (for counts) are provided in the user-defined section, although Probit Analysis is an explicit formulation of a curve-fitting problem with binomial errors.

1.2 Curves and their properties

Functions suitable as models for curve fitting have the following essential properties:

1. Single-valuedness: The function $y = f(x)$ has a unique value for all x within a defined range.
2. Scale-invariance: The function does not depend on choice of units for x or y.
3. Continuity: Vertical asymptotes must not occur within the data range, or in the range into which extrapolation is desired.

There are other desirable properties, such as:

4. Smoothness: Gradients should be no steeper than those expected of a free hand curve drawn to the plotted data.
5. Extrapolability: Extrapolated values should be consistent with reasonable expectation.
6. Origin-invariance: The function does not depend on choice of origin, unless particular physical interpretation applies to the origin (e.g. that y must be non-negative, or that y must be fixed when $x = 0$).
7. Parsimony of parameters: The function should not require more parameters than are justified by the data.
8. Interpretability: The fitted curve gives some insight into possible underlying mechanism that might generate the model.
9. Algebraic simplicity: These are practical advantages in curves that can be computed, differentiated, inverted or integrated on a desk machine.

The following sections describe the most important families of curves in practical use, and some examples are illustrated in Fig. 1.

1.2.1 Polynomials

For most statistical texts, curves are polynomials, or models that can be transformed into polynomials by some change of scale. The advantages of polynomials are that they may always be fitted to data by a unique non-iterative algorithm, and that the appropriate degree to choose may be assessed by a statistical test. They are also algebraically simple. However they are unsatisfactory in practice because of their lack of smoothness and extrapolability with the result that the fitted curves are very sensitive to choice of data range and that they are incapable of fitting data with horizontal asymptotes. The fits often appear satisfactory on the simple criterion of the residual mean squares, i.e., the errors at the data points themselves, but a graph of the fitted curve may show unreasonable fluctuations, particularly at the ends of the range, and the practice of down-weighting the end-points is only a partial palliative.

1.2.2 Splines

The popularity of spline functions can be attributed to the fact that they retain the advantages of polynomials while improving the smoothness and goodness of fit, and thus may be used over a wider class of problem. However the improved fit is obtained at the expense of extra parameters, both explicit (the coefficients of the constituent polynomials) and implicit (the guessed or fitted optimal position of the nodes). Splines are still unsuitable if interpretability or extrapolability are required. If the position of the nodes is to be fitted the model becomes non-linear and uniqueness of solutions is not guaranteed. Therefore whereas they may be satisfactory for fitting calibration curves to accurately determined data points they are not necessarily the best curves for general use with more variable data.

1.2.3 Exponential curves

Exponential curves of the form

$$y = \beta_0 + \sum_{j=1}^{m} \beta_j \exp(-k_j x) \qquad (1)$$

Fig. 1 Popular curves used in Curve Fitting.

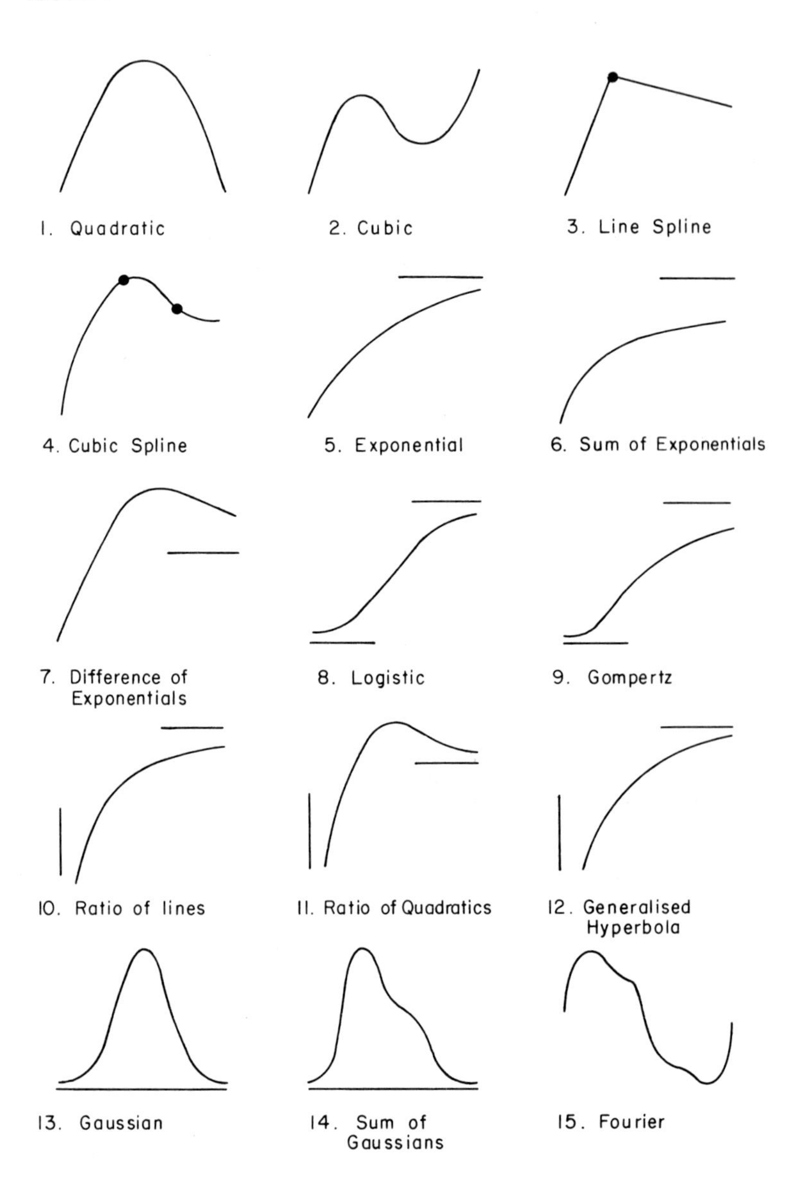

arise as solutions of linear ordinary differential equations whose auxiliary equations have real roots k_j of equal sign, so that the curve has a horizontal asymptote. Models both of growth and of decay lead to curves of this family, as do "compartment models" in which there are flows of material from one compartment to another at rates proportional to the amount in the source compartment. But the simplest cases, m = 1 or 2, are valuable purely as empirical curves wherever a process is approaching a stable state, whatever the underlying model. The curves may be further specialized by imposing constraints such as a fixed asymptote or fixed origin, or a fixed slope at the origin, as in the simplest two compartment model which yields curves such as

$$y = \alpha(k_1 \exp(-k_2 x) - k_2 \exp(-k_1 x))/(k_1 - k_2) \qquad (2)$$

or $$y = \alpha(\exp(-k_2 x) - \exp(-k_1 x)) \qquad (3)$$

depending on which part of the system is observed.

Certain limiting cases must be noted. When k_1 tends to 0 the term

$$\beta_1 \exp(-k_1 x)$$

must be replaced by $\beta_1 x$,

and when k_1 tends to k_2, the term must be replaced by

$$\beta_1 x \exp(-k_1 x).$$

It is not useful to postulate too many terms, but rather to replace the spectrum of discrete k_j values by a continuous spectrum $\beta(k)$ giving rise to a curve

$$y = \beta_0 + \int_0^\infty \beta(k) \exp(-kx)dk \qquad (4)$$

of which a simple example is the generalized hyperbola

$$y = \beta_0 + \beta_1 (1 + \gamma x)^{-\delta} \quad (5)$$

when the spectrum of k has a Gamma distribution. Curves of the form

$$y = \beta_0 + \Sigma\beta_j x^{\theta_j} \quad (6)$$

also belong to this family after a log transformation of the x-axis.

1.2.4 Growth Curves

Growth curves are used in the study of biological growth, in population dynamics and in economics. The differential equation

$$\frac{dy}{dx} = \frac{\beta y}{\theta} \left(1 - \left(\frac{y}{c}\right)^{\theta}\right) \qquad (0 < y < c) \quad (7)$$

provides a range of curves to fit data that grow initially at an exponential rate but tend to an asymptote $y = c$ as the expression in brackets approaches zero, (see, for example, Nelder (1961)). The explicit form of the curve can be written

$$y = c(1 + \theta \exp(-\beta(x-\mu))^{-1/\theta}) \quad (8)$$

and is called the Generalized Logistic Curve. The special case $\theta = 1$ is often called the Logistic Curve, and the limiting case as $\theta \to 0$,

$$y = c \exp(-\exp(-\beta(x-\mu))) \quad (9)$$

is called the Gompertz Curve.

The curves are monotone, since y' is a function of y only, and therefore cannot be used if the growth rises to a maximum and falls again.

1.2.5 Rational Curves

Rational curves are ratios of polynomials,

$$y = P_n(x)/Q_m(x) \quad (10)$$

a polynomial of degree n divided by a polynomial of degree m. Usually we require $n \geq m$ to provide a general form of curve, otherwise the horizontal asymptote is fixed at $y = 0$. There are $n+m+1$ parameters to be estimated, as one coefficient may be cancelled in equation (10).

The curves have vertical asymptotes corresponding to real roots of $Q_m(x)$, and if m is odd there is at least one real root. The curves are therefore only suitable provided there are no vertical asymptotes within the data range or in the range of extrapolation. When these conditions are satisfied the curves form a useful empirical family that tend to require fewer parameters than polynomials or splines fitted to the same data.

Rational curves arise theoretically in such fields as chemical kinetics, and they have been used in many applications in agricultural research and plant physiology (Nelder (1966), Thornley (1976)).

1.2.6 Other useful non-linear curves

Fourier curves are linear in the parameters if the period is known, but sometimes the period is to be estimated which makes the model non-linear.

Gaussian curves or mixtures of Gaussians,

$$y = \beta_0 + \sum_{j=1}^{m} \beta_j (\exp(-\frac{(x-\mu_j)^2}{2\sigma_j^2})/\sqrt{2\pi\sigma_j^2}) \tag{11}$$

are required in spectrography because the coefficients β_j measure directly the contribution of each component. They are easily fitted if the components are well separated, but when two or more overlap, or when the data are very scattered or incomplete it is much more difficult to fit the curve. Fitting is simplified if it can be assumed that all the variances σ_j^2 are equal.

Other curves such as square root or logarithmic curves may arise from theoretical considerations but they tend to be unsuitable as general empirical families if they are undefined over part of the range of x.

1.3 User requirements of a curve-fitting package

The basic user requirements of a curve-fitting package are as follows:

1. the ability to select and compare several different curves to be fitted to the same data,
2. the ability to compare several sets of data fitted by the same model,
3. optional weighting of the observations,
4. estimation of parameters in the model, fitted values and the residual sum of squares.

Desirable extra facilities are as follows:

5. graphical display of the data and fitted curve,
6. standard errors of parameters and correlations between them,
7. predicted values of y at nominated values of x,
8. other functions of parameters such as the slope at any point, maxima and minima, or areas under the curve, and standard errors of prediction,
9. intercepts of the curve with nominated values of y (a root-finding problem),
10. graphical display of the residual sum of squares as a function of parameters,
11. adequate reporting of any difficulties experienced in fitting the curve, with suggestions or alternatives,
12. automatic choice of fitting procedure, initial estimates of parameters and other necessary quantities,
13. partial derivatives of fitted values with respect to the parameters as aids to improved design of the distribution of data points.

1.3.1 User language for curve-fitting using MLP

The curve-fitting facilities within MLP constitute an attempt to provide some or all of the above requirements while maintaining a sufficiently simple user language to encourage the non-specialist to make use of the program in an appropriate manner. The user language makes use of free format directives and mnemonics, and may be used in batch or interactive mode.

To fit a curve one need only supply the data values, for example:

Fig. 2 Output from curve fitting in MLP

```
Y = A + B*R**X    R = EXP(-K)
PASSING THROUGH X=0, Y=   0.000
      PARAMETER            S.E.          CORRELATIONS
 1  R        0.86429     0.02224       1.000
 2  B      -33.74347     3.29650      -0.965   1.000
    A       33.74347
    K        0.14584     0.02573
          X          Y        E(Y)   WTD.RES.      SLOPE
     1.0000     3.4000     4.5792    -1.1792      4.2534
     2.0000     6.1000     8.5370    -2.4370      3.6762
     3.0000    11.7000    11.9577    -0.2577      3.1773
     4.0000    14.9000    14.9142    -0.0142      2.7461
     5.0000    18.8000    17.4695     1.3305      2.3735
     6.0000    20.2000    19.6779     0.5220      2.0514
     8.0000    25.1000    23.2365     1.8635      1.5324
    10.0000    24.8000    25.8947    -1.0947      1.1447
    12.0000    27.2000    27.8804    -0.6804      0.8551
     7.0000               21.5867                 1.7730
     9.0000               24.6624                 1.3244
    11.0000               26.9599                 0.9893
    15.0000               29.9581                 0.5521
    20.0000               31.9178                 0.2663
RESIDUAL M.S.      2.081882    D.F.      7
RESIDUAL S.S.     14.573178
          '.........'.........'.........'.........'
 27.500-                                        ..'X-
       -                                    ..''
       -                           X   .''X
       -                             .''
       -                          ..'
 21.250-                        .'                -
       -                     X'
       -                  X.'
       -                 .'
       -                .
 15.000-             X'                           -
       -            '
       -          .'
       -         .X
       -        .
  8.750-       '                                  -
       -      '
       -     ' X
       -    '
       -   X
```

```
DATA    1    2    3    4    5    6    8    10   12
        3.4  6.1  11.7 14.9 18.8 20.2 25.1 24.8 27.2;
```

and a list of model options

```
CMODEL = EXPONENTIAL  CORIGIN = 0  CVARIANCE = RMS
```

and any additional output options,

```
CPLOT   CSLOPE
```

an optional list of x values for prediction

```
EXTRA  7   9   11   15   20;
```

and an executive instruction

```
FIT CURVE
```

to get the program to fit an exponential curve through the origin, to use the residual mean squares as the estimate of error variance, to obtain a plot, and values of the slope of the curve at each data point and at the extra points listed. The results are shown in Fig. 2. Further curves or data sets may then be specified without altering any of the other options.

Several sets of data may be supplied simultaneously under one DATA statement: they may be separated by solidi (/), or generated automatically from a grouping factor. The analysis then proceeds for each set of data individually, and if the option CPCA is specified a Parallel Curve Analysis is performed in which some of the parameters are made to be the same in each curve while others are allowed to vary. In this way it is possible to test whether the data sets can be fitted by the same curve.

In order to obtain information on the residual sum of squares as a function of parameters directives MAP and DISPLAY provide various forms of systematic evaluation and contour plotting, and from these it can be seen whether there are alternative solutions, and also how the parameters are related in a more general way than that provided by the correlations (Fig. 3). To obtain alternative fits, if they exist, a directive INITIAL provides starting values for the parameters.

Fig. 3 Contours of equal sums of squares for the data of Fig. 2.

The contour interval has been chosen so that the unit contour (surrounding the points marked 0) corresponds to an approximate 95% confidence region for the parameters. Values above 10 are replaced by X and odd values are omitted.

The diagram is derived from MLP output.

```
GRID VALUES OF FUNCTION IN UNITS OF   34.2975

PARAMETER 1    0.7600   TO    0.9200
PARAMETER 2  -20.0000   TO  -52.0000

  101.0722   73.0751   44.2950   17.9782    1.8623
   50.3364   33.6949   17.6039    4.7961    1.3755
   17.5002    9.6513    3.2453    0.5064    6.0662
    2.5636    0.9443    1.2191    5.1091   15.9343
    5.5268    7.5739   11.5253   18.6040   30.9799

       0.76      0.80      0.84      0.88      0.92
-52.00 -XXXXXXXXXXXXXXXXXXXXXXXXXXXXXXXXXX  64  2 -
        XXXXXXXXXXXXXXXXXXXXXXXXXXXXXXXXX  64 22
        XXXXXXXXXXXXXXXXXXXXXXXXXXXXXXXX  64 22
        XXXXXXXXXXXXXXXXXXXXXXXXXXXXXXX8  4 22  0
        XXXXXXXXXXXXXXXXXXXXXXXXXXXXX 86 4 22   0
        XXXXXXXXXXXXXXXXXXXXXXXXXXXX8 6 4 2  00
-44.00 -XXXXXXXXXXXXXXXXXXXXXXXXXX 86 4  2    00  -
        XXXXXXXXXXXXXXXXXXXXXXXX 8 6 4 22  000
        XXXXXXXXXXXXXXXXXXXXXX 8 6 4  2  0000   2
        XXXXXXXXXXXXXXXXXXX8 66 4  2  00000   2
        XXXXXXXXXXXXXXXXX 8 6  4 22  000000   2
        XXXXXXXXXXXXXX 8 66 44 222  0000000  2  4
-36.00 -XXXXXXXXXX 88 66 44  22   00000000  2 446-
        XXXXXX 88  6  44  222   00000000  22 4 6
        X  88  66  44  222   000000000   2  4 6 8
          66  444   22    0000000000   22 4  68 X
          44   2222    00000000000   22  4 6 8 XX
           2222     00000000000    22  4 66 8XXXX
-28.00 -222       000000000     222  4  6 8 XXXXX-
                00000       2222  44 66 8 XXXXXXX
                       22222   44  6  8 XXXXXXXXX
                222222    444  66  8  XXXXXXXXXXX
        2222222     4444   66  88  XXXXXXXXXXXXXX
```

The user does not have to understand the precise details of the methods used to fit the curves, and these methods are discussed in the next section. He does however have to be aware that a solution does not necessarily exist for every set of data, and that the data must resemble in shape at least some members of the parametric family which the model defines. The program attempts to report back helpfully when no solution exists, and to provide the limiting case which is usually a simpler model.

2. METHODS FOR FITTING NON-LINEAR CURVES

The traditional methods of fitting non-linear curves may be classified as follows (Ross, 1975):

1. approximate solutions obtained by reading off points from a free-hand curve and solving simultaneous equations, or by guessing some parameters and fitting the rest by linear regression,
2. compromising the law of error so that by transforming the variables a linear relation may be obtained, as when fitting $y = be^{kx}$ by taking logs of both sides,
3. iterative solutions such as the Newton-Raphson method using explicit derivatives, or general least-squares routines,
4. use of general optimization routines, with or without derivatives, operating on the residual sum of squares.

The first method has much to recommend it, when time permits, as it enables the investigator to think about his data and the adequacy of the model. Indeed much of the difficulty experienced with general optimization routines may be attributed to a failure to recognise the relationship between the model and the data.

The second method produces the wrong answers, but this may not be serious if errors are small, and may provide suitable initial estimates for the correct analysis. For some curves the method of Nelder and Wedderburn (1972) may be used, as embodied in GLIM (NAG (1977)), in which the transformation to a linear model is combined with adjustments to ensure that the correct error distribution is used.

Non-linear least squares algorithms and optimization algorithms can give solutions, but some skill is required in providing suitable initial estimates and step lengths. A more radical approach, as used in MLP, is outlined below.

2.1 Stable parameters

There is no reason why the defining parameters of the model should be used as variables in the optimization algorithms, apart from algebraic simplicity of differentiation and evaluation. From a numerical and statistical point of view they are hardly ever the best variables to use, and much of the difficulty in non-linear optimization may be attributed to unsuitable choice of variables. The concept of temporary parameter transformation should not be unfamiliar as it is used in fitting polynomials and other linear models by orthogonalization. But whereas linear transformations suffice to simplify a linear model-fitting problem, and the optimum transformation may be found automatically either outside or inside the matrix inversion routine, non-linear models require non-linear transformations to be fully effective. Davidon (1978) has experimented with a class of non-linear transformations within the optimization algorithm, with variable results. In MLP non-linear transformations are chosen on statistical grounds, outside the optimization algorithm, so that the objective function is likely to be easy to optimize. Parameters that are easily estimated are called "stable" (Ross (1970)).

For a parameter to be stable its final estimated value should not depend strongly on the value of any other parameter. In the context of curve-fitting any fitted value on the curve corresponding to a data point is potentially a stable parameter, because it contributes a term to the sum of squares regardless of the other parameters, which must increase with increasing discrepancy between the observed and fitted points. Therefore to fit a p-parameter model we need p well spaced ordinates each of which is close to at least one data point. Other quantities such as areas under the curve, slopes or ratios of ordinates are to a greater or lesser extent also stable. The choice of stable parameters is therefore partly a matter of algebra (choice of ordinates so that the equations of transformation may be easily solved) and partly of statistics (choice of ordinates appropriate to the particular data sets).

The choice of ordinates requires first a preliminary analysis of data so that the range of values may be assessed. A simple fit such as a straight line provides sufficiently accurate initial estimates of parameters, indicators of positive or negative trend, and suitable positions for trial ordinates. If these choices are inappropriate for any reason it may be possible to reassess the situation after a certain number of iterations of optimization. (Poor choices usually occur when the data points are very unevenly spaced, or the model is close to the limits of its possible range.)

Even if a full set of stable parameters cannot be conveniently found a partial set is very much more useful than noneat all. The use of stable parameters makes it more difficult to evaluate analytical derivatives, but algorithms using differences instead of derivatives become more reliable because suitable step lengths are more easily found (for instance as a proportion of the expected range of each parameter).

2.1.1 An example of stable parameters

As a simple example, consider the hyperbola through the origin,

$$y = \alpha x/(1+\beta x) \tag{12}$$

fitted to the following data:

x	2	3	4	6	8	10
y	7	10	12	16	19	21

The parameter α is the gradient at the origin, while the ratio α/β estimates the horizontal asymptote. As neither of these quantities may be easily guessed from the data it is not likely that α or β will be stable. If instead we try two stable ordinates, such as η_4 at $x = 4$ and η_8 at $x = 8$, we find that the model may be written

$$y = \eta_4\eta_8 x/(8(\eta_8-\eta_4)+(2\eta_4-\eta_8)x) \tag{13}$$

and that the residual sum of squares surfaces for the two forms of the model are as illustrated in Fig. 4. From initial values $\eta_4 = 12$, $\eta_8 = 19$ the model is fitted in about 2 iterations whereas in (α,β) space from the starting point (3,0) about 7 iterations are required. The correlation between the estimates of α and β is 0.973 while that between η_4 and η_8 is 0.306. We could therefore have chosen other ordinates more nearly uncorrelated, but our empirical guess is sufficiently stable for estimation purposes.

Fig. 4a Contours with respect to defining parameters.

The contours are obtained from the program output of unscaled sums of squares for the defining parameters of the rectangular hyperbola for the data in the text. Even units digits are printed, and values above 20 appear as X.

```
    0.08      0.09      0.10      0.11      0.12
5.10-XXXXXXXXXXXXXXXXXXXXXXXXXXXXXXXXXXXXXXXX 8 -
     XXXXXXXXXXXXXXXXXXXXXXXXXXXXXXXXXXX  6 4
     XXXXXXXXXXXXXXXXXXXXXXXXXXXXXXX 8  4 2 0
     XXXXXXXXXXXXXXXXXXXXXXXXXXXX  6 42 0  8 6
     XXXXXXXXXXXXXXXXXXXXXXXXX 8  4 2 0 8 66 44
     XXXXXXXXXXXXXXXXXXXXXX  6 42 0 8  6 44  22
4.70-XXXXXXXXXXXXXXXXXXX8  4 2 0 8 6  4  222   -
     XXXXXXXXXXXXXX 86 42 0 8 66 44 222   0000
     XXXXXXXXXXX  64 2 0 8 6 44  22   00000000
     XXXXXXXX8  42 0 8 6  4  22   000000000000
     XXXX 86  2 0 8 6 44 22   0000000000000
     X  64 20 8 6  4  22  0000000000000  2222
4.30- 42 0 8 6 44 22   000000000000   222  44 -
      0 8 6 4  22  000000000000   222  44 66 8
      6  4  2   00000000000   22  44 66 8  0
     44 22   0000000000   222  4  6  8 00 2 4
     22   000000000   222  4  6  8 0  2 4 6 8
       0000000   222  4  6  8 0 2 4 6 8 XXXXX
3.90-0000     22  44  6 88 0 2 4 6 8 XXXXXXXXX-
         222  44 66 8  0 2 4 6 8 XXXXXXXXXXXXX
     22  44  6  8 0 22 46 8 XXXXXXXXXXXXXXXXX
     44  6 88 0 2 4 6 8 XXXXXXXXXXXXXXXXXXXXX
     6 8 00 2 4 6  XXXXXXXXXXXXXXXXXXXXXXXXXX
     0 2 4 6 8 XXXXXXXXXXXXXXXXXXXXXXXXXXXXXXX
3.50-4 6  XXXXXXXXXXXXXXXXXXXXXXXXXXXXXXXXXXX-
    0.08      0.09      0.10      0.11      0.12
```

<u>Fig. 4b</u> Contours with respect to stable parameters.

The contours refer to the same function as in Fig 4a, but in a different coordinate system.

The contours are nearly elliptical, and sensible choice of starting values is easily made.

```
    11.00     11.50     12.00     12.50     13.00
20.00-4  2  00   88   6666      4444444444      -
       22  00  88   666    4444444444444444444
      2  00  88  666    44444              4444
       00  88   66   4444        22222        4
      00  88  66   4444     2222222222222
        88  666  444     2222222222222222222
19.50- 88  66   444   222222            22222  -
      88  66  444   22222                 2222
        66   44    2222        0000         2222
       66   44   2222     00000000000        222
      66   44   2222    00000000000000       222
      6   44   222     0000000000000000      222
19.00-6  44   222     000000000000000000     222-
        444  2222    0000000000000000000     222
        44   222    0000000000000000000    2222
       444  2222    0000000000000000000    222
       44   222     000000000000000000    2222
      444   222     00000000000000000    2222
18.50-444   222      00000000000000     2222   -
      444   2222      00000000000      2222   4
      444   22222       000000     2222    444
       444   22222                22222   444
       444    22222             22222    44
        444    22222222    22222222    444    6
18.00-   444    2222222222222222     444    66 -
    11.00     11.50     12.00     12.50     13.00
```

The expression in terms of stable ordinates shows that the model is suitable provided

$$\eta_4 < \eta_8 < 2\eta_4$$

and if these conditions are violated the wrong sort of hyperbola will be fitted with negative β and with the vertical asymptote at some positive value of x.

A very similar diagram to Fig. 4 is obtained if instead we fit the exponential model

$$y = \beta(1-\exp(-kx)) \qquad (14)$$

to the same data. This is not surprising as the interpolated values between x = 0 and x_8 are very similar for the two families of curve, provided the data may be fitted by both models.

2.2 Separability of linear parameters

It has often been noticed that models of the form

$$y = \Sigma_i \beta_i \, z_i(x,\theta)$$

where the functions z_i may involve non-linear functions of θ may be fitted by linear least squares if values of θ are assumed known. The fitted values of β_i may then be substituted back into the sum of squares to obtain a conditionally minimum sum of squares as a function of θ. Richards (1961) showed that not only is the minimum of this reduced sum of squares $R(\theta)$ the minimum of the complete sum of squares in θ and β, but that the variances and covariances of θ may be obtained directly from the second derivatives of $R(\theta)$, and that the variances of the β_i could be obtained with a little extra calculation. The β_i are called linear parameters.

In the context of curve-fitting it is always possible to identify some linear parameters, if only the over-all scale parameter which makes the solution independent of the scale of y. The linear parameters are often less stable than the non-linear parameters, and it is a common experience with optimization algorithms that it takes a long time to recover from a poor initial choice of linear parameters. For example, to fit the model

$$y = \beta_0 + \beta_1 \exp(-kx) \qquad (15)$$

to the following data

x	1	2	3	4
y	3	7	10	16

it is unlikely that the solution, $k = -0.264$, $\beta_0 = 7.211$, $\beta_1 = 8.025$ would be found by optimization from starting values such as $k = 0.1$, $\beta_0 = 51$, $\beta_1 = -53$ because the search has been started in the wrong region of parameter space. The reduced sum of squares as a function of k only is however very easily optimized provided the value $k = 0$ is avoided (in which case a straight line may be fitted).

In MLP all the curve-fitting models so far provided are optimized in a reduced sum-of-squares space of three or fewer dimensions. This also makes it relatively simple to examine graphically the function being optimized, and to check on the existence of alternative solutions.

2.2.1 Separability and stability

The reduced-sums-of-squares function is not necessarily easy to optimize in spite of its lower dimensionality. As it is bounded above by the upper limit Σy^2 there are regions where the function has a non-positive-definite matrix of second derivatives, and local maxima as well as local minima. It is therefore still necessary to attempt to stabilize the space of non-linear parameters. The stable ordinates of Section 2.1 are not suitable in themselves. Instead we may seek functions of the stable ordinates which estimate the linear parameters, namely differences and ratios of stable ordinates. The linear parameters may then be estimated by linear regression. For example the logistic curve

$$y = c/(1 + \exp(-\beta(x-\mu))) \qquad (16)$$

can be fitted through three equally spaced ordinates (η_1,η_2,η_3), but it is possible to transform from (β,μ) space to the space of ratios of ordinates $(\eta_2/\eta_1,\eta_3/\eta_2)$ and to estimate c by linear regression.

The precise method of stabilization depends on the analytical form of the model, combined with an empirical scaling decision made on the basis of the preliminary data analysis. If the transformation is not algebraically simple it is possible to use a transformation appropriate for a simpler curve similar in shape to the required curve.

2.3 Sequential modelling

The complete process of curve fitting in MLP is a sequence of analyses each analysis providing information for the subsequent stage. The first stage is a simple linear analysis from which scaling constants, initial estimates of parameters and step lengths may be obtained, and invalid data may be rejected. For simple models such as the exponential (15) the next stage is optimization of the stabilized non-linear parameters, but for models such as the double exponential

$$y = \beta_0 + \beta_1 \exp(-k_1 x) + \beta_2 \exp(-k_2 x) \qquad (17)$$

some exploratory optimizations of special cases are carried out first, and initial values are chosen close to the best optimum found.

If convergence is satisfactory the program proceeds to the output stage, but if the optimization algorithm terminates on a constraint there are "second chance" procedures whereby the transformation constants are revised and a further optimization is performed. The second chance is needed to distinguish the cases where no valid solution exists from those in which the distribution of data points made it difficult to select a suitable stable parameter system, but from the progress made so far a better transformation may be achieved. For example if most data points for model (15) lie on or near the asymptote the mid-range may not be a very useful place to choose for one of the stable ordinates, but this information is not readily available without a first attempt to fit the model.

2.3.1 <u>Pathological cases</u>

The method of stable ordinates (Section 2.1) explains very clearly why some data will not be fitted by a particular curve. If the stable ordinates summarise the data, then the only data sets that can be fitted are those for which the transformation to defining parameters is valid. Unfortunately the more complex the model becomes the smaller becomes the region of stable parameter space corresponding to valid models. In contrast to polynomials and other linear models it may make matters worse to generalize a form of curve if the generalization does not improve the shape in the required way.

Ambiguous cases arise when the transformation from stable parameters has multiple roots, the usual cause being that too small a portion of curve is being subjected to comparison with the data. The remedy is to simplify the model, but the user should be warned if possible of what has happened.

MLP prints the limiting case if there is no feasible solution, and suggests which model to try next. It prints warnings if the vertical asymptote of a rational curve occurs within the data range. And for all models it is possible to examine graphically the residual sum of squares function to check the uniqueness of the solution, and the user's initial values may be used to override the default procedures.

3. PROGRAM AVAILABILTIY

MLP is written in ANSI FORTRAN IV, and has at the time of writing been implemented on the following machine ranges: ICL System 4-70, IBM 370, ICL 1906B, Burroughs B 6700 and CDC (Cyber) 7600. It may be obtained under licence agreement by application to The Programs Secretary, Statistics Department, Rothamsted Experimental Station, Harpenden, Herts AL5 2JQ. The program may be used in batch or interactive mode, and requires a minimum of 75K bytes on the ICL system 4-70.

4. REFERENCES

Davidon, W.C. (1978) "Optimization by non-linear scaling", (This volume Chapter VI. 2).

Hayes, J. G. (1978) "Data-fitting algorithms available, in preparation, and in prospect, for the NAG library", (This volume Chapter III. 1).

N.A.G. (1977) GLIM Manual.

Nelder, J.A. (1961) "The fitting of a generalisation of the logistic curve", *Biometrics* **17**, 89-110.

Nelder, J.A. (1966) "Inverse polynomials, a useful group of

multifactor response functions", *Biometrics* **22**, 128-141.

Nelder, J.A. and Wedderburn, R.W.M. (1972) "Generalised linear models", *J.R.Statist.Soc. A* **135**, 370-384.

Richards, F.S.G. (1961) "A method of maximum likelihood estimation", *J.R. Statist.Soc. B* **23**, 469-476.

Ross, G.J.S. (1970) "The efficient use of function minimization in non-linear maximum likelihood estimation", *J.R.Statist.Soc. C* **19**, 205-221.

Ross, G.J.S. (1975) "Simple non-linear modelling for the general user", Proc.40th Session Int.Statist.Inst. Warsaw, Vol. 2, 585-593.

Ross, G.J.S., Kempton, R.A. and Payne, R.W. (1978) "MLP Manual" (in preparation), Rothamsted Experimental Station.

Thornley, J.M. (1976) "Mathematical models in plant physiology", Academic Press, London.

PART IV

COMPUTER AIDED DESIGN AND SIMULATION

IV. 1

INTERACTIVE COMPUTING: A NEW OPPORTUNITY

H.H. Rosenbrock

(University of Manchester Institute of Science and Technology)

1. INTRODUCTION

The title of this paper has given me some difficulty, because the current terminology has no short and simple word for what I want to discuss. "Interactive computing" includes a great deal which will not concern me - for example the on-line editing of programs. An alternative title would have included "computer-aided design", but this is widely identified with geometric design, and this is a rather small part of what I want to discuss.

My subject will be the use of a computer for problem-solving, in such a way that a significant part of the solution is generated in the user's mind, and another significant part is handled by the computer. Specifically, constraints on the solution, and decisions about whether a solution is satisfactory, and if not how it is to be modified, will largely be left to the operator. The computer will analyse specific configurations, and produce from them a concise and informative report to the user to guide his decisions. This report, if it is to provide enough bandwidth for rapid communication, will almost inevitably be in some graphical form.

I contrast this with many other ways in which computers can be used. In batch computing, even with fast turn-round, there is not usually any attempt to set up a dialogue of the type described. Problems are formulated so that the computations give, if possible, the complete solution in one run. The interactive editing of programs is also outside my scope, since again there is no continuing dialogue of the kind I have described.

Computer-aided design of mechanical components is certainly part of the area which I mean to include. Economically, it may be very significant, particularly when it is linked to production. Yet as will appear, I regard this and most of the other existing examples of my theme as only the most elementary beginnings of something of the greatest potential significance. In the next ten years, I would expect interactive computing, in the full sense in which I am using the term, to be one of the most important areas of growth in engineering: and possibly also in other areas where the formulation of problems is at least as difficult as their solution.

2. INTERACTIVE COMPUTING

The ideas of interactive computing and computer-aided design are not new. In the early 1960's an interactive graphical facility was developed in the Electronic Systems Laborator at MIT. The first version of this used largely analogue equipment for doing such tasks as rotation of objects on the screen. Subsequently a fully digital version was developed in association with Project MAC. Many will have seen the "sketchpad" film which demonstrated this early work.

Subsequently computer-aided design facilities were developed for a number of geometrical design problems - that is, design problems concerned with the geometrical shape of objects For example, systems exist for the design of mechanical components, for architectural design and for ship design. These systems can be connected to the production system - for example a mechanical part may be designed, the cutter path may be defined, and a tape may be produced for the numerically-controlle milling machine which will make the part.

Design problems of this kind have many of the features which make interactive computing appropriate, and among them we may list the following.

(i) The design has to satisfy a large number of constraints. These may be connected, for example, with stresses and stress concentration, material to be used, machining methods, assembly, maintenance, etc. The constraints wi be complicated, and therefore difficult to formulate. I their number is large, say 50 or more, it will be very difficult for the designer to list them all. The human mind finds it difficult to produce comprehensive lists of this sort. Some possibility will be overlooked until an example actually arises. It will usually be easier and more efficient for the designer to look at tentative solutions to the problem - if these transgress constrain then he can recognize it at once. The design can then be modified, and a new solution inspected. By proceeding in this way only a few of the possible constraints will be met, and there is no need to produce a full list.

(ii) No standard approach to the problem can be suggested which will deal with all possible difficulties. New difficulties may always arise which have not been met before and for which new methods of solution have to be found.

(iii) The meaning of a "satisfactory design" is complex and ill-defined. There are many desirable properties - cheapness, reliability, high performance, ease of maintenance, etc. - but these are competing aims. Too much of one will usually mean less of others. A compromise has to be sought, and this depends on the trade-offs between different advantages. The definition of what is satisfactory has to proceed in parallel with the search for a solution.

(iv) Often the basic data are uncertain and allowance has to be made for this in the design. Estimates of the uncertainty have to be made in the light of their likely importance, the extra cost entailed, and so on.

These are the characteristics of a task which cannot easily be reduced to an accurate formulation and solved by algorithmic means. It is also the kind of task for which the human mind has a characteristic ability. The designer envisages the object he is designing and relates it to its function, its assembly, and its manufacture. He will carry out some standard checks on its performance - for example its strength. But if the object is unusual, he may suspect new possibilities of failure, and make special calculations to see whether these will occur. Such a use of human imagination, founded in experience, is the only way of guarding against types of failure which have not yet been experienced.

What I wish to suggest is that these characteristics, which make a task suitable for interactive computing, are not in any way confined to geometrical design. There are many other kinds of engineering design problems which have the same characteristics. For example, the layout of printed-circuit boards is a geometrical problem: the design of the circuit is not, but it shares the characteristics I have described. The design of an automobile body is geometrical, but the design of the suspension system is a problem in dynamics having similar features.

Nor is it only engineering which gives rise to such loosel
defined problems. They arise in many other areas too. If interactive computing is the appropriate way of attacking such problems, then its scope is a very wide one.

3. AN EXAMPLE

I would like to illustrate these remarks by an example from my own work, not because illustrations could not be drawn just as well from other fields but simply because it is the area I know best.

Control systems are used very widely in industry: to keep the temperature constant in a room or in a chemical reactor, to control the speed of an electrical generator or an aircraft engine, to amplify the forces in the brakes or steering of a car, etc. Up to 1950 or so, systems like these were designed largely by trial and error. Then graphical methods were introduced, which gave great insight into the behaviour of the syste
The calculations were made with a slide-rule, and graphs were plotted with pencil and paper. These methods developed until by 1960 they were highly refined. Indeed they could hardly be taken any further at that time because of the limitations on computing, and some control engineers felt that the area was worked out.

About 1960, two things coincided to bring about a revival of interest. The first was a new problem, rocket guidance, demanding new methods. The second was the increasing availabilit
of computers. It happened that the guidance problem was amenab
to algorithmic solution, while early batch computers were just about powerful enough to make the calculations. So from about 1960 to 1975, a great effort was expended on this problem.

Very powerful algorithms were developed, and solutions to the guidance problem can now be obtained in a routine fashic
Yet attempts to apply similar techniques to industrial problems were not very successful. This is partly because the industria
problems are much more difficult to define, and cannot effectively be reduced to an algorithmic solution.

Starting about 1967, my own group in Manchester went back to the early graphical methods, and began to extend them. We used a computer with graphical output, and were able to develop displays (Fig. 1) from which the designer can assess stability, speed of response, sensitivity to disturbances, and other

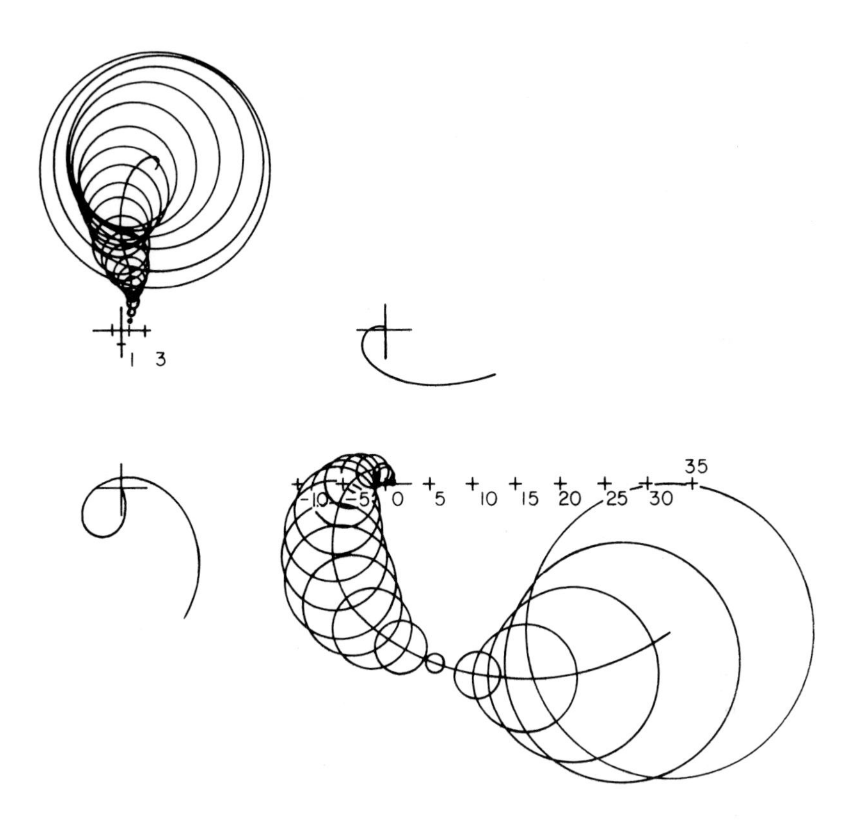

Fig. 1 Inverse Nyquist array for a control system

properties of the system in which he is interested. At the same time, if the performance of the system is not satisfactory, the displays suggest how it may be improved. In these respects the displays carry on the tradition of the early pencil-and-paper methods, but of course with a much greater computing power available. Problems of much greater complexity can therefore be attacked.

This line of work has been highly successful, whether judged by the ability to solve industrial problems or the number of other research groups who are developing analogous methods. My belief is that the methods we have will be greatly improved in the next five years. Nothing in this seems to be special to control engineering, and I would expect to see very similar developments in many other fields.

4. SOFTWARE PROBLEMS

In a conference of computer specialists, there is a natural interest in the needs for software development. This is not my own special interest, but experience suggests some conclusions which may be useful.

(i) There is a need for software which allows the user to communicate naturally and easily with graphical input or output devices. It should as far as possible be hardware independent, so that for example displays can be shown on a Tektronix 4010 or 4014 or a Digital GT40 at will. The usual graphical functions are needed: drawing points or lines, moving pictures, changing scale, adding text, etc. But the software should allow special features to be added easily: for example in my own work we needed to display several hundred circles. It was inefficient in the use of storage to record every point on each circumference in a display file, so we store the origin and centre and generate the points (between which line segments are drawn) as the circles are displayed.

The display software therefore needs to be simple, basic, but flexible. Some of the better-known graphics packages should be regarded as applications packages for special problems (engineering draughting, architecture, etc.) rather than as basic graphical communication programs of the kind I have described.

(ii) Similarly, the user needs basic software which allows him to generate dialogue with the computer in an easy way. For example, everyone who writes an interactive program has to face the problem of checking the validity of commands, recognising inputs as commands or data, allowing freedom of format for entries, etc. Standard solutions for these and similar problems would be a great help.

(iii) Some quidance from computer specialists on program structure as perceived by the user would be helpful. For example, should one permit the user to go from one branch directly to another, or should one make him descend and re-ascend? Or is it immaterial which one allows? Most interactive programs seem to use the same internal structure for the program as is perceived by the user, but there appears to be no logical connection between the two. Is one structure better for the user and another for the programmer?

(iv) The ideal interactive system would allow the user great freedom to change the facilities, for example by generating macros for sequences of operations which he wishes to perform frequently. At the same time, the system should help the user by picking up his errors before he has engaged himself deeply in a particular erroneous sequence of calculations. These two requirements, for freedom and for guidance, are in conflict; how can this best be resolved?

5. CULTURAL DIFFICULTIES

When I have talked in the past about the joint use of computers and the human mind for solving design problems, a common reaction has been "Ah yes, you are talking about the unscientific part of engineering". It has to be recognized that within our culture, science has come to be regarded as something opposed to, and superior to, human skill. So it is much more natural for us to regard the computer as central, rather than the man, To suggest that computers should be used to assist and enhance human skills is to oppose deep-seated attitudes in our society.

This is unfortunate, for two reasons. First, it is likely to mean that a promising line of development is less vigorously pursued than it deserves. Secondly, it is likely to produce computer systems which are destructive of human skills, and contribute to a worsening of conditions for those who have to work with them. It is no accident that Taylor called his system "scientific management" and described it (1906) in these words.

"Under our system the workman is told minutely just what he is to do and how he is to do it; and any improvement which h makes upon the orders given him is fatal to success."

Such "science" leaves no scope for the creative use of human intelligence, nor for that interplay between experience and the imagination out of which science and technology were born

There are, as I have said elswhere (1977) two paths that could be pursued:

"The first is to accept the skill and knowledge of designers, and to attempt to give them improved techniques and improved facilities for exercising their knowledge and skill. This demands a truly interactive use of computers in a way that allows the very different capabilities of the computer and the human mind to be used to the full. Such a route will lead to a continually evolving ability in the designer, and a continued interaction between his ability and the techniques and equipment he uses.

"The alternative is to subdivide and codify the design process, incorporating the knowledge of existing designers, so that it is reduced to a sequence of simple choices. Thus "de-skilled" the job can be done by men with much less trainin and much less experience. In the areas affected, design skill will gradually die and there will be no effective dialogue between the operator of the system and the man (in the Research Department?) who devised and will modify it.

"Given equal effort applied to the two possibilities it seems most likely, for reasons given above, that the first would be more effective, cheaper, and more satisfying to the user. If, contrariwise, almost all research effort is devoted to the second approach, then however inherently less promising that will eventually become the only effective method."

In this I am saying, for a particular case, what Weizenbaum (1976) has said in a more general context. There is a strong danger that we may be led, by the current ideology, to uses of computers which are both inefficient and offensive to human values. There is still the opportunity to avoid this, and I hope it will be taken. If it is, then there is much work of great interest and value to be done both by computer scientists and be engineers.

6. REFERENCES

Rosenbrock, H.H. 1977 "The futre of control", *Automatica Vol.* **13**, 389-392.

Taylor, F.W. 1906 "On the art of cutting metals", Third edition, revised (American Society of Mechanical Engineers).

Weizenbaum, J., 1976 "Computer power and human reason", W.H. Freeman, San Francisco.

IV. 2

THE USE OF NUMERICAL SOFTWARE IN THE DIGITAL SIMULATION LANGUAGE PMSP

T. Chambers

(Central Electricity Generating Board, Computing Centre, London)

1. INTRODUCTION

This paper describes the use of numerical software in the context of digital simulation. The motivation for using digital simulation at the CEGB is first discussed and an effective method of applying simulation techniques via a high level language (PMSP) is also described. User requirements from simulation studies are given in terms of numerical software and also the extent to which algorithms in PMSP satisfy these requirements. The algorithms themselves are then described and finally, an important application of PMSP to a very large simulation study is given.

2. THE USE OF DIGITAL SIMULATION

A digital model of an engineering system offers many advantages during the design and commissioning stages of new power plant. It can obviate the necessity for expensive prototype equipment or the engineer may use the model to supplement physical means for finding the solution to a practical problem. Physical experimentation can, of course, be especially expensive when it contains an element of destructive testing and it is then much more economic to perform instead the same experiments on an accurate simulation model.

One of the major uses of digital models in the CEGB is in the design, commissioning and operation of power station automatic control schemes. Until the plant is built and operated, its response characteristics will be known mainly from the performance of existing similar plant (if any). The power station control systems engineer (and later the plant operator) needs information about plant responses which is available at the design stage only from a computer simulation. This information is needed because sufficient flexibility has to be built into the control system design to cope with departures from designed performance characteristics and digital models are invaluable in determining the sensitivity of the total plant response to such departures.

As another example of the use of digital models, consider the problem of satisfying the temperature constraints imposed by material limitations in a nuclear power station. The operating and commissioning engineers must know what disturbances to the plant may cause these constraints to be exceeded. The simulation model will produce transient records for such disturbances and these give the engineer a valuable insight into the interactive working of the various parts of the station and the effect of operating procedures on the safety limits.

CEGB engineers apply the techniques of digital simulation via the use of a simulation language called PMSP.

3. PMSP - THE PLANT MODELLING SYSTEM PROGRAM

A very effective means of applying the techniques of simulation is via a high level simulation language, and the Computing Centre has developed such a language, PMSP, to facilitate the simulation of dynamic systems. This language is FORTRAN based and uses IBM double precision arithmetic where necessary, which in practice turns out to be nearly everywhere. It is used extensively on a regular basis by about three hundred users within the CEGB Computing Centre, by the laboratories at Marchwood, Leatherhead and Berkeley and in the Scientific Services departments of several regions.

The nuclear power industry also makes use of PMSP at the Nuclear Power Company (Risley and Whetstone) and the United Kingdom Atomic Energy Authority (Winfrith, Dounreay and Risley). PMSP is intended for use by engineers who use digital simulation mainly as a design tool. The aim is to make simulation for such people as easy as possible. The terminology is simple and readily understandable to the engineer so that the input usually looks very much like the original mathematical definition of the problem. No great programming knowledge is required and the language offers a great many simply-invoked functions which commonly occur in engineering terminology. Among these are functions such as rate and position limits, dead spaces, dead bands, first and second order lags, function generators, hysteresis, steps, ramps and many others. The main appeal of the language, however, lies in the range of mathematical and analytical techniques available to the user and these will be described later. The input to PMSP is basically in the form of systems of first order ordinary differential and algebraic equations. It may be asked at this stage why we do not attempt to solve the partial differential equations from which the ordinary differential equations usually come as a result of spatial discretization. Apart from historical reasons, involving the use in the past of analogue computing equipment, there are three main reasons.

(a) A very common use of simulation is in the design of control systems. Crudely speaking, the input to a piece of plant is varied in a controlled manner, and the consequent variation in the output is analyzed. The details of the internal structure of the plant may be of little interest as long as the response reflects adequately the true behaviour. Hence in the model, an averaging process can be used to simplify calculation and the system is then sometimes known as a "lumped" parameter system. This averaging process often removes sufficient physical dimensions from the model so that a system of o.d.e.s naturally results.
(b) An engineer, although usually concerned with non-linear models, may nonetheless require access to the techniques of linear systems analysis such as frequency responses, and o.d.e.s lead very naturally to this type of calculation. This is a very common application of digital simulation in control system design.
(c) It seems likely that no general p.d.e. solving package exists which is as flexible and can offer as many facilities as PMSP.

Given, then, a presented system of o.d.e.s, PMSP will convert these equations from PMSP code into a FORTRAN subroutine and will interface the latter with a wide range of numerical algorithms, also coded in FORTRAN. Many other features of the language will not be described here, such as automatic sorting, checking for sequencing errors and free format data input. The model is exhaustively checked by PMSP before numerical calculations are allowed to proceed.

4. PROPERTIES OF PMSP MODELS

The models consist of large sets of non-linear ordinary differential and algebraic equations. Typically there will be between 30 and 400 o.d.e.s and possibly several thousand associated algebraic equations. The equations are non-linear because nature is non-linear and realistic models always take account of this. The o.d.e.s are usually "stiff" with a stiffness ratio of about 10^6 and are usually quite strongly cross-coupled, making the automatic separation of "stiff" and "non-stiff" components very difficult. There is an important minority of problems which are not "stiff" in the usual sense but have solutions of approximately fixed frequency - usually 50 cycles per second. As often occurs in many types of simulation model, the Jacobians associated with the differential equations are sparse - typically 5% dense - but unfortunately exhibit no symmetry numerically or positionally. The equations are almost always discontinuous - in the first derivative because of discontinuous forcing functions and in higher derivatives because of the use of function fits to empirical or experimentally obtained data.

5. USER REQUIREMENTS FROM PMSP

In terms only of numerical algorithms, users require the following facilities, all of which are available in PMSP:

(a) Numerical solution of ordinary differential equations, both stiff and non-stiff.
(b) Steady state solutions By this is meant the quiescent state of the model, i.e. that state at which time derivatives are zero or acceptably small. Almost all studies, at some stage, require the steady state to be found, the model solved for a short while at this state, and a disturbance then applied. Accurate steady state solutions are essential so that cause and effect can be distinguished.
(c) Access to linear analysis techniques Although models are almost always non-linear, valuable insight can be gained through the application of linear techniques.

In this context, PMSP provides:

(i) Numerical linearization of the model about an operating point (usually the steady state condition).
(ii) Eigenvalue analysis of the linearized system. A knowledge of the eigenvalues can answer certain questions concerning the dynamic stability of the model.
(iii) Frequency response calculations. This facility is very widely used by engineers in control system design. Output from frequency response calculations is usually requested in the form of Nyquist, Inverse Nyquist or Bodé plots.

(iv) Steady state sensitivity analysis. This provides an estimate of the condition of the steady state solution with respect to the model data. This device is especially useful when the error tolerance of the data is known.

(v) Origins of instability, stiffness or oscillatory modes. It sometimes happens that the user's model exhibits, for example, dynamic instability as indicated by at least one eigenvalue of the linearized system having a positive real part. This knowledge is of limited use unless that part of the model which is, in some sense, the "cause" of the instability can be detected. A partial answer to this problem is to examine the associated eigenvector, due allowance having been made for the effect of different physical units. This eigenvector then "points", sometimes with remarkable accuracy, to the required area of the model.

PMSP can perform all the above tasks on simple commands from the user and using exactly the same definition of the model. The fact that the model need not be changed according to algorithmic requirements is of great importance to the user.

In terms of performance of numerical algorithms, it is found that users usually have the following descending order of priority:

(a) Robustness This is easily the most important requirement. Answers must be obtained if possible.

(b) Minimum intervention in respect of provision of such items as error criteria (see Section 6.8), scaling of problems, number of iterations allowed, etc.

(c) Comprehensive error diagnostics All such diagnostics must be expressed in terms of properties of the simulation model. As a simple example, an error message which says "INTEGRATOR VARIABLE *FLUX* IS MISSING" is far more valuable to the user than the equivalent "COLUMN 12 OF THE JACOBIAN IS NULL - FACTORISATION IMPOSSIBLE".

(d) Efficiency This comes last in order of priority, though it is of course essential to see that simulation is as cheap as possible, consistent with (a), (b), (c) above.

With regard to (a) above, experience over a number of years indicates that the algorithms used are very robust for the models presented to PMSP, though not necessarily in any other context.

Regarding (b), PMSP requires no provision of error criteria with the single exception of the numerical solution of differential equations. Here it seems unavoidable that users should state the accuracy required. For all other applications, answers are computed to the limiting accuracy for the machine and algorithms used. For practical applications, the extra work involved in proceeding from a "reasonable accuracy to the limiting accuracy is not great and provides greater confidence in the solutions obtained (see Section 6.8).

Where needed, PMSP scales a problem automatically. For example, in solving non-linear simultaneous equations (steady state problems), the equations are scaled to remove the effect of different physical units. When eigenvalues are needed, the matrix in question is balanced beforehand and whenever sparse LU factorizations are to be performed, the matrix is equilibrated.

Finally, regarding (d), it is difficult to assess the efficiency of the algorithms provided. For a new algorithm to be considered as a replacement for an existing one, it would have to be shown, over a period of time, to be overwhelmingly superior. Examples of replacements which have taken place are the use of sparse instead of full matrix techniques and the inclusion of several "stiff" integration algorithms.

6. DESCRIPTION OF NUMERICAL ALGORITHMS USED

Whenever, in the following text, reference is made to sparse matrix algorithms, it will imply the use of the algorithms described in Curtis and Reid (1971).

6.1 Numerical solution of ordinary differential equations

As described earlier, almost all systems presented fall into the category known as "stiff", and here three different variable step algorithms are available.

The first is an implementation of Gear's algorithm (Gear, 1971). The strategy as proposed in that reference is all but useless for applications to large, sparse systems such as those for which PMSP is used. The following is a brief summary of the changes made to that scheme:

(a) All matrix operations are replaced by sparse matrix code. The need for this change is self-evident for systems of hundreds of differential equations, known in advance to be sparse.

(b) Gear's policy of recalculating the Jacobian numerically whenever the order of method being used changes or there are problems with convergence, is disastrous for large systems. In general, we adopt the policy that if the current iteration matrix solves the non-linear algebraic problem at each time step in a reasonable number of iterations, then we retain it. When there is a problem with convergence, we have found that this is almost always because the current steplength and/or order has changed greatly from that in use when the interation matrix was last calculated. It is rarely because the Jacobian elements themselves have changed too much. We therefore keep a copy of the last Jacobian calculated in core, in sparse form, and whenever the rate of convergence is too slow or divergence occurs, the iteration matrix is simply reconstructed by combining the stored Jacobian, the _current_ steplength and _current_ order. In practice, little more is required than a re-scaling of the diagonal elements of the iteration matrix followed by a "fast" sparse LU decomposition. The decomposition is a "fast" one because it assumes that the Jacobian structure has remained unaltered and that the pivotal sequence found to be best at the start of the integration is still adequate.

The iteration matrix, referred to above, is of the form:

$$D = I - \alpha hJ \tag{6.1.1}$$

where h is the steplength, α depends on the order and J is the Jacobian of the system of differential equations. I is the identity matrix.

Thus when divergence occurs, α and h are updated but not J. D is stored in core as:

$$D = (I/\alpha h - J) \tag{6.1.2}$$

so that only the diagonal terms of D need be changed. D is then decomposed into its triangular factors, maintaining sparsity, and these triangles over-write the old ones. This procedure usually produces sufficiently improved convergence but other recovery procedures are then tried if this is not so. Ultimately, J may have to be recomputed and if this fails to produce convergence, then usually it implies that the structure of J has changed and needs to be recomputed, a very expensive process. Whenever LU factorization takes place, a check is kept on the pivotal growth of errors and if this is too great, a new pivotal sequence is calculated but this necessity is found only rarely. It is commonly found that during a transient solution, only one Jacobian need be calculated over many hundreds of seconds, re-scaling being adequate to produce convergence. It has been found too, that successive corrector iterations respond well to an Aitken δ^2 acceleration process.

The second method available for stiff systems is one due to Fowler and Warten (1967). This is a non-linear explicit scheme which fits a simple exponential function locally to the solution. It is designed for systems having widely spaced eigenvalues of the local Jacobian. Even when the eigenvalues are clustered, the scheme works well but in the presence of oscillatory modes, efficiency seems to decrease rapidly.

Finally, the last and most generally used algorithm available for stiff systems is an unusual implementation of the implicit trapezoidal rule. As is well known, the simplest method of solving the non-linear algebra at each time step is the method of successive substitution so that the iterative sequence to generate the solution at the (n+1)th time step is, with obvious notation:

$$\underline{y}_{(n+1)}^{(i+1)} = \underline{y}_n + \frac{h}{2} (\dot{\underline{y}}_n + \dot{\underline{y}}_{n+1}^{(i)}) . \qquad (6.1.3)$$

This method, of course, diverges rapidly for stiff problems for usable values of h. If, however, three successive members of the sequence (6.1.3) are saved and an Aitken δ^2 process applied, componentwise, to these three vectors, the "accelerate vector so produced is, under certain circumstances, a good approximation to the solution of the algebraic problem. If this vector is then used as the first of a new triad obtained from (6.1.3) and the acceleration applied again, convergence of the accelerated vectors is often rapid. The method works best when the "stiffest" eigenvalue lies on the real axis, well separated from the others. This kind of distribution is, fortunately, frequently found in models presented to PMSP. The method is a variable step method possessing the great advantage of requiring no storage of Jacobians and thus makes light demands on core and has very low overheads. It is robus in the presence of discontinuities, presumably because of its low order (2) and its being a one-step method.

For non-stiff problems, users can choose between a 4th order variable step Runge-Kutta scheme or Gear's non-stiff option (Gear 1971). Little effort has gone into the incorporation of good non-stiff methods because very few systems encountered are non-stiff.

Lastly, there is a small minority of problems having an oscillatory solution of approximately fixed frequency. This frequency is always known in advance but this cannot be exploited with conventional methods, which prove extremely expensive. An attempt is being made to answer this problem by implementing an adaptation of the methods of Gautschi (1961). These methods have the property that if the solution of the differential equations has the form:

$$y = a + b\sin(\omega t) + c\cos(\omega t) \tag{6.1.4}$$

then these methods yield exact answers for any steplength h. The algorithm being investigated at the moment combines the three algorithms:

$$y_{n+1} = y_n \cos(\nu) + \frac{h\sin(\nu)}{\nu}\dot{y}_n \tag{6.1.5}$$

$$y_{n+1} = y_n + \frac{2h\sin(\nu)}{\nu}\dot{y}_{n-1} \tag{6.1.6}$$

$$y_{n+1} = y_n + h\left[\frac{\sin(3\nu/2)}{\cos(\nu/2)}\dot{y}_n - \tan(\nu/2)\dot{y}_{n-1}\right] \tag{6.1.7}$$

where $\nu = h\omega$, ω the known frequency. Formulae (6.1.6) and (6.1.7) are 2nd order. Formula (6.1.5) is used initially to generate one extra point, apart from the initial point and (6.1.7) is then used to generate the solution. The solutions from (6.1.6) and (6.1.7) are compared to estimate truncation error and so enable steplength doubling and halving to be done. By restricting steplength changing to halving and doubling, the new values for the trigonometric terms in (6.1.5) to (6.1.7) can be generated very economically. A trigonometric Lagrangian interpolation formula is used to generate extra points for step halving or print-out purposes.

6.2 Steady state problems

These require the calculation of the quiescent state, i.e. that state at which time derivatives are acceptably small. This implies the solution of a square system of non-linear algebraic equations. For less than about 50 equations, an implementation of a Broyden rank-1 update scheme (Broyden, 1965) is used. In conjunction with an automatic scaling procedure to remove the effect of physical units, this scheme has proved remarkably robust over many years of use. For very large systems, however, it is essential to exploit sparseness and for this an implementation of Schubert's method (Schubert, 1970) coupled with sparse matrix techniques, is proving equally effective. Up to 420 non-linear equations have been solved with this method. It too includes automatic scaling and requires the storage in core of a Jacobian approximation and its triangular decomposition, both in sparse form. For both these algorithms, solutions are found to the limiting accuracy using the technique described in Section 6.8.

6.3 Numerical linearization of non-linear equations

For access to linear systems techniques, the Jacobian of the system must be computed accurately. In computing accurate numerical partial derivatives, the usual problems of conflict between heavy cancellation of leading digits and truncation error is solved by using central difference approximations applied iteratively to achieve the limiting accuracy. The process is accelerated by Richardson extrapolation.

6.4 Computation of eigenvalues of the linearized system

To obtain information concerning the stability and/or stiffness of the dynamic system, some of the eigenvalues of the Jacobian (calculated as in Section 6.3) are needed. Very few eigenvalues are required from the matrix, these being at either end of the whole spectrum. Those with most positive real part give information concerning stability, those with most negative real part give information concerning stiffness. For this problem, an effective method is the following procedure:

(a) Scan the matrix and remove physically ignorable elements.
(b) Balance the matrix.
(c) Reduce the matrix to upper-Hessenberg form using stabilized elementary similarity transformations. Call this matrix H.

(d) Scan the sub-diagonal elements of H for zeros and, if any, split the matrix into groups of smaller matrices.
(e) Using Hyman's method, find the determinant and its first and second derivatives with respect to λ, of:

$$D = (H-\lambda I) \tag{6.4.1}$$

where λ is the current approximation to an eigenvalue. Use these in Laguerre's method for computing the roots of polynomials.
(f) Use root suppression (see Wilkinson, 1963) to remove the possibility of converging more than once to an eigenvalue already found.

We have found that this method has the great advantage of allowing eigenvalues to be extracted at will from either end of the spectrum. This can be achieved simply by making the initial estimate for the root lie to the left ("stiffness" or right ("instability") in the complex plane of all eigenvalues. The algorithm described is extremely robust and has been used successfully on very many problems over a number of years. Its main disadvantage is its comparative slowness. Interestingly, although convergence with Laguerre's method is guaranteed only when all roots are real, we have never detected any difficulty in obtaining convergence for complex roots, only in the rate of convergence when these roots lie close to others.

6.5 Frequency response calculations

The frequency response of a stable system of linear differential equations is the state which the system attains after an infinite time in response to a sinusoidal forcing function of fixed frequency. To calculate the frequency response we have to find the LU decomposition of matrices of the form:

$$\left[\begin{array}{c|c} A & \omega I \\ \hline -\omega I & A \end{array}\right] \tag{6.5.1}$$

where A is square and sparse, I the identity matrix and ω the frequency in question. Often A will be nearly singular and ω very small and in that case, iterative refinement of the solution is essential. Since it is not known in advance how badly-conditioned A is with respect to inversion, iterative refinement is always applied for all A and ω. Sparseness is exploited by performing a sparse factorization of (6.5.1). The matrix has to be factorized for possibly hundreds of values of ω for the same A, because users usually require the frequency responses to construct Nyquist, Inverse Nyquist or Bodé plots. We have found it best to solve for values of ω in decreasing orders of magnitude. The pivotal sequence determined for the first value of ω is used for all subsequent values of ω. If iterative refinement fails, then a new pivotal sequence is calculated and the calculation proceeds. Hence, almost all decompositions are "fast" in the sense of Curtis and Reid (1971). It is rare for more than one pivotal sequence to be needed.

6.6 *Sensitivity analysis*

It is important to know the effect on numerical solutions of perturbations in the model data, especially when the error tolerance of the data is known. We have found that it is sufficient to do this for steady state calculations, since sensitivity here usually indicates that most other calculations which may be requested (e.g. numerical integration) will probably also be sensitive. If numerical calculations are shown to be highly sensitive over the known tolerance, then the validity of the whole simulation study must be called into question. The steady state sensitivity analysis may be found as a frequency response (see previous section) for ω very small.

6.7 *Origins of instability, stiffness and oscillatory modes*

As mentioned in Section 5, the knowledge that a model is, for example, dynamically unstable is of little use unless the cause of the instability can be detected. Almost always, dynamic instability results from a user coding error but we have found that the eigenvector corresponding to the "unstable" eigenvalue indicates, often with remarkable accuracy, that area of the model in which the coding error lies. The effect of physical units is removed from this eigenvector and its non-zero components are given to the user as the physical names of the offending differential equations. The eigenvector is found by performing inverse iteration on the matrix:

$$(H-\lambda I) \qquad (6.7.1)$$

where H is the accurately computed Jacobian of the system (Section 6.3) and λ the "unstable", "stiff", or "oscillatory" eigenvalue, as required. If λ is complex, real arithmetic is still used by performing inverse iteration on a matrix twice as large as (6.7.1). Sparse LU factorizations are performed on (6.7.1) and using the sparse triangles so formed, a series of forward and back substitutions is performed, starting with an arbitrary vector, in the usual way. The effect of physical units can be removed by performing a simple diagonal similarity transformation on H where the diagonal terms are simply the steady state values found as in Section 6.2.

6.8 Convergence criteria used to terminate iterative sequences

In Section 5 it was mentioned that PMSP users usually proved unwilling to provide termination and accuracy criteria for iterative numerical calculations. A general technique is used by PMSP to terminate all iterative sequences. This method provides:

(a) the limiting accuracy in numerical procedures for the given algorithm and machine precision;
(b) independence of machine precision;
(c) greater confidence in numerical answers;
(d) no user intervention.

The technique has been described by Wilkinson (1965) and is based upon the idea that if an iterative sequence of approximations Z_i, i the iteration number, is converging then:

$$|Z_{i+2} - Z_{i+1}| < |Z_{i+1} - Z_i| \qquad (6.8.1)$$

until the limiting precision for the algorithm and machine has been reached. Initially (6.8.1) may not be true until the sequence Z_i "settles down" and so the convergence criteria used are:

$$|Z_{i+1} - Z_i| < \varepsilon \text{ and } |Z_{i+2} - Z_{i+1}| < |Z_{i+1} - Z_i| \qquad (6.8.2)$$

ε is a "generous" error tolerance, dependent on the algorithm used. Iteration continues until (6.8.2) is satisfied.

There is a possible danger in using (6.8.2) that excessive computer time may be used if convergence is very slow. However, in view of the criteria described in Section 5, the risk is in general worth taking. The criteria (6.8.2) are used in PMSP to terminate iterative sequences for steady state solutions, numerical linearization, eigenvalue calculations, frequency response calculations (iterative refinement), and "origin" calculations (inverse iteration).

6.9 An application of PMSP to a large simulation study - the AGR simulator

The CEGB operates a Nuclear Power Training Centre at Oldbury-on-Severn in Gloucestershire, about 100 miles from the Computing Centre. It is proposed to provide real time hands-on training for nuclear power station staff and to use a digital simulator for this purpose.

The basic idea is to provide exact replicas of the control desks of each of three Advanced Gas-cooled Reactor stations - Hinkley Point "B", Dungeness "B" and Hartlepool. A digital model of the selected station resides in the Computing Centre in an IBM 370 computer and a tutor, at a separate desk, creates typical fault conditions in the model to which the pupil, seated at the "station" desk, must respond. In this way we simulate real fault conditions in the AGR stations.

Two PDP11/34 computers will scan the status of the pupil's desk five times per second and transmit this information by telephone line to the Computing Centre. Here the model equations are solved continuously at real time and the solution transmitted back to Oldbury where the PDP11s receive the information and update the pupil's desk, i.e. his meters, recorders, alarm lamps and VDUs. There is a wide spectrum of fault conditions available to the tutor over the whol range of operating conditions, from start-up to full load running. Because of this and the severe transient conditions which result, a very detailed non-linear PMSP model is used consisting of about 400 differential equations. This model consists of many separate PMSP modules, each representing in detail a distinct part of the station, e.g. boiler region, turbine, reactor and feed system. Each is tested thoroughly "off-line" and they are finally assembled to represent the whole station. Steady state calculations are crucial here and the methods described in Section 6.2 are used on very large systems of equations on a day-to-day basis.

Real-time operation is obviously essential for realistic training, i.e. when t seconds of time have elapsed in the training session, the model equations must have been solved up to t seconds in the independent variable. It is this requirement of matching real time to model time that presents severe problems for the numerical integration algorithm used. Whenever it is possible for the algorithm to overtake real time, the CPU is relinquished to other time sharing users but when severe transient conditions are met and real time operation is threatened, the CPU is taken over again by the simulator. This part of the algorithm is complicated and will not be described more fully here.

The algorithm used to compute the real time solution is Gear's method strategically revised, and combined with sparse matrix techniques. Gear's algorithm was chosen, firstly, because the equations are stiff, and, secondly, because well-tested code already existed for this algorithm in PMSP (see Section 6.1). There are several strategic differences in the Gear real-time algorithm from that described in Section 6.1. First, we restrict the order of method to be no more than 3 and demand a relative accuracy of 1%. The most fundamental changes necessary occur, not surprisingly, in the method used to solve the non-linear algebraic equations at each time step. Because the algorithm must not get "stuck" and thus ruin real-time operations, we make the important assumption that the system Jacobian J does not vary much from that obtained at the start of the training session. This means that numerical approximations to J are never calculated during real-time running. However, the steplength h and order, characterized by α, will change so that we will need sparse factorizations of equation (6.1.1) for arbitrary values of α and h. However, we have restricted the order to be less than 4 and we similarly restrict h to take discrete values in a fixed range, say 0.01 to 2.0, each intermediate value of h being a constant multiple (say $\sqrt{2}$) of the previous value. Hence for each value of h in the range 0.01 to 2.0 and for the three values of α we need sparse LU factors of (6.1.1) at various stages in the real-time session. Since J in (6.1.1) has been assumed constant numerically and structurally, all these sparse factors are computed and stored on disc before training begins and at any stage in the session, the relevant triangular factors are read into core to be used for the solution of the algebraic problem. We read in such factors only when there is difficulty with convergence or when steplength is increased.

Other features of the algorithm are that all known discontinuities are numerically smoothed out. This is simple for such characteristics as step functions, ramp functions, limits, etc. and is crucial for an algorithm like Gear's, which is acutely sensitive to discontinuities. The commonest source of discontinuity is a change in status of the pupil's or tutor's desk. However, when this information is sent to the Centre, the time at which the change is status occurred is sent with it and we force the algorithm to "land" on these times. This does not remove the effect of the discontinuity but does remove the chattering effect often seen in these circumstances.

Finally, if real-time is threatened and convergence in the corrector step is taking too long, the last converged solution is sent down to the training centre. This produces a slewing effect on the solution in the direction of increasing time which corrects itself as real-time computation is achieved again.

7. ACKNOWLEDGEMENT

I am grateful to the Central Electricity Generating Board for permission to publish this paper.

8. REFERENCES

Broyden, C.G. (1965) "A class of methods for solving non-linear simultaneous equations", *Maths. Comp.*, **19**, 577-593.

Curtis, A.R. and Reid, J.K. (1971) "Fortran subroutines for the solution of sparse sets of linear equations", AERE Report R.6844, HMSO, London.

Fowler, M.E. and Warten, R.M. (1967) "A numerical integration technique for ordinary differential equations with widely separated eigenvalues", *IBM J. Res. Develop.*, **11**, 537-543.

Gautschi, W. (1961) "Numerical integration of ordinary differential equations based on trigonometric polynomials", *Numerische Mathematik*, **3**, 381-397.

Gear, C.W. (1971) "Numerical initial value problems in

ordinary differential equations", Prentice-Hall Series in Automatic Computation, New Jersey.

Schubert, L.K. (1970) "Modification of a quasi-Newton method for non-linear equations with a sparse Jacobian", *Maths. Comp.*, **25**, 27-30.

Wilkinson, J.H. (1963) "Rounding errors in algebraic processes", NPL Notes on Applied Science No. 32, HMSO, London.

Wilkinson, J.H. (1965) "The Algebraic Eigenvalue Problem", Clarendon Press, Oxford.

PART V

DIFFERENTIAL AND INTEGRAL EQUATIONS

V. 1

SOLUTION OF LARGE, STIFF INITIAL VALUE PROBLEMS - THE STATE OF THE ART

A.R. Curtis

(Computer Science and Systems Division, AERE Harwell, Oxfordshire)

1. INTRODUCTION

Ten years ago, stiffness was a serious difficulty in ODE problems, whose nature was imperfectly understood by most people who met it. Today, largely because of the introduction of backward difference methods, even extremely stiff problems, both large and small, are routinely solved. Understanding of the phenomenon has also progressed greatly; there are still many details of the automatic solution of such problems where substantial improvements can be made. The practitioner's skill can still affect his success; today's sophisticated software tools offer scope for learning to use them well; so the process of solving stiff problems, especially large ones, is still an art - or, at least, a craft.

In section 2 of this paper we discuss stiff initial value problems in terms of their size, stiffness and source. In section 3 we discuss software design from the user and technical viewpoints, beginning to pay some regard to the fact that many of the largest and stiffest problems arise from mass action kinetics, and illustrating by practical examples the importance of careful thought about error control; we also discuss numerical methods. Section 4 contains a brief account of the simplied problem specification facilities provided by the author's program FACSIMILE. Section 5 gives an account of experience at Harwell and elsewhere in using FACSIMILE, especially on large or extremely stiff problems.

2. PROBLEM CLASSIFICATION AND SOURCES

2.1 Mathematical statement of problem

The equations to be solved can be written in the form

$$\dot{y} = f(y), \quad y(0) = y_0 \tag{2.1.1}$$

where y,f are n-vectors. We are interested in the stiff case, in which the Jacobian matrix

$$J = \partial f/\partial y \tag{2.1.2}$$

has at least one eigenvalue λ whose real part is large and negative in relation to the time scale T of evolution of the solution. That is, the non-dimensional product T Re($-\lambda$) is large for this eigenvalue; thus perturbations of the solution in the direction of the corresponding eigenvector would be extremely rapidly damped in relation to T. This definition of stiffness does not require a wide spread of eigenvalues, or even that $n > 1$. The time-scale may be determined by relatively slowly-varying forcing terms, e.g. diurnal and seasonal variations in wind, temperature and insolation in atmospheric chemistry studies.

It is now well-known that many classical methods for numerical solution of the initial-value problem (2.1.1) fail dismally, or at best are extremely inefficient, in the stiff case, because the step size h is limited by $|h\lambda| \leq c$, a constant, typical of the method, of order unity. Successful methods are implicit; that is, the solution vector y_r at the end of integration step r is obtained by solving one or more sets of equations, whose Jacobian matrix is related to J. Newton iteration is necessary in solving these (generally non-linear) equations, to avoid re-introducing the severe step-size limitation of non-stiff methods.

We shall be mainly concerned with the case when n, the dimension of the problem, is large.

2.2 Classification by size and stiffness

We identify the non-dimensional product $|\lambda h|$, where λ is an eigenvalue (with negative real part) of the Jacobian matri J and h is the integration step-size, as some kind of measure of stiffness. For this parameter, what matters is that the software should allow it to become extremely large (of order the reciprocal of a computer rounding error) without getting into difficulty, just as the stiff numerical method on which the software is based permits arbitrarily large values. This is a matter of robustness which needs to be tested, independently of performance and reliability tests at more moderate stiffness; it would be wise to use test problems with realistic non-linearities for the prupose. Unfortunately, since h is method-dependent, such a definition implies use of an adequately good method in order to assess stiffness.

In the sequel, we shall find it helpful for some purposes to use a rough classification of problems by size, even though it is necessarily rather subjective. We shall call problems small if n, the number of variables, does not exceed about 10; medium if $n \lesssim 100$; large if $n \lesssim 1000$. Even larger problems are solved, but do not seem to introduce new concepts.

2.3 Sources of stiff problems

It is genuinely difficult to give an exhaustive account of problem sources, but the following rough classification may be helpful:

A. Artificial test problems.

B. Parabolic partial differential equations - method of lines.

C. Overall kinetics of nuclear or chemical reactors.

D. Electronic circuit design.

E. Mass action kinetics.

Problems of class A tend to be small (some medium ones are constructed by idealising problems of type B), and surprisingly are rarely extremely stiff. The product $|h\lambda|$, which is a measure of the degree of stiffness, often will not exceed 10^3 or 10^4 in solving such problems; this may not provide a sufficiently searching test of the robustness of programs in the face of practical problems where values of 10^{10} or even higher are not uncommon.

It is common to include a component in (2.1.1) of the form

$$\dot{y} = \lambda y \qquad (2.3.1)$$

for a large negative λ, whereas a more testing form might be

$$\dot{y} = \lambda y + g(t) \qquad (2.3.2)$$

where g(t) is a slowly-varying transcendental forcing function, e.g.

$$g(t) = (\alpha-\lambda) \exp(\alpha t), \quad |\alpha| << (-\lambda). \qquad (2.3.3)$$

For the solution $y=\exp(\alpha t)$, severe cancellation occurs in (2.1.2), illustrating our point about robustness; a non-linear test problem would be even more searching.

Class B tends to produce problems which are only mildly stiff (unless they include phenomena, such as mass action kinetics, which fall under another heading). It is a good source of large problems (one-dimensional with up to ~ 50 variables per mesh point, or two-dimensional), and of medium ones (one-dimensional with $\lesssim$ 5 variables per mesh point).

There are some small test problems of class C in the literature (e.g. Enright and Hull 1976); they do not seem to be particularly demanding, although clearly they can be more realistic than those of class A. In practice, they often arise as simplifications of problems which fall into both class B and class E.

Class D is capable of providing quite large problems which are extremely stiff. Considerations of user convenience in problem description often lead to mixed algebraic and differential systems; although in principle these can be reduced to the form (2.1.1) by solving the algebraic equations, this is neither easy nor desirable in practice. Fortunately, a very simple extension of stiff methods enables them to handle such generalised problems.

The author has been mainly concerned with problems of class E, which often involve consideration of spatial homogeneity and thus introduce also features of class B. Small reaction systems, treated as spatially homogeneous, give small problems and large homogeneous systems give medium problems (polymerisation processes can give large problems). Consideration of spatial inhomogeneity (with diffusion and/or advection) requires one copy of the reaction system in each spatial mesh, thus generating a large problem. Simultaneous solution of several replicates of a reaction system (e.g. with different starting conditions) is often needed when choosing parameter values to give best fit to results of kinetic experiments, thus generating a medium problem. In both these cases, arrays of variables and of reaction systems are desirable.

Chemical reaction kinetics is the most obvious source of such problems, but they have also risen in metallurgy, genetics, aerosol physics, astrophysics and solid state physics. Extreme stiffness can result when certain of the species modelled are very reactive, but their concentrations remain close to values giving balance between production and removal rates. The time derivative of the concentration y of such a species is the difference between two nearly equal terms, a production term p which is independent of its concentration and a removal term ry which us (nearly) proportional to the concentration, i.e.

$$\dot{y} = p - ry \ . \qquad (2.3.4)$$

Thus J has an eigenvalue close to $-r$, while the time-scale T of evolution of the solution may be of order $|y/\dot{y}|$. Thus we have

$$-\lambda\, T \simeq |ry/(p-ry)| >> 1 \qquad (2.3.5)$$

since p and ry are nearly equal. In our experience, the right-hand side of (2.3.4) can be dominated by rounding error (double-precision on IBM 370, $\sim 10^{-16}$), and the software must be able to cope.

Most of the remainder of this chapter will be based on experience of problems of class E.

3. SOFTWARE DESIGN CONSIDERATIONS

3.1 User considerations

The user of software for solving stiff ODE problems is unlikely to be a numerical analyst, or indeed a mathematician: he may be a chemist, biochemist, or metallurgist; a chemical, electronic or heat transfer engineer; or one of several kinds of theoretical physicist. It follows that there are a number of matters, which need consideration in the software design, but in which the user is typically not interested. Ideally, the software should be designed so that the user is not called on to be aware of these matters; if that is impractical within the state of the art, care should be taken to minimise the demands made on him, and to guide him in making the choices which the software is unable to make for itself. In this section we discuss some of these aspects, first from the user's viewpoint and then the designer's. Many of the problems become more severe as one goes up the size spectrum of section 2.4, and where this is the case we indicate this dependence.

3.2 Error control and estimation

Typically, the user wants answers to, say, 1% accurac but 1% relative to what? Often he has no idea in advance how large individual variables will become, and indeed individual components of the solution may vary through many orders of magnitude in a single run. It is rare, in our experience, for all components to be of order unity; see for example problem CHM1 below. Only where absolutely necessary, therefore, should the user be asked to indicate the magnitude of his variables in order to assist the software with error control; and where it is necessary, care must be taken to ask him the right questions in terms he can understand.

We have found that a good solution, in general terms, to the question asked above is to interpret the user's accuracy tolerance as a relative one for each component of the solution vector y.

Control of step-length and, sometimes, order of integration method is usually done on the basis of some weighted norm, e.g.

$$||\varepsilon_r|| = \max_i |\varepsilon_{ri}/Y_{ri}| \qquad (3.2.1)$$

of an estimate ε_r of the vector of local truncation errors at step r, and the question at issue is what "test numbers" Y_{ri} to use. Unfortunately, many tests of stiff ODE solvers take all the Y_{ri} to be unity, using the term "absolute errors" which tends to conceal the fact that an assumption is being made. The discussion below of Enright and Hull's (1976) test problem CHM1 shows how inappropriate this can be. In our program FACSIMILE we make $Y_{ri} \simeq |y_{ri}|$, while safe-guarding against zero values, by updating the Y_{ri} each step according to the formula

$$Y_{ri} = \alpha \, Y_{r-1,i} + (1-\alpha) \, |y_{r-1,i}|. \qquad (3.2.2)$$

Here α is small enough to allow exponentially-decreasing components to be followed; we take α=0.3, although it could appropriately be made to depend on error tolerance and order of integration formula.

Ideally, the user would like a guarantee that the computed solution lies within his specified error tolerance; it is beyond the present state of the art to achieve this, and he should normally be advised to apply a safety factor (10 is reasonable) in specifying the tolerance required.

Failing a guarantee, the user might like an *a posteriori* global error estimate. Such estimates can be produced relatively cheaply, by numerical solution of the variational equation, and are reasonably accurate in estimating the errors at the end of each integration step. Unfortunately, as we show later (section 3.2.4), on stiff problems such errors can be far smaller than the error tolerance; interpolation within the step to provide output values at a specified point normally introduces interpolation errors comparable to the tolerance. Therefore caution is advisable in providing such estimates to the user, who may interpret them over-otpimistically.

3.2.1 Need for relative error criterion

We illustrate the weakness of "absolute errors" by examining test problem CHM1 of Enright and Hull (1976), see Table 1. All the methods tested by Enright and Hull (using "absolute errors") seemed quite efficient on this problem, many "solving" it in fewer than 20 integration steps even at a tolerance of 10^{-6}. However, to specify an "absolute error" tolerance in the range of 10^{-8} to 10^{-2} for this problem means that almost arbitrarily large relative errors are acceptable, certainly on components y_2, y_3 and y_4 which determine the course of the reaction. Because the very relaxed tolerance demanded is met in general (at least at the end of each integration step, which is where error tests were done), the poor error control does not give disastrous global errors, by the criteria used in the test program.

Table 1: Test problem CHM1 (Enright and Hull)

$$\dot{y}_1 = -7.89.10^{-10} y_1 - 1.1.10^{7} y_1 y_3, \qquad y_1(0) = 1.76.10^{-3}$$

$$\dot{y}_2 = 7.89.10^{-10} y_1 - 1.13.10^{9} y_2 y_3, \qquad y_2(0) = 0$$

$$\dot{y}_3 = 7.89.10^{-10} y_1 - 1.13.10^{9} y_2 y_3$$

$$-1.1.10^{7} y_1 y_3 + 1.13.10^{3} y_4, \qquad y_3(0) = 0$$

$$\dot{y}_4 = 1.1.10^{7} y_1 y_3 - 1.13.10^{3} y_4, \qquad y_4(0) = 0$$

$0 \le t \le 1000$:

$1.76.10^{-3} \ge y_1 \ge 1.60.10^{-3}$

$0 \le y_2, y_4 \le 1.5.10^{-10}$

$0 \le y_3 \le 10^{-11}$ (Note: $y_2 = y_3 + y_4$)

Steadily-progressing consumption of y_1, with y_2, y_3, y_4 reaching maximum values and then slowly decreasing.

We show in Table 2 the results of solving this problem with (a) relative error tolerance, (b) "absolute error" tolerance, (c) error tolerance relative to fixed values of 2.10^{-3}, 10^{-10}, 10^{-11}, 10^{-10} for the four components. In each case we give the number of integration steps and the maximum relative error on y_2, y_3, y_4 or $(y_1(0) - y_1)$, including errors due to interpolation at equally-spaced time points in the range, the "true solution" being computed with a relative error tolerance of 10^{-8}. This shows how poor the results are, using "absolute errors". They are virtually unresponsive to the tolerance specified, because even the smallest tolerance is far larger than the important components y_2, y_3, y_4. The error e_1 indicates the relative accuracy with which the progress of the reaction (change in y_1) is represented, while e_2 is an ordinary relative error. Using method (c) the results are considerably better; the error e_1 was as large as shown only near the beginning of the run, showing clearly

Table 2: Accuracy and cost on problem CHM1 for various error control methods

Method	RELATIVE(a)			ABSOLUTE(b)			SCALED ABSOLUTE(c)		
τ	NS(d)	e_1(e)	e_2(f)	NS	e_1	e_2	NS	e_1	e_2
10^{-2}	62	5.4E-3	1.05E-2	11	4.7E-1	2.03E-1	29	3.0E-1	8.86E-3
10^{-4}	108	6.0E-4	2.99E-4	11	4.7E-1	2.03E-1	56	1.9E-2	6.71E-4
10^{-6}	207	4.2E-5	3.14E-6	15	4.7E-1	1.45E-1	110	2.6E-3	8.19E-6

Notes: (a) Standard FACSIMILE "test numbers" Y_i.

(b) $Y_i = 1.0$.

(c) Y_i = 2.0E-3, 1.0E-10, 1.0E-11, 1.0E-10 for i=1,2,3,4.

(d) Number of integration steps taken, $0 \le t \le 10^3$.

(e) Max(|error on $y_1|/(y_1(0)-y_1)$) at $t=20,40,60,\ldots,10^3$.

(f) Max(|error on $y_i|/y_i$), i=1 to 4, at $t=20,40,60,\ldots,10^3$.

(g) All errors calculated using RELATIVE with $\tau=10^{-8}$ for "true values".

the need for tighter control of errors on y_2, y_3 and y_4 where they are small. Controlling relative errors achieves this, and we see that specified error tolerances are nearly achieved (e_2), although the progress of the reaction still has a rather large relative error (e_1) at $\tau=10^{-6}$ This problem, far from being easy to solve as it appeared, thus turns out to be quite difficult and interesting on deeper study.

3.2.2 Study of a more demanding problem

As a further illustration of special error control problems which can arise, we consider test problem CHM9 (Table 3) of Enright and Hull; this is the most demanding of their ten test problems, typically accounting for about half the total computer time for those programs able to solve it. It is a model of the Zhabotinsky reaction, with a solution of period 303 sec., containing a short (3 sec.) pulse during which the

Table 3: Test problem CHM9 (Enright and Hull)

$\dot{y}_1 = 77.27(y_2-y_1y_2+y_1-8.375.10^{-6}y_1^2)$,	$y_1(0) = 4$
$\dot{y}_2 = (-y_2-y_1y_2+y_3)/77.27$,	$y_2(0) = 1.1$
$\dot{y}_3 = 0.161(y_1-y_3)$,	$y_3(0) = 4$
$0 \le t \le 300$; limit cycle period = 302.9,	
$1 \le y_1 \le 1.25.10^5$, $3.10^{-3} \le y_2 \le 2.10^3$,	
$1 \le y_3 \le 3.2.10^4$.	
For our solution, $0 \le t \le 605$, so as to see the second cycle	

variables range through factors of order 10^5, followed by slow recovery. The model time in the test problem was 300 sec., covering the first pulse and most of the subsequent slow recovery, but not the second pulse. It seemed important to verify that subsequent pulses could also be simulated accurately, and interesting to investigate any cumulative timing errors, so we extended the model time to 605 sec., virtually two complete cycles.

Because of the large range covered by each variable, we used only our standard relative error criterion (3.2.2). A "true" solution was generated with a tolerance of 10^{-8}, and was stored whenever $\log_{10}y_i$, i=1 to 3, passed through a half-integer value (excluding a few near-extreme values which might have been missed at relaxed tolerance). The reason for this, rather than recording the result at fixed times, was to avoid what could have been large apparent relative errors of order $\Delta t\ d(\ln y_i)/dt$ arising from quite small timing errors Δt. The value of this choice is well illustrated by a typical result at tolerance 10^{-6}, where $\Delta t = -1.207.10^{-3}$, $d(\ln y_2)dt = 40.79$, which would have given an artificial relative error on y_2 of -0.05 approximately, whereas the actual relative error (at a fixed value of $\log_{10}y_1$) was $-2.8.10^{-6}$.

The effect of varying the error tolerance τ is illustrated in Figure 1, which shows phase-plane plots ($\log_{10}y_3$ vs. $\log_{10}y_1$ and $\log_{10}y_3$ vs. $\log_{10}y_2$) of the limit cycle at a tolerance of 10^{-8}. The curves at tolerances $\tau=10^{-6}, 10^{-5}, 10^{-4}, 10^{-3}$ are not distinguishable from these. Even at $\tau=10^{-2}$ the curves were followed fairly accurately the first time round, but the start of the second pulse gave a large departure (shown by dashed lines); at $\log_{10}y_3=0.5$, the error in $\log_{10}y_1$ is about 1.55, corresponding to about a factor 35. This very serious error (from which this model recovers, but which in another non-linear problem could switch to an unstable solution), indicates the unwisdom of trying to solve demanding non-linear problems at relaxed tolerances.

Figure 2 shows histograms of the frequency of occurrence of $\tau^{-1}|\delta y/y|$ for various values of τ. If a safety factor of 10 had been applied in choosing τ because of the difficulty of the problem, good results would have been obtained for τ in the range 10^{-5} to 10^{-3}, and fair results at $\tau=10^{-6}$. As in the case of CHM1, all errors quoted here include effects of interpolation within a step. Table 4 shows errors in period of oscillation as a function of τ. We see that specified relative tolerance is approximately achieved.

Table 4: Errors in period T as a function of tolerance τ

τ	10^{-3}	10^{-4}	10^{-5}	10^{-6}
ΔT	$-2.48.10^{-1}$	$-3.77.10^{-2}$	$-5.09.10^{-3}$	$-1.19.10^{-3}$
$\Delta T/T\tau$	-0.82	-1.24	-1.68	-3.93

3.2.3 Initial choice of step size

It is usual, but unnecessary, to ask the user to suggest an initial trial step-size. If we use a first-order method to begin with, the error estimate has the form

$$\varepsilon = Ch^2\ddot{y} = Ch^2J\dot{y} = Ch^2 J\, f, \qquad (3.2.3)$$

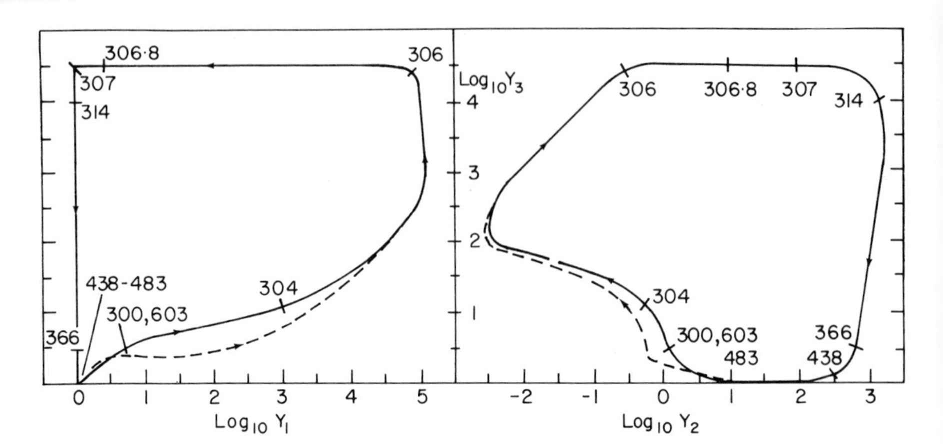

Figure 1: $\text{Log}_{10}y_3$ vs. (left) $\log_{10}y_1$ and (right) $\log_{10}y_2$.
Solid curve: $\tau=10^{-8}$ to 10^{-3}; dashed curve: $\tau=10^{-2}$ (second time round cycle). Numbers indicate times at which points are passed.

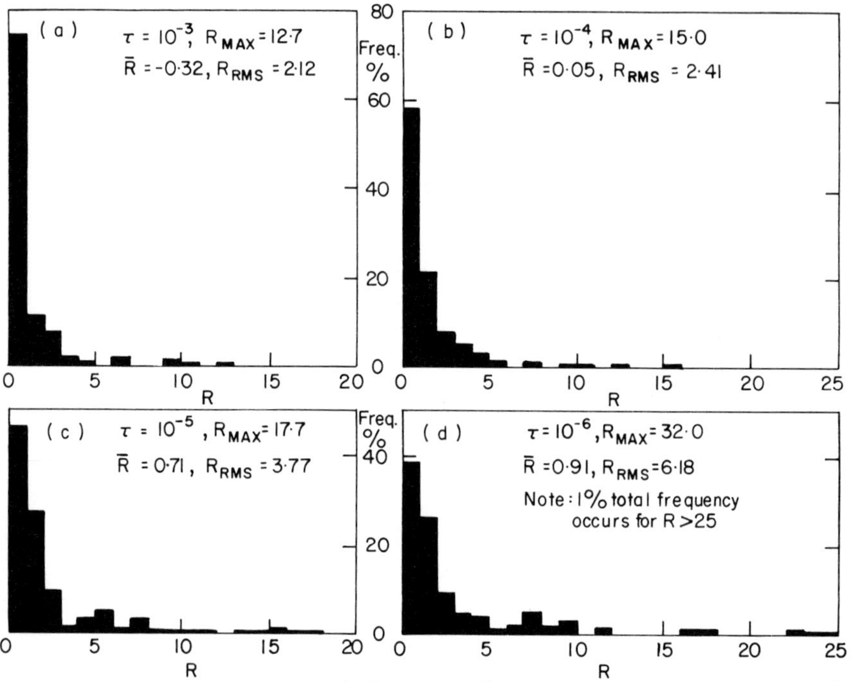

Figure 2: Histograms of $|R|$, i.e. $|$Relative error/tolerance$|$.
Note: $R = [y_\tau - y_\varepsilon]/[\tau y_\varepsilon]$, where $\varepsilon=10^{-8}$ and y_τ is value computed tolerance τ.

where C is a known constant, by differentiating (2.1.1). Since J is needed anyway during the course of the integration it costs little more to compute it before starting and use it to choose h so that

$$||C\, h^2\, J\, f|| < \tau \qquad (3.2.4)$$

where τ is the specified tolerance. This requires initial "test numbers" Y_{oi}: no difficulty arises if $y_i(0)\neq 0$, we take $Y_{oi}=|y_{oi}(0)|$; if $y_i(0) = 0$ but $f_i(y(0))\neq 0$, we take $Y_{oi}=h|f_i(y(0))|$, so the corresponding term on the left of (3.2.4) is proportional to h rather than h^2; for components whose initial value and derivative are both zero, the user is asked to supply as Y_{oi} a "threshold" value for the component, below which errors are unimportant; if he does not do so, a default value can be used, which may be quite large implying a "don't care" action, or very small for safety (FACSIMILE uses 1.0).

3.2.4 Behaviour of global error estimate

The global errors η may be estimated by integrating the variational equation, and we may investigate the effect of this by considering the simple form (2.3.2),(2.3.3). When many steps of the same size h have been taken, the calculated solution at step r, and other entries in the history array (for a multi-step method), are proportional to ρ^{-r}, where $\rho = \exp(-\alpha h)$, with η independent of r in the limit. Writing $a = \alpha h$, $b = -\lambda h$, we find for the Kth order backward difference method

$$\eta = (a-\phi_K(\rho))/(b+\phi_K(\rho)) \qquad (3.2.5)$$

where $\phi_K(\rho)$ is the polynomial of degree K in ρ which satisfies

$$\phi_K(\rho) \equiv a + O(a^{K+1}). \qquad (3.2.6)$$

For example for $K = 1$, $\phi_1(\rho) = 1-\rho=1-e^{-a}$, so if $b \gg 1 \gg |a|$ we have

$$\eta \simeq a^2/2b\ . \qquad (3.2.7)$$

This is smaller by a factor b^{-1} than the local truncation error estimate $a^2/2$, so that if step-end values are used to get global error estimates very small estimates will result. A similar result holds for other values of K. However, on using values y_{r-k}, k=0 to K, to interpolate within the step to give output at off-step points, the interpolation errors are the same order as the local truncation error estimate.

3.3 Jacobian matrix elements and sparsity pattern

It is not uncommon when testing stiff ODE solvers to supply them with a subroutine which evaluates the Jacobian matrix J(y) for a given vector y, as well as one to evaluate f(y). The test problems are often small, so coding of this subroutine is not difficult; and moreover no harm is done in such cases by treating it as a full matrix. In the author's view, this is not a realistic way of testing how methods are likely to perform in practice. Even on small problems, it is not reasonable to ask the user to code the evaluation of J; and if he is asked to do so he is likely to make mistakes. On medium and large problems it becomes increasingly important to exploit the sparsity which commonly occurs in J (most of its elements being zero); but it is not reasonable in general to expect the user to specify a list of which components of f depend on each component of y. From the point of view of the user, J is not part of the problem specification, but of the means to solve it; so it should be the concern of the software. We discuss some techniques for achieving this.

There seem to be two alternatives: (a) automatic generation of a Jacobian evaluation subroutine, following symbolic differentiation of the functions f(y), or (b) numerical estimation of the matrix elements by finite differencing. Method (a) is used in FAST (Stutzman *et al* 1976). FACSIMILE uses method (b); it uses a technique similar to method (a) for finding the sparsity pattern.

Use of method (b) apparently needs (n+1) evaluations of f to generate a finite difference approximation to J. However, Curtis *et al* (1974) showed that in sparse cases far fewer are often needed. As an example of the savings which this technique can bring, a problem with 541 variables (including t: 10 variables in each of 54 cells of a two-dimensional PDE) needed only 16 evaluations of f to calculate J. It is no disadvantage with this technique if a few columns of J are nearly full, but a nearly full row (corresponding to a component of y whose derivative depends on nearly all components) can be expensive, and it may be worth while in such a case to try omitting the off-diagonal components in such a row from J; often there is little or no deleterious effect.

Finite-difference increments for the components of y can be chosen to be proportional to the "test numbers" Y_i. A suitable constant of proportionality is $\sigma = (c\tau)^{\frac{1}{2}}$ where τ is the relative error tolerance and c is a constant taken as 10^{-3} in FACSIMILE.

An alternative method for initial detection of the Jacobian sparsity is to change one component of y at a time, by a "large" amount, to see which components of f are affected. Since this apparently need be done only once, the cost does not at first sight seem unduly high. However, this approach caused so many difficulties, especially with undetected non-zeros becoming important later in the run, that we abandoned it. In the current version of FACSIMILE the pattern is found by a single preliminary evaluation of $f(y)$ in a special dependence-recording mode; this takes quite a lot of code, and is only practicable because FACSIMILE executes interpretively the user's code for evaluation of $f(y)$, which may be quite complex.

Many stiff ODE solvers recompute J unnecessarily often. Typically, sets of linear equations of the form

$$(I - \gamma hJ)\, z = r \qquad (3.3.1)$$

have to be solved, where h is the step-size and γ a numerical coefficient depending on the order of integration formula. Commonly, an LU decomposition of the coefficient matrix of (3.3.1) is retained, and may have to be updated for one of two reasons: (a) if γh changes, a fairly frequent occurrence, or (b) if the elements of J currently in use have become so outdated that convergence of corrector iterations is no longer obtained. In many problems J changes only slowly, and case (b) is relatively rare. Most solvers treat both cases alike, recomputing J and then over-writing it first with $(I-\gamma hJ)$ and then with the LU decomposition. Advantage is often obtained by storing J as well as the LU decomposition; in case (a), the current J is used to recompute the LU decomposition, and only in case (b) is J itself recomputed. The extra storage cost is not large in relation to total problem requirements.

3.4 Problem definition

In the case of small problems, no great hardship is caused by requiring the user to code a Fortran subroutine, say, to evaluate the components f(y) for values of y, supplied by the software. The same is true of some medium problems, especially those arising from PDEs with only a few variables per mesh point. But in the case of large reaction systems, which may give rise to medium or large problems, conversion from a set of reactions to a set of derivative functions f is laborious and error-prone, and the user will welcome help in the form of special problem-description facilities. This is especially true in a research environment, where the model of a process is frequently changed. The same is true in simulating large electronic circuit kinetics. If the software tries to meet the user's wishes in this respect, it ceases to have the character of a library subroutine and becomes a complete program package. There is then scope for deciding how far to integrate the problem specification into a general simulation language, with a much wider range of user-oriented facilities.

Some program packages (e.g. Dickinson and Gelinas 1976; Curtis *et al* 1977) accept data in the form of chemical reactions as a (full or partial) definition of the right hand side of (2.1.1). The problem is to accept specifications of reactions in a form acceptable to the user, and to generate the components of f by assembling, for each component, terms arising from the rates of those reactions in which the corresponding component of y occurs, the rates being computed from the law of mass action.

SETKIN (Dickinson and Gelinas 1976) is a preprocessor for EPISODE; thus solving a problem becomes a two-step process - first SETKIN translates the problem description, producing an output file which is used as input to a solver program incorporating EPISODE. SETKIN incorporates a library of reaction data for atmospheric chemistry applications, and it seems that it offers less direct help (although it could be extended) to other reaction kinetics applications. It generates Fortran subroutines to evaluate f and J from given y, from references to its library of reactions; it is not clear whether it can do so from user-specified reaction data as an alternative.

FACSIMILE is so far defined only by a series of data preparation documents (Curtis 1975; Curtis and Kirby 1977; Curtis 1977), but a full user's manual is now being prepared (Curtis *et al* 1977). We give a brief account of its facilities in §4.

3.5 Available solution methods

Two kinds of numerical method came out well in the tests by Enright and Hull, (a) backward difference methods and (b) second derivative multistep methods. For reasons explained by Lindberg (1974), one may have to consider (c) trapezoidal rule methods for problems in which the object is to discover whether a system is stable or not, since then the stability boundary in the complex $h\lambda$ plane must be the imaginary axis; but they do not seem competitive for simulating the behaviour of a stiff system known to be stable.

Backward difference methods for stiff problems were popularised by Gear (1968), and have been much used since. Several implementations (DIFSUB,GEAR.REV3 and EPISODE) performed very well on Enright and Hull's (1976) tests, with GEAR.REV3 marginally the best; but because the error control method used ("absolute errors") was inadequate on some at least of the tests, not too much attention should be paid to details of the placing. All three programs have been in existence for some time, so they are well established and available (through the Argonne Code Center). They can be recommended for small and for some medium sized problems; being more or less standard ODE subroutines they do not provide powerful problem description features, and do not exploit sparsity in the Jacobian matrix. Some slight intervention would be needed to achieve reliable relative error control.

Second derivative multistep methods (Enright 1974) introduce second derivative terms into the usual multistep formula, using $\ddot{y}=J\dot{y}$ to compute them. This gives the Jacobian matrix J an enhanced rôle, possibly explaining inefficiency reported by Enright and Hull, and raising doubts about reliability if J was evaluated by finite differences. Moreover, since J^2 enters into the Newton iteration method, deleterious effects on sparsity can be expected in medium or large problems. This method is not recommended for large or very stiff problems.

4. FACSIMILE - A PROBLEM-ORIENTED BACKWARD-DIFFERENCE PROGRAM

FACSIMILE is a backward-difference ODE solver developed at Harwell with powerful problem-description features. It provides a command language and a high level programming language for problem description with user-defined VARIABLEs (which are controlled by differential equations) and PARAMETERs (which are constants, or functions of variables and/or other parameters).

The programming language is used to write code blocks with two kinds of name: system-defined names (implying execution at specific stages, e.g. to compute the functions f(y)) or user-defined names (which can be called as subroutines or to generate output). Arrays of variables or parameters are permitted, and code can be made to operate on whole arrays (of similar dimensions), on an element-by-element basis; or matrix operations on arrays may be specified; or terms resulting from discretisation of spatial PDE operators may be added into the time derivatives of an array of variables (one element for each spatial mesh division). Powerful indexing and looping facilities are available in the programming language, and Fortran-like expressions are allowed. There is an application-oriented syntax for specifying reaction kinetics equations, exemplified by

```
|K1|K2: A+B=C;
```

which is interpreted as: form the expression

```
R=K1.A.B-K2.C
```

and subtract R from the time derivatives of A and B and add it to that of C. Here K1,K2,A,B,C are user-defined names (which can be up to 10 characters long). Fortran-like expressions may be substituted for K1 or K2, and all the quantities may be arrays, thus specifying similar reactions involving corresponding array elements.

The flavour of the command language can perhaps be given by an example of a command to produce output:

```
WHEN TIME = 1 2 5 10 20 | CALL PRINT;
```

Here, PRINT is the name of a user-defined code block, TIME is the independent variable; the vertical bar indicates that the integration process is to be stopped when the value 20 is passed, and control returned to the command processor.

FACSIMILE has been widely used for reaction kinetics and other problems, at Harwell and elsewhere, for about 18 months. It is available on commercial terms (including a substantial discount to universities and other educational establishments). At present, it is restricted to large IBM 360 or 370 computers, but it is hoped soon to have it available on large ICL 2900 machines. Internally, it uses a Harwell implementation of backward difference methods, Harwell sparse matrix handling subroutines, automatic sparsity pattern determination, and automatic initial step size selection. It also has parameter-fitting facilities, controlled by a Harwell non-linear least squares subroutine.

5. EXPERIENCE WITH FACSIMILE

FACSIMILE is frequently used on small problems, where its high level programming and command languages make it attractive to the user; but it is on medium and, even more, on large problems that its benefits are greatest, so we concentrate on experience of such problems.

Our first atmospheric pollution problem (Derwent *et al* 1976) involved over 50 chemical and 20 photolytic reactions among 35 species in each of 10 atmospheric layers. In addition, diffusion between layers of the 20 stable species was simulated. The stiffness was extreme: one excited species had a lifetime of only 10^{-9} sec. at ground level, due to collisional de-activation; on the other hand, there was a 30-year (10^{9} sec.) time-scale for the removal from the atmosphere of chlorine, the active species resulting from the pollutant studied, necessitating simulation of 5.10^{9} sec. for the model to come to equilibrium. The computation was done using an earlier program, CHEK, from which FACSIMILE was developed; it has since been repeated using FACSIMILE (which permits great condensation of the problem data), and took about 10 minutes computer time. A simplified model was run at tolerances of 10^{-3} and 10^{-4} to check accuracy; differences were generally within about 3 times the larger tolerance, a very satisfactory result.

A two-dimensional (latitude and height) model (Derwent and Curtis 1977) is now in use to study annual and diurnal variations in the troposphere. In the basic model, 10 species in each of 54 cells (18 latitude times 3 height) are modelled, with diffusion and advection of 6 species. Further species are added as necessary for diurnal studies, up to a maximum of 14. Simulation of one year, with diurnal variations smoothed out but annual variations of wind pattern, temperature and insolation included, takes about 6 minutes. Simulation of diurnal variation (sunset, sunrise and sunset) takes about one minute. A version with 6 layers, to give improved vertical resolution is being prepared. In these two-dimensional studies we have recently successfully tried omitting the effects of diffusion and advection from the Jacobian matrix (using high-level language facilities for user over-ride), while retaining them in the functions f, to economise in matrix operations; maximum step-size was reduced, but so was total computer time. Originally, 4285 non-zeros in J produced 12575 in its LU decomposition; ignoring transport terms reduce these figures to 3241 and 3349.

The carbon-burning phase of stellar nuclear synthesis has been modelled, both in hydrostatic and explosive conditions. The same model is used, but temperature dependence of rate coefficients changes the time-scale dramatically, from 10^7 sec. to 0.1 sec. About 40 nuclides are modelled, undergoing over 100 nuclear reactions, most of them reversible. Each run takes about one minute of computer time, and we use them as test problems after major program updates.

Transmission of nerve impulses along an axon according to the Hodgkin-Huxley model has been simulated, with four variables at each of 25 or 100 mesh points. This problem is highly non-linear, and is sometimes used as a test problem.

Enhanced diffusion (Marwick and Piller 1977; Marwick 1977) of solute atoms in a metal, induced by point defects caused by radiation damage, has been simulated, and this has helped in understanding of the processes involved. Four species were modelled at each of either 25 or 100 mesh points. The parameter-fitting facilities are being used to choose rate coefficients for best agreement with experimental data.

6. ACKNOWLEDGEMENTS

In all the above problems (and many others) problem description data for FACSIMILE is prepared by the scientists concerned, with only general advice on use of the facilities from those responsible for the program (plus specific mathematical analysis, where necessary, about such matters as discretisation of PDE operators). The users have always been appreciative, responsive and constructive in discussions about program enhancements, and their comments are valued. I should like to conclude, therefore, by expressing my indebtedness to them, to my colleagues more directly concerned with FACSIMILE development, and to other colleagues whose numerical analysis has been incorporated, for their invaluable help.

7. REFERENCES

Lapidus, L. and Schiesser, W.E. (editors)(1976). "Numerical Methods for Differential Systems", Academic Press Inc., New York.

Individual references:

Curtis, A.R. (1975). "Data preparation for FACSIMILE 1", Report C.S.S.26, Computer Science and Systems Division, A.E.R.E. Harwell.

Curtis, A.R. (1977). "Further extensions to FACSIMILE 3", Report C.S.S.50, Computer Science and Systems Division, A.E.R.E. Harwell.

Curtis, A.R. and Kirby, C.R. (1977). "New features for computer program FACSIMILE", Report C.S.S.46, Computer Science and Systems Division, A.E.R.E. Harwell.

Curtis, A.K., Powell, M.J.D. and Reid, J.K. (1974) "The estimation of sparse Jacobian matrices", *J. Maths. Applics.* **13**, 117-119.

Curtis, A.R., Chance, E.M., Jones, I.P. and Kirby, C.R. (1977). "FACSIMILE: A program for modelling reaction kinetics, transport processes, and other stiff initial value problems", A.E.R.E.-R.8775, A.E.R.E. Harwell.

Derwent, R.G., Eggleton, A.E.J. and Curtis, A.R. (1976). "A computer model of the photochemistry of halogen - containing trace gases on the troposphere and stratosphere", A.E.R.E.-R.8325, A.E.R.E. Harwell.

Derwent, R.G. and Curtis, A.R. (1977). "Two dimensional model studies of some trace gases and free radicals in the troposphere" A.E.R.E.-R.8853, A.E.R.E. Harwell.

Dickinson, R.P. and Gelinas, R.J. (1976). "SETKIN: A Chemical Kinetics Preprocessor Code", in Lapidus and Schiesser (1976).

Enright, W.H. (1974). "Second derivative multistep methods for stiff ordinary differential equations", SIAM J. Num. Anal. 11, pp.321-331.

Enright, W.H. and Hull, T.E. (1976). "Comparing Numerical Methods for the Solution of Stiff Systems of ODEs Arising in Chemistry", in Lapidus and Schiesser (1976).

Gear, C.W. (1968). "The automatic integration of stiff ordinary differential equations", in Information Processing 68, Proceedings of the IFIP Congress, Edinburgh, 1968, North Holland (Amsterdam).

Hindmarsh, A.C. (1976). Discussion contribution, Lapidus and Schiesser (1976), p.274.

Lindberg, B. (1974). "On a Dangerous Property of Methods for Stiff Differential Equations", BIT 14, pp.430-436.

Marwick, A.D. and Piller, R.C. (1977). "The effect of point defect fluxes on radiation-enhanced diffusion in nickel", A.E.R.E.-R.8682, A.E.R.E. Harwell.

Marwick, A.D. (1977). "Modelling radiation-enhanced diffusion using FACSIMILE", A.E.R.E.-R.8741, A.E.R.E. Harwell.

Stutzman, L.F., Deschard, F., Morgan, R. and Koup, T. (1976). "FAST: A Translator for the Solution of Stiff and Nonlinear Differential and Algebraic Equations", in Lapidus and Schuesser (1976).

V. 2

SOME FACTORS AFFECTING THE EFFICIENCY OF STIFF INTEGRATION ROUTINES

H.H. Robertson

(Imperial Chemical Industries Limited and Department of Mathematics, University of Manchester)

1. INTRODUCTION

We are interested in the efficient numerical integration of the initial value problem for systems of ordinary differential equations of the form

$$\frac{dy}{dx} = f(y), \; y(0) = y_0, \; y \in R^n \qquad (1.1)$$

where the Jacobian f' has eigenvalues λ_i, $i=1,2,\ldots,n$, such that $Re(\lambda_i)<0$ and $\max_{i,j} Re(\lambda_i)/Re(\lambda_j)$ is large. In general, the solution exhibits a fast transient followed by increasingly smooth behaviour. In the region of fast transients a small step-length h has to be used to obtain good accuracy. As the transients decay, the step-length h may be increased, provided the integration procedure remains stable. Highly stable integration procedures are implicit in character and involve the solution of sets of non-linear algebraic equations at each step of the integration. These may be written in the form

$$F(y) = y - h\gamma f(y) - \Psi = 0. \qquad (1.2)$$

where γ is a formula dependent constant and ψ involves past values of y and f.

In Section 2 we review briefly the stability properties of multistep methods and discuss some interactions of stability and accuracy.

The equations (1.2) may be solved by the Newton iteration or more usually by the Chord method;

$$y^{k+1} = y^k - M^{-1} F(y^k), \qquad (1.3)$$

where M^{-1} is an approximate inverse for $F'(y)$. For rapid convergence of the iteration we need a good initial approximation y^0. The usual predictors involving past values of function and derivatives are not very suitable for this purpose and their use in error estimation and control also raises certain snags. These two topics are discussed in Section 3.

In Section 4 we introduce problems where the Jacobian has more detailed structure associated with the fast and slow motions of the system. We mention some special methods and in Section 5 we examine some implications of structure for conditions for local convergence of the iteration (1.3). Finally, we give some examples of stiff systems with structure and illustrate the benefits of using the structure of the Jacobian to update the iteration matrix M^{-1} of (1.3).

2. NUMERICAL STABILITY

For systems of stiff equations it is desirable to increase the step-length h progressively as the fast transients decay. The method must therefore be numerically stable for large values of the step-length. Euler's backward difference formula for the derivative and the trapezoidal rule have long been recognised as suitable procedures.

Dahlquist (1963) studied the problem and defined the property of A-stability for multi-step methods applied to the test equation.

$$\frac{dy}{dx} = \lambda y, \tag{2.1}$$

where λ is a complex quantity with negative real part.

A linear k-step method is written in the form

$$\sum_{j=0}^{k} \alpha_j y_{n+j} = h \sum_{j=0}^{k} \beta_j f_{n+j} \tag{2.2}$$

and for the test equation we have the associated stability polynomial:

$$\Pi(r) = \sum_{j=0}^{k} \alpha_j r^j - h\lambda \sum_{j=0}^{k} \beta_j r^j. \tag{2.3}$$

A method is said to be absolutely stable in a region R of the complex plane if, for all $h\lambda \in R$, all roots of the stability polynomial satisfy

$$|r_s| < 1, \quad s=1,2,\ldots,k. \tag{2.4}$$

A k-step method is A-stable if its region of absolute stability contains the whole of the left half-plane $Re(h\lambda) < 0$.

Dahlquist (1963) showed that

(i) an explicit linear multistep method cannot be A-stable;
(ii) the order of an A-stable method cannot exceed two;
(iii) the second-order A-stable method with smallest error constant is the trapezoidal rule.

In the search for higher order methods of suitable stability Widlund (1967) proposed the property of $A(\alpha)$-stability. A numerical method is $A(\alpha)$-stable, $\alpha \in (0, \pi/2)$ if its region of absolute stability contains the infinite wedge $|arg(-h\lambda)| < \alpha$. Widlund showed that implicit linear multistep methods of order three and four exist which are $A(\alpha)$-stable for any $\alpha < \pi/2$.

Gear (1969,1971) combines stability properties and accuracy in his definition of stiffly stable. A numerical method is said to be stiffly stable if it is absolutely stable in the region R_1 $(Re(h\lambda) \leq D)$ and accurate in the region R_2 $(D < Re(h\lambda) < \alpha, |Im(h\lambda)| < \theta)$. This property permits step-length increases which keep the large $(h\lambda)$ in R_2 but difficulties occur in dealing with systems with highly oscillatory components.

In further stability studies a number of authors (Ehle (1969), Axelsson (1969) and Chipman (1971)) have proposed more stringent requirements than A-stability. When a one-step method is applied to the test equation (2.1) it yields

$$y_{n+1} = Q(h\lambda)y_n. \tag{2.5}$$

A-stability ensures that

$$|Q(h\lambda)| < 1, \quad Re(h\lambda) < 0.$$

For many A-stable one-step methods, however, $Q(h\lambda)$ is such that

$$|Q(h\lambda)| \to 1 \text{ as } \mathrm{Re}(h\lambda) \to -\infty,$$

so that numerical approximation to the rapidly decaying transients with large $|\lambda|$ may decay very slowly. The trapezoidal rule behaves in this fashion (Rosenbrock (1963)). To overcome this difficulty a stronger stability property is required.

A one-step method is L-stable if it is A-stable and $|Q(h\lambda)| \to 0$ as $\mathrm{Re}\,(h\lambda) \to -\infty$.

L-stable methods ensure that fast transients can be damped out of the numerical solution as the step-length h is rapidly increased.

In the application of A-stable one-step methods to large systems of stiff non-linear equations Prothero and Robinson (1974) have found that

a) some A-stable methods give highly unstable solutions and,
b) the accuracy of the solutions often appears unrelated to the order of the method used.

This led them to study the stability and accuracy of numerical solutions to the more general test equation

$$y' = g'(x) + \lambda(y - g(x)). \tag{2.6}$$

Solution of this equation by one-step methods gives a difference equation of the form

$$e_{n+1} = y_{n+1} - g(x_{n+1}) = \alpha(h\lambda)\, e_n + \beta(h, h\lambda, g(x)) \tag{2.7}$$

Prothero and Robinson (1974) define a stability property, termed <u>S-stability</u>, which generalizes the A-stability property to equation (2.6) and they examine the asymptotic accuracy of one-step methods in the limit $|h\lambda| \to \infty$. Thus the error in the trapezoidal rule in this limit is of the form

$$\beta \simeq -1/6\, h^3 (h\lambda)^{-1} g_n'''$$

while for the closely related implicit midpoint rule

$$\beta \simeq -1/4 h^2 g_n''\ .$$

One-step methods for which $\beta \to 0$ as $|h\lambda| \to \infty$ are said to be stiffly accurate.

Prothero and Robinson demonstrate that for one-step methods which are not stiffly accurate there is considerable deterioration in the effective order as $Re(h\lambda)$ is increased, in contrast to errors of stiffly accurate methods which tend to zero as $Re(-\lambda) \to \infty$.

For a fuller perspective of the subject of numerical stability we refer to Lambert (1973) and to Hall and Watt (1976). Comparisons of various methods for stiff systems have been made by Enright, Hull and Lindberg (1975), Ehle and Lawson (1975) who devised generalized Runge-Kutta processes for stiff systems and more recently by Alexander (1977) who makes some favourable comments on the use of diagonally implicit Runge-Kutta methods. The coefficient matrix of these methods is lower triangular with all diagonal elements equal so that the implicit equations at each stage are solved sequentially and moreover the same iteration matrix may be used. Strongly S-stable methods are derived and are shown to confirm Prothero and Robinson's observations on the importance of stiff stability. The methods have advantages for highly oscillatory functions and low accuracy requirements.

3. PREDICTION AND ERROR ESTIMATES

The iteration matrix M of the Chord iteration (1.3) may be employed for a number of integration steps. It is taken to be an approximation to the derivative $F'(y^*)$ at the solution point y^* of $F(y) = 0$. Thus

$$M = F'(y^*) + E, \qquad (3.1)$$

for some E which in practice is usually given by:

$$E = M - F'(y^*) = F'(y^{\dagger}) - F'(y^*), \qquad (3.2)$$

where $y^{\dagger}$ is some previously determined point on the solution. To ensure rapid convergence it is desirable to have an accurate initial approximation y^0 which may be obtained by prediction. The usual prediction formulae associated with prediction/correction methods are not very suitable for this purpose. By definition, the derivative $f(y)$ in stiff problems is sensitive to errors in y. Similar difficulties are encountered with the accuracy of error estimates based on the difference between predicted and corrected values. The basic difficulty here is the lack of stability of explicit predictors. The problem has been discussed by Robertson and Williams (1975) who illustrate the difficulties of accurate prediction by solution of the set of equations

$$\frac{dy}{dx} = Ay$$

where

$$A = -\frac{1}{2}\begin{pmatrix} 1001 & 999 \\ 999 & 1001 \end{pmatrix},$$

and with initial conditions

$$y(0) = \begin{pmatrix} \sqrt{2} \\ 0 \end{pmatrix}.$$

The solution of these equations is

$$y(x) = \frac{1}{\sqrt{2}}\begin{pmatrix} e^{-x} + e^{-1000x} \\ -e^{-x} + e^{-1000x} \end{pmatrix}$$

These equations were integrated by the trapezoidal rule which has the local truncation error

$$\tau(x_{n+1}) = y(x_{n+1}) - y(x_n) - 1/2\, h(y'(x_{n+1}) + y'(x_n)).$$

This is compared with Est = $(\tilde{y} - y^p)/5$, where y^p is the prediction from the mid-ordinate rule and $\tilde{y}$ are computed values. Table I shows the results and we note the discrepancy of 10^3 in the values of τ and Est which is entirely due to magnification of small errors in y in the computation of the derivative Ay. The sensitivity of derivatives to error in y follows from the definition of stiffness.

When predictors involving derivative values are employed to calculate initial approximations for the corrector equations, highly accurate values of y are required to ensure sufficiently accurate derivatives. This requirement for additional accuracy in y values may be avoided by the use of extrapolation on past values of y. Alternatively, we may construct the series

$$\pi^s = (y^s - \psi)/h\gamma \tag{3.3}$$

and obtain an estimate of $f(y^*)$ from the limit of this series which is insensitive to error in y^s. It can also be shown

Table I

y_1, y_2 are solutions of the differential equations computed from the trapezoidal rule. Est_1 and Est_2 are the estimates of the local truncation error while τ_1 and τ_2 are the corresponding exact local truncation errors.

x	y_1	$10^5\tau_1$	10^5 Est_1
0.01	0.700 103 053 7	-	-
0.02	0.693 083 669 1	-	-
0.03	0.686 222 770 7	0.00633	8.92433
	(0.686 668 987 2)		
0.04	0.679 371 045 5	0.00569	5.93807
	(0.679 074 142 0)		
x	y_2	$10^5\tau_2$	10^5 Est_2
0.01	-0.700 038 848 5	-	-
0.02	-0.693 126 472 6	-	-
0.03	-0.686 194 235 0	-0.00516	8.91046
	(-0.685 748 712 0		
0.04	-0.679 390 069 3	-0.00569	-5.95180
	(-0.679 687 659 5)		

that the chord iteration applied to the equations for the reduced derivatives k,

$$k - hf(\gamma k + \psi) = 0 \qquad (3.4)$$

which is used in the Nordsieck form of integration formulae, also produces accurate derivative values directly. The linear transformation $y = \gamma k + \psi$ ensures that the chord iteration (1.3) and the corresponding chord iteration for k on (3.4) are equivalent. Corresponding y values from the iteration on (3.4) are obtained by $y^s = \gamma\, k^s + \psi$ and we note that, as previously discussed, k^s is very different from $hf(y^s)$.

4. STRUCTURE

In addition to the basic requirement of a large Lipschitz constant many stiff systems have significant structure in the

Jacobian f'. An important feature for large systems is sparsity. Curtis and Reid (1974) discuss the numerical evaluation of the Jacobian and minimization of truncation and round-off errors. Curtis, Powell and Reid (1974) evaluate the implications of sparsity for the computation.

In Chapter II.3 Reid reviews efficient algorithms for the identification and manipulation of sparse systems and in Chapter V.1 Curtis describes techniques for exploiting sparsity patterns in the context of our problem.

Another aspect of structure is the magnitude and rates of change of individual elements of the Jacobian. There is a close relationship with singular perturbations of the initial value problem. For the purposes of preliminary orientation we consider the simplest form of such problems consisting of a pair of scalar equations of the form

$$\frac{dx}{dt} = f(x,y) \text{ and } \varepsilon\frac{dy}{dt} = g(x,y),$$

where ε is small and positive and where the initial conditions are given. For a qualitative description of the behaviour of this system we refer to Wasow (1965). Suffice to say that under appropriate circumstances the solution rapidly approaches an epsilon-neighbourhood of the curve $g(x,y)=0$. We identify the x and y variables with the slow and fast motions, respectively, and observe that the Jacobian of the system,

$$\begin{pmatrix} f_x & f_y \\ \varepsilon^{-1}g_x & \varepsilon^{-1}g_y \end{pmatrix}$$

has in general, much larger elements in the bottom row. This structure of the Jacobian is typical of the type we wish to consider. We are not, however, concerned with the analytical properties of the system for limiting values of the parameters. We note that this matrix can be transformed by scaling of rows and columns to one with a much larger second column. A linear transformation of the dependent variables of the differential equation produces a new system of equations whose Jacobian is similar to the original and with effects which may correspond to scaling operations on the rows and columns. In fact, we prefer to take the column form of structure.

For our purposes, a more appropriate description of the systems of interest is in terms of the Jacobian and higher derivatives since we are concerned essentially with local properties in the neighbourhood of the solution in the asymptotic region where the fast transients have died away. Our problem may therefore be written as the numerical integration of the equations

$$\frac{dy}{dx} = f(y), \quad y(0) = y_0, \text{ and } y \in R^n, \qquad (4.1)$$

where the Jacobian f' has eigenvalues λ_i, $i = 1,2,\ldots,n$ such that $Re(\lambda_i)<0$ and $\max_{i,j} Re(\lambda_i)/Re(\lambda_j)$ is large. We suppose further that y can be partitioned into

$$\begin{pmatrix} y_1 \\ y_2 \end{pmatrix}, \quad y_1 \in R^{n_1}, \; y_2 \in R^{n_2}, \; (n_1 + n_2 = n),$$

such that the conformal partition of f' by columns into

$$f' = \begin{pmatrix} f_{11} & f_{12} \\ f_{21} & f_{22} \end{pmatrix} = (f'_1, f'_2) \text{ satisfies } \|(f'_1, 0)\| << \|(0, f'_2)\|. \qquad (4.2)$$

Further conditions on f_{22} ensure the stability of the fast motions associated with y_2. When f' varies slowly there is little need to adjust its value for computational purposes but when the changes in f' along the solution affect convergence properties and when the structure described by (4.2) is maintained along the solution there is the possibility of increased efficiency by updating the partition f'_2. We can express this property by requiring that df'/dx or $f''f$ has the same structure as f'.

In a recent paper Hofer (1976) describes a partially implicit method suitable for large stiff systems having a few equations associated with small time constants. The modified midpoint rule is applied to the slow motions and the trapezoidal rule is combined for the fast motions according to the following formulae:

$$y_{1,j+1/2} = y_{1,j} + 1/2hf_1(y_j)$$

$$y_{2,j+1/2} = y_{2,j} + 1/2h(f_2(y_j) + f_2(y_{j+1/2}))$$

$$y^*_{1,j+1} = y_{1,j} + hf_1(y_{j+1/2})$$

cont.

$$y_{2,j+1} = y_{2,j+1/2} + 1/4\, h\, (f_2(y_{j+1/2}) + f_2 \begin{bmatrix} y^*_{1,j+1} \\ y_{2,j+1} \end{bmatrix})$$

$$y_{1,j+1} = 1/2(y_{1,j+1/2} + y^*_{1,j+1} + 1/2\, hf_1 \begin{bmatrix} y^*_{1,j+1} \\ y_{2,j+1} \end{bmatrix}).$$

When the method is applicable it provides useful computational savings although there is no *a priori* guarantee of stability which depends on the coupling between the fast and slow motions.

Dahlquist (1969) introduced the SAPS method which has been analyzed by Oden (1971). The equations are written in the form.

$$\frac{dy_1}{dx} = f_1(y) \tag{4.3}$$

$$\frac{dy_2}{dx} = Ay_2 + \phi \tag{4.4}$$

where A is a piecewise constant matrix with all its eigenvalues in the left half-plane and where f_1 and ϕ have Lipschitz constants which are small compared with $\|A\|$. Smooth approximate particular solutions of (4.4) are obtained by a semi-analytical approach, while a conventional method is used to integrate the non-stiff system (4.3). The method has advantages for highly oscillatory functions and is applicable when linearization about the particular solution is sufficiently accurate.

Miranker (1973) uses numerical methods of boundary layer type and develops these to cope with parameterless systems. Comparison with conventional numerical methods for general stiff systems is unfavourable but there are useful pointers for the automatic identification of fast motions. Robertson (1976) shows how structure can be used to produce an approximate inverse M^{-1} for the iterative solution of implicit procedures (1.3). Given a partition $\begin{pmatrix} y_1 \\ y_2 \end{pmatrix}$ of the dependent variables y we take conformal partitions of the Jacobian F' and write

$$F' = \begin{pmatrix} F_{11} & F_{12} \\ F_{21} & F_{22} \end{pmatrix} = \begin{pmatrix} a & b \\ c & d \end{pmatrix} \tag{4.5}$$

and formally write the inverse as

$$(F')^{-1} = \begin{pmatrix} (a-bd^{-1}c)^{-1} & -(a-bd^{-1}c)^{-1}bd^{-1} \\ -(d-ca^{-1}b)^{-1}ca^{-1} & (d-ca^{-1}b)^{-1} \end{pmatrix}$$

$$= \begin{pmatrix} \alpha^{-1} & -\alpha^{-1}bd^{-1} \\ -\delta^{-1}ca^{-1} & \delta^{-1} \end{pmatrix}. \qquad (4.6)$$

Recall that the structure introduced in (4.2) requires that b and d are much "larger" than a and c and that b and d are more sensitive to changes along the solution. This motivates consideration of the iteration matrix

$$M = \begin{pmatrix} I & b \\ c & d \end{pmatrix}. \qquad (4.7)$$

Formally we have

$$M^{-1} = \begin{pmatrix} I+b\delta_I^{-1}c & -b\delta_I^{-1} \\ -\delta_I^{-1}c & \delta_I^{-1} \end{pmatrix}, \quad \delta_I = d - cb \qquad (4.8)$$

The corresponding iteration in terms of the partition of y and F becomes

$$\begin{aligned} y_1^{k+1} &= y_1^k - F_1(y^k) - b(y_2^{k+1} - y_2^k) \\ y_2^{k+1} &= y_2^k + \delta_I^{-1}(cF_1(y^k) - F_2(y^k)) \\ y_1^{k+1} &= h\gamma f_1(y^k) + \psi_1 - b(y_2^{k+1} - y_2^k) \\ y_2^{k+1} &= y_2^k + \delta_I^{-1}(cF_1(y^k) - F_2(y^k)), \end{aligned} \qquad (4.10)$$

in terms of the integration equations (1.2). This iteration has a Newton-like character for fast motions and resembles repeated substitution with correction terms for the slow motions. This is a very plausible algorithm when we recall that the slow motions have much longer time constants so that repeated substitution of $F_1 = 0$ is rapidly convergent whereas the fast motions, corresponding to $F_2 = 0$ require local derivative information to be incorporated in the iteration for rapid convergence. An example will be presented later which shows that this iteration can be successfully employed. If the dimensions of δ_I are much less than those of the full matrix $M(n_2 \ll n_1)$ considerable computational savings accrue. This is however only one stage in the integration, for as the step-length h is further increased, it will become expedient to take account of the structure of the partition a. If a remains diagonally dominant we may take advantage of this. If $a=a_d+e_a$ where a_d is a diagonal matrix, the corresponding modifications to (4.8) may provide adequate convergence at the expense of a minimum of additional computation. Again, we may consider the partition y_1 to be further partitioned into fast and slow motions when an approximation of the form (4.8) will be appropriate for a^{-1}. In this way we can introduce a spectrum of motions into the problem. Robertson (1976) also considers the problem of inverse updating.

5. LOCAL CONVERGENCE OF THE PARALLEL CHORD ITERATION

The parallel chord iteration for solution of the equation $F(y) = 0$ may be written in the form

$$M(y^{k+1} - y^k) = -F(y^k), \quad k = 0,1,2,\ldots, \tag{5.1}$$

where y^0 is given and where the matrix M may be taken to be an approximation to the derivative at y^* so that

$$M = F'(y^*) + E \tag{5.2}$$

for some perturbation E.

Sufficient conditions for the local convergence of the iteration $y^{k+1} = G(y^k)$ are given by Ostrowski's theorem (Ortega and Rheinboldt, 1970). In particular, this includes the essentially necessary condition $\rho(G'(y^*))<1$, where ρ is the spectral radius and y^* is a fixed point of G.

Robertson (1976) examines the convergence of (5.1) and obtains the sufficient condition for local convergence

(i) $\| F'(y^*)^{-1} \| \cdot \|E\| < 1/2$

where $F(y^*) = 0$.
An estimate of the spectral radius of $G'(y^*)$ is

$$\rho(G(y^*)) \leqslant \|M^{-1}E\| \leqslant \frac{\|F'(y^*)^{-1}E\|}{1-\|F'(y^*)^{-1}E\|} \leqslant \frac{\|F'(y^*)^{-1}\| \ \|E\|}{1-\|F'(y^*)^{-1}\| \ \|E\|} \qquad (5.3)$$

If P and Q are non-singular matrices and the iteration

$$QMP(z^{k+1} - z^k) = - QF(Pz^k), \quad k = 0,1,2,... \qquad (5.4)$$

is locally convergent, where $z^0 = P^{-1}y^0$ is given, then the iteration (5.1) is locally convergent. It can be shown that condition (i) for local convergence of (5.1) may be replaced by

(ii) $\|P^{-1}F'(y^*)^{-1}Q^{-1}\| \ \|QEP\| < 1/2$,

where P and Q are non-singular matrices.

The corresponding estimate of the spectral radius is

$$\rho(G'(y^*)) \leqslant \frac{\|P^{-1}F'(y^*)^{-1}Q^{-1}\| \ \|QEP\|}{1-\|P^{-1}F'(y^*)^{-1}Q^{-1}\| \ \|QEP\|}. \qquad (5.5)$$

Condition (ii) and the estimate (5.5) show the importance of transformations in dealing with our type of application where the Jacobian F' and the perturbation E have common features of structure.

When the matrix M^{-1} of the parallel chord method is taken to be a perturbation of the inverse Jacobian so that

$$M^{-1} = F'(y^*)^{-1} + W \qquad (5.6)$$

a sufficient condition for convergence of (5.1) is

(iii) $\|F'(y^*)\| \ \|W\| < 1$.

An estimate of the spectral radius of $G'(y^*)$ is

$$\rho(G'(y^*)) \leqslant \|F'(y^*)\| \ \|W\|. \qquad (5.7)$$

From consideration of the equivalent iteration (5.4) it can be shown that condition (iii) may be replaced by

(iv) $\|QF'(y^*)P\| \; \|P^{-1}WQ^{-1}\| < 1.$

The corresponding estimate of the spectral radius is

$$\rho(G'(y^*)) < \|QF'(y^*)P\| \; \|P^{-1}WQ^{-1}\| \tag{5.8}$$

Finally we remark that the practical usefulness of the transformations introduced in (5.4) is in obtaining a readily computable and accurate estimate of the spectral radius $\rho(G'(y^*))$. Also by considering partitions of E which are set to zero it enables us to determine accurately the convergence effects of updating the matrix M in a variety of ways.

6. APPLICATIONS

We now give a number of examples of systems of stiff differential equations which exhibit the structure defined in (4.2) of a Jacobian which has column partitions of widely separated "magnitudes". All of the examples discussed are in spaces of small dimensions for reasons which are partly historic. and partly due to the amount of work and space required to analyze more realistic cases. They are therefore somewhat artificial if they are judged from the viewpoint of computational savings to be gained in taking account of structure. Large systems with structure are of practical importance and in suitable cases there is a significant increase in efficiency by making use of appropriate inverse updating procedures. For further details of some of the examples discussed here and a comparative review of methods for the solution of stiff systems see Enright *et al.* (1975).

Example 1

Gear (1969) discusses the application of stiffly stable integration formulae based on backward difference approximations of the derivative and uses the following example (with a sign change for y_3) to illustrate the method

$$\frac{dy}{dx} = f(y), \quad y \in R^3$$

where

$$f_1 = -0.013y_1 + 10^3 y_1 y_3$$

$$f_2 = 2.5 \times 10^3 y_2 y_3$$

$$f_3 = 0.013y_1 - 10^3 y_1 y_3 - 2.5 \times 10^3 y_2 y_3,$$

with initial conditions

$$y(0) = \begin{pmatrix} 1 \\ 1 \\ 0 \end{pmatrix}.$$

This application is from chemical reaction kinetics and y_3 represents the concentration of a very reactive species which is an intermediate in the course of the reaction and always stays small. y_1 and y_2 are monotonically decreasing and increasing respectively. y_3 increases to a maximum and thereafter is monotonically decreasing. It is not difficult to show that $y_3 < 1.3 \times 10^{-5}$. The Jacobian of the time derivative f is given by

$$J = \begin{pmatrix} -0.013 + 10^3 y_3 & 0 & +10^3 y_1 \\ 0 & +2.5 \times 10^3 y_3 & +2.5 \times 10^3 y_2 \\ 0.013 - 10^3 y_3 & -2.5 \times 10^3 y_3 & -10^3 y_1 - 2.5 \times 10^3 y_2 \end{pmatrix}$$

Due to the smallness of y_3, the last column of this matrix has elements which are very much larger than the remainder. y_1 and y_2 are clearly associated with slow motions whilst y_3 is identified with the fast motion. In terms of the structure of (4.1) $n_1 = 2$ and $n_2 = 1$. The matrix updating procedure associated with the partitioned approximate inverse (4.8) no longer requires matrix inversions. This is most clearly seen from (4.10) where δ_I, from being a matrix requiring inversion, has degenerated to a scalar. This example illustrates well the features of the Jacobian which are typical of chemical reaction kinetic systems with very reactive intermediates. Another common feature of reaction kinetic systems is the polynomial character of f. Here f is quadratic in the elements of y so that f'' is a constant operator and $f'(y^\dagger)-f'(y^*)$ is linear in $(y^\dagger - y^*)$. It will readily be seen that this implies that the perturbation $E=F'(y^\dagger)-F'(y^*)$ has the same structure as the Jacobian so that the matrix M of the iteration and the corresponding inverse are updated mostly as required by change in y_1 and y_2 and less frequently by change in y_3. Robertson (1966) discusses a similar system from reaction kinetics in three dimensions and also having quadratic derivatives. This system has an invariant but is otherwise similar to the above.

Example 2

Lindberg (1974) considers a system of dimension 4, where the derivatives are given by

$$f_1 = -k_1 y_1 y_2 + k_3 y_3$$

$$f_2 = k_3 y_3 - 2k_2 y_2^2 - k_1 y_1 y_2$$

$$f_3 = -k_3 y_3 + k_1 y_1 y_2$$

$$f_4 = -k_4 y_4 + k_2 y_2^2$$

and where

$$k = \begin{pmatrix} 10^2 \\ 10^4 \\ 1 \\ 1 \end{pmatrix} \text{ and } y(0) = \begin{pmatrix} 1 \\ 1 \\ 0 \\ 0 \end{pmatrix}.$$

If we make the change of variable $\eta = Py$ where P is the permutation matrix interchanging 1 and 3 and 2 and 4, the Jacobian becomes

$$\mathcal{J} = P^{-1}JP = \begin{pmatrix} -1 & 0 & k_1\eta_4 & k_1\eta_3 \\ 0 & -1 & 0 & 2k_2\eta_4 \\ 1 & 0 & -k_1\eta_4 & -k_1\eta_3 \\ 1 & 0 & -k_1\eta_4 & -k_1\eta_3 - 4k_2\eta_4 \end{pmatrix}.$$

We observe that η_3 and η_4 are associated with the fast motions, so that matrix updating may be confined to the last two columns. Eventually, η_4 becomes sufficiently small so that the last column alone requires frequent updating when the associated matrix inversion degenerates to scalar operations.

Example 3

Liniger and Willoughby (1967) have solved a problem from nuclear reaction theory where

$$\begin{aligned} f_1 &= 0.2(y_2 - y_1) \\ f_2 &= 10y_1 - (60 - 0.125y_3)y_2 + 0.125y_3 \\ f_3 &= 1. \end{aligned}$$

y_2 is associated with the fast motion. As in the previous example a permutation of the variables produces the standard form. This system has only one non-linear derivative so that rank one updates by the Sherman-Morrison formula suffice for all updating except change of step-length.

7. NUMERICAL COMPARISON OF PARTITIONED UPDATING

The system is taken from mass action kinetics and has been used as a test example by Enright *et al.* (1975). It is very stiff and elsewhere in this Part Curtis makes some penetrating comments on the effect of different accuracy criteria in solving effectively the same system. A further description of the system is found in Robertson (1976). The equations are written

$$\frac{dy}{dt} = \alpha r, \quad \alpha = \begin{pmatrix} -1 & -1 & 0 & 0 \\ 1 & 0 & 0 & -1 \\ 1 & -1 & 1 & -1 \\ 0 & 1 & -1 & 0 \end{pmatrix}, \; r = \begin{pmatrix} k_1y_1 \\ k_2y_1y_3 \\ k_3y_4 \\ k_4y_2y_3 \end{pmatrix}. \qquad (7.1)$$

The columns of the matrix α are the stoicheiometric vectors for the individual reactions and the components of r are the corresponding rates. The k_i are the reaction rate constants. We study the system with

$k^T = (10^{-9}, 10^7, 10^3, 10^9)$ and $y^T(0) = (2 \times 10^{-3}, 0, 0, 0)$.

The Jacobian of the system is

$$J = \begin{pmatrix} -(k_1 + k_2 y_3) & 0 & -k_2 y_1 & 0 \\ k_1 & -k_4 y_3 & -k_4 y_2 & 0 \\ k_1 - k_2 y_3 & -k_4 y_3 & -(k_2 y_1 + k_4 y_2) & k_3 \\ k_2 y_3 & 0 & k_2 y_1 & -k_3 \end{pmatrix}$$

which is singular, corresponding to the invariant $y_2 - y_3 - y_4$. This does not affect the computations to any great extent although we could have chosen to reduce the order of the system at the expense of complicating the right-hand sides. For the standard integration formulae which are linear in function and derivative values Robertson and McCann (1969) have pointed out that the approximate solutions also satisfy the invariant relation apart from round-off.

We identify the fast motions with y_3 and to a lesser extent with y_4. Typical values of the variables are

$y^T = (2 \times 10^{-3}, 0.4 \times 10^{-10}, 0.2 \times 10^{-11}, 0.3 \times 10^{-10})$.

If the Jacobian of the non-linear equations F = 0 is partitioned in the form $\begin{pmatrix} a & b \\ c & d \end{pmatrix}$ we observe that a is almost diagonal, has much smaller elements and is much more slowly varying than b, c or d. We therefore assume that a can be updated less frequently than b, c, d. Four separate runs of the program at tabular interval 10^3 and over the range $0\text{-}10^5$ have been carried out with different updating strategies. A corresponds to use of a = I throughout and therefore gives a Newton-like iteration for the fast motions and repeated substitution for the slow motions as in (4.10). In B, a = I initially and thereafter a is updated and a full inversion of the Jacobian is carried out whenever the perturbation E_{11} corresponding to a satisfies $\|E_{11}\|_\infty > 1/2$. C is similar to B but the criterion for updating is $\|E_{11}\|_\infty > 1/2.10^{-1}$. In D the full Jacobian is updated and inverted when convergence is unsatisfactory. Thus whenever we refer to an inversion in A, B and C it means an update of the inverse with respect to changes in the fast motions only, the recomputation of a full inverse only being called for whenever a is updated.

Table II

Table of Inversion and Iteration Counts

$10^{-3}x$	A NIT	A NINV	B NIT	B NINV	B UPD	C NIT	C NINV	C UPD	D NIT	D NINV
1	816	27	816	27	0	812	27	1	796	27
2	26	2	33	3	1	23	2	0	22	2
3	19	1	10	0	0	18	1	0	18	1
4	23	2	8	0	0	10	1	0	8	1
5	23	2	8	0	0	8	0	0	8	0
6	23	2	8	0	0	8	0	0	8	0
7	23	2	10	0	0	8	0	0	8	0
8	23	2	8	0	0	10	0	0	8	0
9	23	2	8	0	0	8	0	0	8	0
10	23	2	8	0	0	8	0	0	8	0
Total	1022	44	917	30	1	913	31	1	892	31

We have taken iteration (NIT), inverse (NINV) and a-update counts (UPD) for the tabular intervals $x = 0(10^3)10^4$ and the results are shown in Table II. The programme uses an explicit integration method to begin with and then switches to the trapezoidal rule. Automatic step adjustment takes place in order to satisfy a local relative accuracy criterion of 10^{-3}. In all cases a large number of iterations and inversions take place in the first interval where the step-length is initially small and builds up by several magnitudes to $h \simeq 10^2$. Clearly, in this region a = I is almost as good an approximation as using the full Jacobian. Over the first three intervals, methods A and B are very similar and methods C and D are very similar in performance. Thereafter the step-length in A becomes limited by our convergence restriction which allows a maximum of four iterations per step. The inversions here are associated with attempts to increase the interval, an unsuccessful doubling of the interval being followed immediately by halving. In B, C and D convergence remains satisfactory and only one inversion takes place for C and D associated with interval increase. The approximate inverses are all different constants over this range.

The table gives the numbers of iterations, inversions and full matrix updates performed in the tabular intervals as shown. In cases A, B and C, NINV refers to the number of partial inversions whereas in case D it refers to inversion of the full matrix. UPD gives the corresponding number of updates of the partition of the Jacobian associated with the slow motions (the matrix a of (4.5)).

Further computation over the range $10^4 < x < 10^5$ shows that the step-length of A continues to be restricted by slow convergence. About ten more inversions are required for C and D, each of them full inversions. A similar number of inversions is required for B but only three of them are full inversions.

To summarize, we observe that the convergence properties of A are satisfactory up to $x = 3 \times 10^3$ and thereafter B gives practically the same iteration count as D. If A is used over the first part of the range and then we switch to B, only one full inversion and 30 "partial" inversions are required compared with 31 full inversions for D. In fact, due to the special structure of a, this partition may be adequately represented by a diagonal matrix throughout so that a full inverse need never be computed.

This example illustrates the potential improvements in computational efficiency to be gained by taking acount of the structure of the perturbation E in matrix updating strategies. The improvements will depend in general on the range of integration, the stiffness and structure of the differential equations, convergence and step adjustment criteria. This system of equations is very stiff with a good separation between the eigenvalues associated with slow and fast motions. Under these circumstances the approximation a = I is well worthwhile. This leads to the iteration (4.10) corresponding to a Newton-like iteration for the fast motions and a repeated substitution with corrections for the slow motion. Periodic updating of a (reduction of E_{11} to zero) leads to further efficiency.

8. CONCLUDING REMARKS

The art of solving stiff systems is now in a highly

developed state but new developments, particularly in the field of Runge-Kutta formulae, stiff stability and structured equations may lead to still further advances in efficiency.

9. REFERENCES

Alexander, R. (1977) "Diagonally Implicit Methods for Stiff ODE's", *SIAM J. Num. Anal.* in press.

Aris, R. (1969) "Elementary Chemical Reactor Analysis", Englewood Cliffs, N.J.: Prentice-Hall.

Axelsson, O. (1969) "A Class of A-Stable Methods", *BIT* **9**, 185-199.

Butcher, J.C. (1964a) "Implicit Runge-Kutta Processes", *Math. Comp.* **18**, 50-64.

Butcher, J.C. (1964b) "Integration Processes Based on Radau Quadrature Formulas", *Math. Comp.* **18**, 233-244.

Chipman, F.H. (1971) "A-Stable Runge-Kutta Processes", *BIT* **11**, 384-388.

Croujiex, M. (1975) "Sur L'approximation des Équations Différentielles Operationelles Linaires par des Methodes de Runge-Kutta". Thése présentée a L'Université Paris VI.

Curtis, A.R., Powell, M.J.D. and Reid, J.K. (1974) "On the Estimation of Sparse Jacobian Matrices", *J. Inst. Maths. Applics.* **13**, 117-119.

Curtis, A.R. and Reid, J.K. (1974) "The Choice of Step Lengths When Using Differences to Approximate Jacobian Matrices", *J. Inst. Maths. Applics.* **13**, 121-126.

Dahlquist, G. (1963) "A Special Stability Problem for Linear Multistep Methods" *BIT* **3**, 27-43.

Dahlquist, G. (1969) "A Numerical Method for Some Ordinary Differential Equations with Large Lipsihtz Constants" in Information Processing 68 (Proceedings of the IFIP Congress 1968), North Holland Publishing Co.

Ehle, B.L. (1969) "On Padé Approximations to the Exponential Function and A-Stable methods for the Numerical Solution of Initial Value Problems" University of Waterloo Department of Applied Analysis and Computer Science, Research Rep. No. CSRR 2010.

Ehle, B.L. and Lawson, J.D. (1975) "Generalized Runge-Kutta Processes for Stiff Initial-value Problems", *J. Inst. Maths. Applics.* **16**, 11-21.

Enright, W.H., Hull, T.E. and Lindberg, G. (1975) "Comparing Numerical Methods for Stiff Systems of Ordinary Differential Equations", *BIT* **15**, 10-48.

Gear, C.W. (1969) "The Automatic Integration of Stiff Ordinary Differential Equations" in Information Processing 68 (Proceedings of the IFIP Congress 1968), North Holland Publishing Co. 187-193.

Gear, C.W. (1971) "Numerical Initial Value Problems in Ordinary Differential Equations", Prentice-Hall, Englewood Cliffs, N.J.

Hall, G. and Watt, J.M. *ed.*(1976) "Modern Numerical Methods for Ordinary Differential Equations", Oxford University Press, Ely House, London W1.

Hindmarsh, A.C. (1974) "GEAR: Ordinary Differential Equation System Solver", LRL Report UCID - 30001 Revision 3.

Hofer, E. (1976) "A Partially Implicit Method for Large Stiff Systems of ODEs with only Few Equations Introducing Small Time-Constants", *SIAM J. Numer. Anal.* **13**, 645-663.

Lambert, J.D. (1973) "Computational Methods in Ordinary Differential Equations", John Wiley and Sons, London.

Lindberg, B. (1974) "Optimal Stepsize Sequences and Requirements for the Local Error for Methods for (stiff) Differential Equations", Tech. Report No. 67, Department of Computer Science, University of Toronto.

Liniger, W. and Willoughby, R. (1967) IBM Res. Rep. RC-1970.

Miranker, W.L. (1973) "Numerical Methods of Boundary Layer Type for Stiff Systems of Differential Equations", *Computing* **11**, 221-234.

Norsett, S.P. (1974) "Semi-explicit Runge-Kutta Methods", University of Trondheim, Mathematics and Computation No. 6/74.

Oden, L. (1971) "An Experimental and Theoretical Analysis of the SAPS-method for Stiff Ordinary Differential Equations", Royal Institute of Technology, Stockholm, Department of Information Processing, Tech. Report No. NA 71.28.

Ortega, J.M. and Rheinboldt, W.L., (1970) "Iterative Solution of Non-linear Equations in Several Variables", Academic Press, New York and London.

Prothero, A. and Robinson, A. (1974) "On the Stability and Accuracy of One-step Methods for Solving Stiff Systems of Ordinary Differential Equations", *Math. Comp.* **28**, 145-162.

Robertson, H.H. (1966) "The Solution of a Set of Reaction Rate Equations" in Numerical Analysis: An Introduction (J. Walsh *ed.*), Academic Press, London and New York.

Robertson, H.H. (1967) "The Approximate Inverse Algorithm for the Solution of Stiff Differential Equations", Research Report MSDH/67/86, Imperial Chemical Industries.

Robertson, H.H. (1976) "Numerical Integration of Systems of Stiff Ordinary Differential Equations with Special Structure", *J. Inst. Maths. Applics.* **18**, 249-263.

Robertson, H.H. and McCann, M.J. (1969) "A Note on the Numerical Integration of Conservative Systems of First-order Ordinary Differential Equations", *Comput. J.* **12**, 81-82.

Robertson, H.H. and Williams, J. (1975) "Some Properties of Algorithms for Stiff Differential Equations", *J. Inst. Maths. Applics.* **16**, 23-34.

Rosenbrock, H.H. (1963) "Some General Implicit Processes for the Numerical Solution of Differential Equations", *Comp. J.* **5**, 329-330.

Wei, J. and Prater, C.D. (1962) "Analysis of Complex Reaction Systems", *Adv. in Catalysis* **13**, 203-391.

Wasow, W., (1965) "Asymptotic Expansions for Ordinary Differential Equations", Interscience Publishers, New York.

Widlund, O.B. (1967) "A Note on Unconditionally Stable Linear Multistep Methods", *BIT* **7**, 65-70.

V. 3

NUMERICAL SOFTWARE FOR INTEGRAL EQUATIONS

L.M. Delves

(University of Liverpool)

1. INTRODUCTION

Integral equations arise in a wide variety of application areas, and increasing effort has been expended in recent years in providing standard numerical techniques for their solution; readily available algorithms now exist for solving all of the "standard" forms of integral equations, at least in one dimension. Unfortunately, most of the equations arising in practice seem to be non-standard, and providing a widely useful set of routines is less easy than for, say, initial value ordinary differential equations.

We describe here the best currently available methods for solving integral equations of various types, and ways in which these methods can be used to provide a flexible tool for solving a wide variety of (one-dimensional) problems. This account is necessarily very brief; readers interested in details of either the mathematical background or of the numerical methods are referred to Delves & Walsh (1974).

The integral equations most frequently discussed in the mathematical or numerical analysis literature are those which come under the classical heading of:

a) Volterra equations of the first kind

$$0 = g(x) + \int_0^x K(x, y, f(y))\, dy \qquad (1.1)$$

b) Volterra equations of the second kind

$$f(x) = g(x) + \int_0^x K(x, y, f(y))\, dy \qquad (1.2)$$

c) Linear Fredholm equations of the first kind

$$0 = g(x) + \int_a^b K(x,y)\, f(y)\, dy \qquad (1.3)$$

d) Linear Fredholm equations of the second kind

$$f(x) = g(x) + \int_a^b K(x,y)\ f(y)\ dy \qquad (1.4)$$

e) Linear Fredholm equations of the third kind (eigenvalue equations)

$$f(x) = 0 + \lambda \int_a^b K(x,y)\ f(y)\ dy\ . \qquad (1.5)$$

Here, $f(x)$ is the unknown function; $K(x,y)$ or $K(x,y,f(y))$ is the kernel of the equation; and $g(x)$ is a known driving term. In the following sections, we consider each of these standard types of equation, and attempt to answer the questions:
a) What methods are available (as tested software?)
b) How do we compare methods and routines?
c) Which appears "best" as the result of this comparison?
Sections 2 - 5 are devoted to question (a), but also contain for convenience the current answers to question (c). Most of these answers are based on subjective judgements, or the (reasonable) belief that theoretical improvements in methods will be rewarded by improved performance in practice. The controlled and objective testing of integral equation routines is still at an early stage; the current position is reported in Section 6.

Finally, in Section 7, we discuss in more detail the types of equation which arise in pratical problems, and the way in which implementations of currently available methods can be extended to give a more flexible software tool for solving these problems.

2. VOLTERRA EQUATIONS OF THE SECOND KIND

2.1 Numerical Methods

There is a strong link between Volterra equations and initial value ordinary differential equation problems: the latter can always be expressed as Volterra equations; and an equally strong link between the solution methods used. Equation (1.2) can be rewritten in the form

$$f(x+h) = g(x+h) + \int_0^x K(x+h,\ y,\ f(y))\ dy + \int_x^{x+h} K(x+h,\ y,\ f(y))\ dy \qquad (2.1.1)$$

which suggests a step-by-step solution based on a "suitable"

grid and a discretization of the integrals. Such methods are almost universally used for Volterra equations; available methods all have direct analogues for ordinary differential equations:
a) Runge-Kutta Methods. Methods of Runge-Kutta type are simple to implement, self-starting, and as for ordinary differential equations have no stability problems. The original development of such methods was carried out by Pouzet (1963), who also gave a published algorithm; an improved version of this algorithm, which avoids function re-evaluation at the boundary between steps, is given by Phillips (1977a, 1977b).
b) Multistep (predictor-corrector) Methods. These methods also retain the characteristics which they show for differential equations; that is
(i) they seem potentially more efficient than Runge-Kutta processes, though this is not easy to demonstrate in practice.
(ii) they require special starting methods for the first few grid points. Runge-Kutta methods can be used as starters; and special starting methods have been developed by Day (1968).
(iii) they may be unstable. Instability of methods for Volterra equations has been analysed by Noble (1969) and is reasonably well understood.

The major difference in procedure, compared with differential equations, lies in the error (and step size control) techniques needed. As the independent variable x increases, the kernel $K(x,y)$, $0 < y < x$ may change character completely, with the result that a mesh which seemed adequate for small x is seen to be inadequate for large x. To retain the accuracy of the solution it is then necessary to start again; or better to go back and fill in values at places where the accuracy is now inadequate (Phillips (1977a)). Because of the complications this causes, variable step algorithms, and indeed error estimates, are much less well developed than for ODE's.

2.2 *Algorithms*

Published or widely available algorithms are also less well developed. For a single Volterra equation with non-singular kernel, the best available algorithm appears to be one based on a seventh order, variable step size (with fill-in) multistep method; both method and algorithm were developed by Phillips (1977a), and the algorithm is expected to appear in NAG in due course. For systems of Volterra equations, and for weakly singular kernels, variable step (but without fill-in) methods based on Simpsons rule are given, with algorithms, by Logan (1976); these algorithms are not yet widely available. No algorithms exist for "stiff" Volterra problems.

3 FREDHOLM EQUATIONS OF THE SECOND KIND

3.1 Numerical Methods

Fredholm equations have as counterpart two-point (second order) boundary value differential equations. This analogue suggests, correctly, that non-linear Fredholm equations will be significantly harder to solve than linear equations; unlike Volterra equations, for which non-linearity plays a relatively minor rôle. Existing algorithms all restrict themselves to the linear case, which is fortunately relatively common in mathematical physics; and those for which tested algorithms exist are of two distinct types

a) Nystrom or quadrature methods The integral in (1.4) is discretized with an N-term quadrature rule with points and weights $(x_i, w_i, i = 1, 2, \ldots, N)$, leading to a set of approximate equations for the solution at the quadrature points:

$$f_N(x_i) = g(x_i) + \sum_{j=1}^{N} K(x_i,x_j)\, w_j\, f_N(x_j). \qquad (3.1)$$

The performance of such a method depends on the quadrature rule used.

b) Expansion methods . The solution $f(x)$ is approximated by a linear expansion of the form

$$f(x) \simeq f_N(x) = \sum_{i=1}^{N} a_i\, h_i(x) \qquad (3.2)$$

where $\{h_i(x),\ i=1,2,\ldots,N\}$ is a known set of functions and one of several algorithms used to determine the coefficients a_i, $i=1,2,\ldots,N$. The most common algorithms used are:

(i)	Galerkin or variational	(Delves and Walsh (1974)	ch. 7,9)
(ii)	Least squares	(Ibid	ch. 7)
(iii)	Collocation	(Ibid	ch. 7)

In practice, the accuracy achieved depends almost entirel on the choice of expansion set $\{h_i\}$, rather than on the algorithm used to determine the coefficients; and there is in any cas a strong connection between these three algorithms: the least squares method is formally a variant of Galerkin, while if for an N-term expansion we consistently use an N-point quadrature rule, all three methods give identical results apart from round off errors (see Delves (1977)). Further, these results are identical (at the quadrature points) with those obtained from the Nystrom method using the same quadrature rule.

The methods differ, however, in two important respects: their speed; and the ease of providing as part of the calculations a cheap and usable error estimate.

Operation counts and the availability of error estimates, for various methods are shown in Table I. This Table suggests that the Fast Galerkin method of Delves (1977) should prove attractive for large N at least; for comparative tests, see Section 6 below.

Table I

A comparison of methods for solving linear Fredholm equations of the second kind

Method	Set up Equations	Solve Equations	Error Est.?	Reference
Nystrom	N^2	$\frac{1}{3}N^3$	NO	Delves & Walsh (1974) Ch. 6
Collocation	$2\ N^3$	$\frac{1}{3}N^3$	NO	Delves & Walsh (1974) Ch. 7
Mod. Nystrom (El Gendi)	N^2	$\frac{1}{3}N^3$	NO	El Gendi (1969)
Galerkin	$3\ N^3$	$\frac{1}{3}N^3$	YES	Delves & Walsh (1974) Ch. 9
Fast Galerkin	$\phi(N^2 \ell n N)$	$\phi(N^2)$	YES	Delves (1977)

3.2 Algorithms

Algorithms based on these methods are of two types:
1) Non-automatic routines, which ask the user to supply a value of N, and return f_N and perhaps an error estimate. The most widely available are:
a) An implementation of the El Gendi (1969) method by Miller and Symm (1975). A version for Green's function kernels is

in the NAG FORTRAN MK 5 library; a faster version for smooth kernels should appear at MK6.

b) The Nystrom method provides a simple write-your-own approach, though in my experience most do-it-yourself writers stick to the (usually expensive) repeated trapezoid or Simpson rules. A version in which the user can specify the quadrature rule freely will enter the NAG Algol 68 library at release 2 (due November 1977).

c) A non-automatic implementation of the Fast Galerkin algorithm (Delves and Abd-Elal (1977)).

2) Automatic routines, which ask the user what accuracy he requires and attempt to choose N to satisfy this request. Such routines are of course much harder to write, and usually costlier to run. They have obvious attractions for the user. The best known are three related routines due to Atkinson (1976); these use the Nystrom method with iterative solution of the equations and successive doubling of N, and are based on, respectively, Simpson, Boole, and Gauss quadrature rules. Section 6 describes briefly comparative tests of these routines, and of a crude automatic version of Fast Galerkin; these tests show that the routines do behave fairly well, and that there is little to choose between them over-all. However, the development of automatic routines is at an early stage, and they are much less sophisticated than any of the automatic quadrature routines for which comparable test results are reported in Lyness and Kagonove (1977).

4. EQUATIONS OF THE FIRST KIND

Formally, Fredholm and Volterra equations of the first kind can be solved by the same methods as their second kind counterparts. In practice, the equations which result may be very ill-conditioned, and this difficulty reflects the underlying properties of first kind equations. These properties are outlined in chapters (12) and (13) of Delves and Walsh (1974); the major theoretical difficulties are:

(i) Existence and uniqueness: equations (1.1), (1.3) may have no solutions, or a non-unique solution.

(ii) Ill-posedness: Worse (since most users know that a solution exists to their equation), first kind problems are ill-posed in general: given that a change δ_g to the driving term involves a change δ_f in the solution, it is possible for the ratio $\|\delta_f\|/\|\delta_g\|$ to be arbitrarily large.

The ill-posed nature of the problem requires special

attention numerically, and we briefly discuss proposed palliatives.

4.1 First Kind Linear Fredholm Equations

a) Singular function expansions (Baker *et al.*, (1964); Hanson (1971)). For every integrable kernel a set of functions $u_i(x)$, $v_i(y)$ exists such that the expansion

$$K(x,y) = \sum_{i=1}^{\infty} K_i \; u_i(x) \; v_i(y) \qquad (4.1)$$

converges in the mean. The functions $u_i(x)$, $v_i(y)$ are the singular functions of K, and the K_i are the singular values. Equation (4.1) yields the basis of a set of necessary and sufficient existence and uniqueness conditions for first kind Fredholm equations, and a truncated form of (4.1) can be used to yield approximate solutions. Unfortunately, finding (approximations to) the singular functions, a necessary first step, is expensive.

b) Regularization Methods (Tihonov (1964), Phillips (1962)) Since a direct solution of the ill-posed problem $K f = g$ leads typically to wild oscillations in the computed f, we seek to impose smoothness constraints on the solution. The method of regularization chooses an operator L, and attempts to minimize the residual $\|K f - g\|$ while keeping $\|L f\|$ bounded; in practice, we introduce a small parameter α and minimize

$$\|K f - g\|^2 + \alpha \|Lf\|^2 . \qquad (4.2)$$

Typical choices of L are $L = I$, $L = d/dx$, $L = d^2/dx^2$. In the case $L = I$, and with an L_2 norm in (4.2), the approach reduces to the solution of the second kind equation

$$(K^T K + \alpha I) f = K^T g \qquad (4.3)$$

which is certainly well-posed for nonzero α.

c) Augmented Expansion Methods. As an alternative to expansion (4.1) we can expand the solution as in (3.2), and seek to impose smoothness constraints on $f_N(x)$ directly. Many successful calculations rely on the implicit smoothing imposed by minimizing $\|K f_N - g\|$ with a least squares (Turchin *et al.* (1971)) or better, an L_1 norm. However, this is not satisfactory for large N, and Babolian (1977) discusses the imposition of additional explicit constraints on the expansion coefficients a_i in (3.2).

The regularization and augmented expansion methods seem to work reasonably well in practice; both contain arbitrary parameters to be adjusted by the user, and no widely available implementations of these methods seem to exist as yet.

4.2 First Kind Volterra Equations

These are less troublesome in practice than first kind Fredholm equations, essentially because there usually exists a related second kind equation with the same solution. Differentiating (1.1) formally we obtain

$$K(x,x,f(x)) + \int_0^x \frac{\partial K}{\partial x}(x,y,f(y))\,dy = \frac{dg(x)}{dx} \qquad (4.4)$$

which is a non-linear second kind equation. In the linear case $K(x,x,f(x)) \equiv K(x,x)f(x)$, and if $K(x,x) \neq 0$ over the range of interest (and if the derivatives in (4.4) exist) equation (4.4) is straightforward to solve numerically. Theoretical difficulties therefore may only arise when these conditions are <u>not</u> satisfied; and it is usually quite possible to use methods developed for second kind equations directly. For example, the algorithm of Pouzet (1963) makes no distinction between first and second kind equations. Alternatives are:

(i) The direction solution of the reduced form (4.4), or of an alternative reduction to a second kind equation which follows on integrating (1.1) by parts.

(ii) The use of methods analogous to those outlined above for Fredholm first kind equations. For a discussion of regularization methods for Volterra equations, see Schmaedeke (1968).

5. EIGENVALUE EQUATIONS

Since it is easy to show that the (linear) eigenvalue Volterra equation

$$f(x) = \lambda \int_0^x K(x,y)\, f(y)\, dy \qquad (5.1)$$

has no non-trivial square integrable solutions for any finite λ and all square integrable kernels K, only linear Fredholm eigenvalue problems arise in the standard classification. These problems are very closely related numerically to the inhomogeneous second kind equations (1.4), and any method used to discretize (1.4) can be used to discretize (1.5), yielding a linear algebraic eigenvalue problem which can be solved by standard means. Two features of the eigenvalue problem should be kept in mind however:

(i) A symmetric kernel ($K(x,y) \equiv K(y,x)$) has only real eigenvalues. It is obviously convenient if the discretized equations retain this relation, and direct applications of the Nystrom, least squares, and (symmetric) Galerkin procedures do so (but the least squares procedure leads to a non-linear algebraic eigenvalue problem, and so is not recommended).

(ii) Perhaps more important is the detection of spurious eigenvalues. An N-term, or N-point, discretization of (1.4) leads to an $N \times N$ linear algebraic set of equations which will always yield N eigenvalues. The original equation may have no eigenvalues; one or finite number; or an infinity of eigenvalues. It is therefore possible that some or all of the computed eigenvalues are spurious and it is important to be able to detect and flag these; for a method which does so, see Spence (1975, 1976).

A routine for the eigenvalue problem should be available in NAG fairly shortly.

6. COMPARING INTEGRAL EQUATION ROUTINES

Test Methodology

Given two routines which purport to solve the same problem, it is natural to ask which is the better, and to construct objective tests to (try to) find out. Test packages are familiar sights in other fields: numerical quadrature, non-linear optimization, ordinary differential equations, for example; and are likely to become so in this field too, although so far only limited testing has been carried out.

Test packages in other fields have been of two main types:

a) Traditional, or Battery tests: a large number (20-50) of different problems graded from easy to very difficult, are fired at the routines, and their response recorded.

b) Statistical or Problem Family packages The defects of battery tests are discussed at length by Lyness & Kaganove (1977); but the main defect can be summarized briefly: the performance of the routine under test may be a rapidly varying (even discontinuous) function of the parameters in the problem on which it is tested. As an example consider a problem having some kind of discontinuity at a point x_0. Varying x_0 does not change the innate difficulty of the problem; but the response of a given routine may be quite different according as it does or does not happen to use x_0 as a mesh point. Problem Family test packages seek to average over this behaviour; they contain a number of problem families with different types of "difficult" features. Each family has one or more parameters which affect the position but not the degree of the difficulty; and the average response of the competing routines is measured, the average being taken over these parameter values within some chosen limits. A test package of this type was first introduced for quadrature routines by Lyness & Kaganove (1977); a similar package for integral equation routines is described in Riddell & Delves (1977), together with test results for several automatic and non-automatic Fredholm routines. The performance parameters measured are speed, and accuracy (for a non-automatic routine), accuracy control (for an automatic routine). For automatic routines it is tempting but misleading to measure <u>speed</u> in terms of the number of kernel evaluations used; typical Fredholm routines, with simple kernels, appear to spend less than 5% of their time in kernel evaluations, and 95% or more in thinking about them.

Figures 1 to 5 show the results obtained for one problem family, defined as follows:

$$f(x) = g(x) + \int_{-1}^{1} K(x,y)\, f(y)\, dy$$

$$K(x,y) = C^2/\{(x - P_2)^2 + C^2\}\{(y - P_3)^2 + C^2\} \qquad (6.1)$$

$$g(x) = P_1[1-C\{\arctan(\frac{1-P_3}{C}) + \arctan(\frac{1+P_3}{C})\}]/[(x-P_2)^2+C^2].$$

Solution: $f(x) = P_1$

This problem family has a (degenerate) kernel with a hyperbolic spike. P_1, P_2 and P_3 are parameters of the family and are varied randomly by the package; they affect the position of the spike in the (x,y) plane. The constant C determines the width, and hence difficulty, of the spike, and has the value C = 0.2 in Figs. 1 to 5.

For the precise averages being taken, and for additional results, see Riddell & Delves (1977). These results show the large difference between runtime and kernel evaluations as a figure of merit: the Simpson-rule based automatic routine of Atkinson (1976) is wildly extravagent of kernel evaluations (and the higher-order Boole routine is worse!) but runs fast by virtue of its low overheads. Fig. 1 also shows that the "Fast Galerkin" algorithm tested fulfills its theoretical promise of being faster than Nystrom only for very large N. It is however reasonably competitive, and is much faster than standard Galerkin (see Fig. 5).

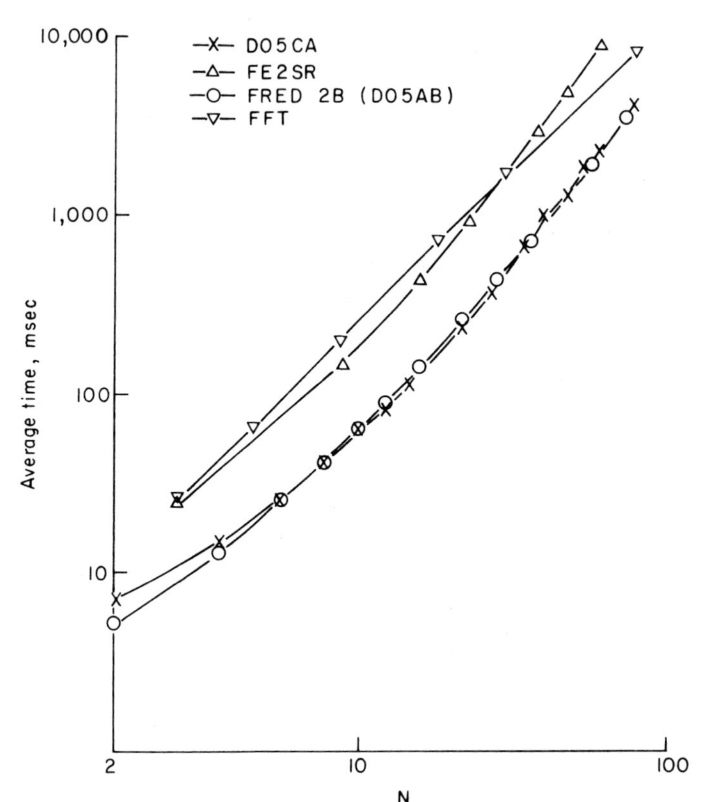

Fig.1 Relative speeds of several non-automatic routines for Fredholm equations of the second kind. The ordinate N is the number of quadrature points or of expansion terms used; times are in milliseconds on an ICL 1906S computer. The routines are as follows:

FRED2B Modified Nystrom routine using the Gauss-Chebyshev quadrature rule, based on the El-Gendi (1969) algorithm and implemented by Miller and Simpson (1975). To appear in NAG MK6 as routine DO5AB.

FFT An implementation of the Fast Galerkin technique (Delves and Abd-Elal (1977))

DO5CA NAG Algol 68 routine DO5CA, calling an N-point Gauss-Legendre rule

FE2S A Nystrom routine by Thomas (1977)based on Simpson's rule but producing also a higher-order corrected solution. The timing difference between this routine and DO5CA reflects essentially the cost of computing this correction term

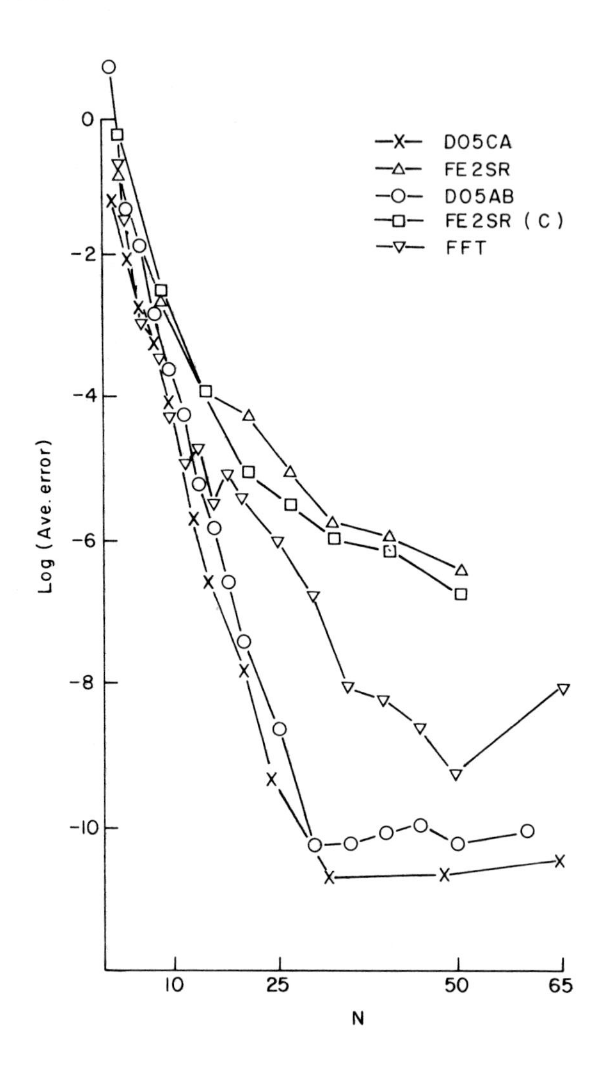

Fig. 2 Relative accuracy of several non-automatic routines for Fredholm equations of the second kind, for the problem family described by equation (5.1) with C = 0.2. Reported errors are averaged over parameter values for P_1, P_2, P_3 between ± 1. Routines as in Fig. 1; FE2SR and FE2SR(C)refer to the Simpson and Corrected Simpson solution, respectively, from routine FE2SR

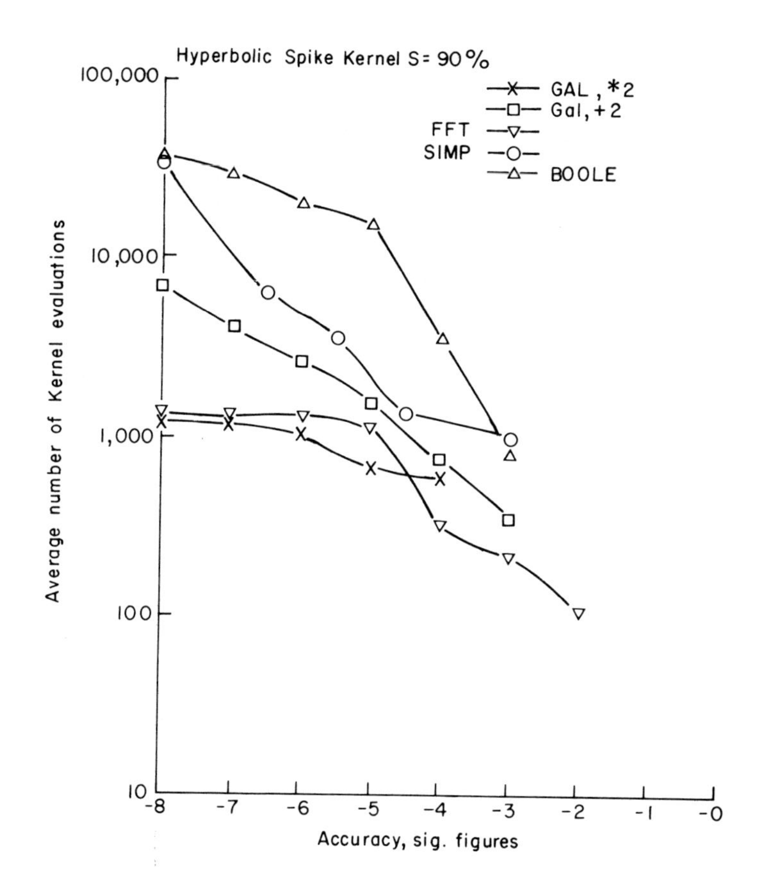

Fig. 3 Average number of kernel evaluations used by several automatic routines for Fredholm second kind equations, on the problem family (5.1). The averages taken in this and Figs. 4 and 5 are defined more precisely in Riddell and Delves (1977). The routines are as follows:

SIMP, BOOLE Routines based on the Nystrom method using the Simpson and Boole quadrature rules, with successive doubling of N and iterative solution of the linear equations (Atkinson (1976))

FFT A crudely automated version of the Fast Galerkin algorithm (Delves (1977)) using this algorithm's built in error estimate with successive doubling of N.

Galerkin * 2, + 2 Similar routines to FFT, but using the standard Galerkin algorithm and two different strategies

*2 Double N (as in FFT)

+2 Add 2 to N: Clearly not a good idea.

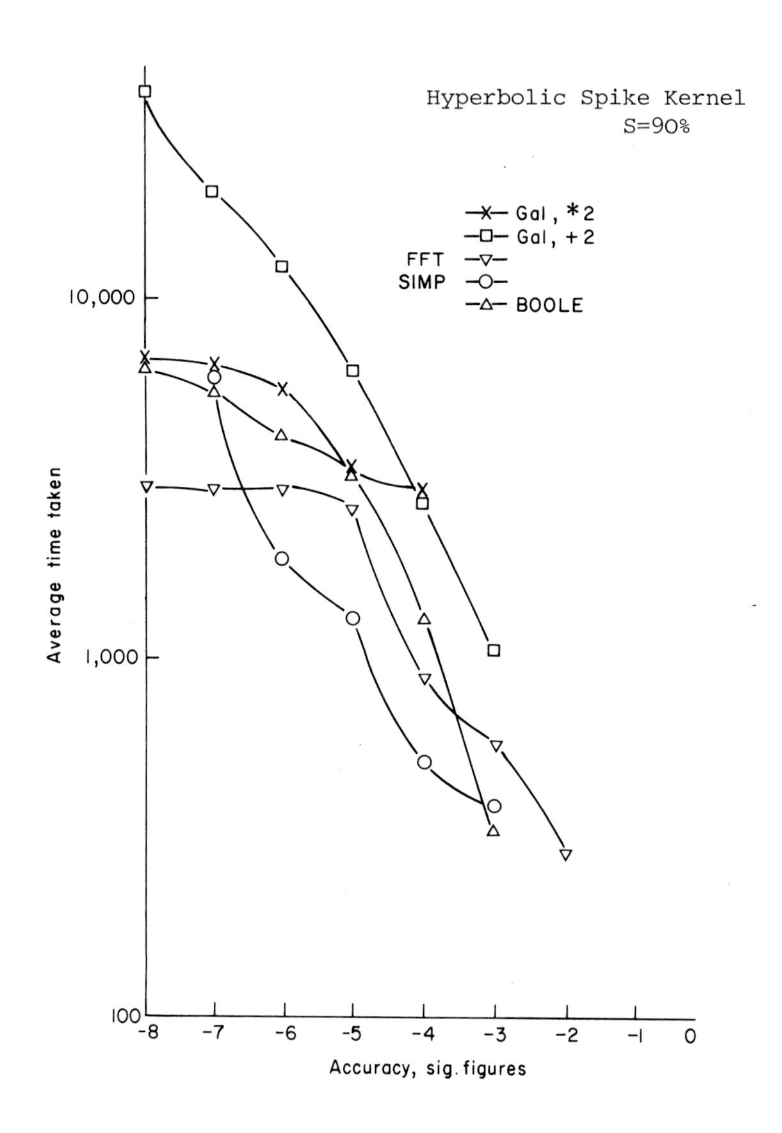

Fig. 4 Average time taken by the automatic routines of Fig. 3, on the same problem family. Note that the most economical routine (in function evaluations) is not necessarily the fastest

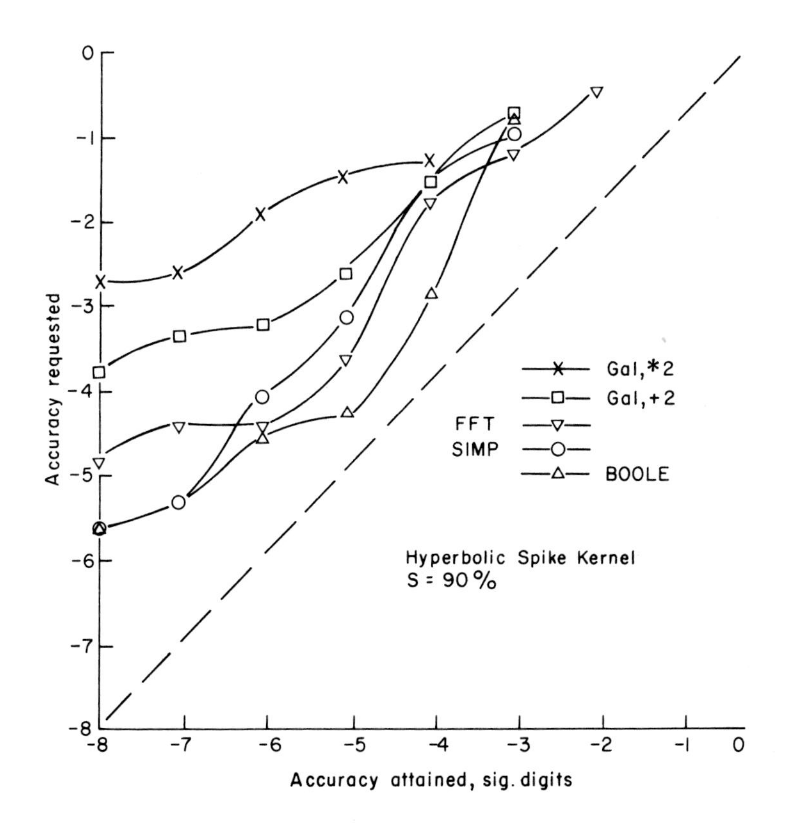

Fig. 5 A test of the error control in the automatic routines of Fig. 3, on problem family (5.1). This figure shows the accuracy request which must be input to the routine to produce an actual error $\leq \varepsilon$ requ exactly 90% of the time. All the routines err on the side of caution on this criterion

7. DEALING WITH PRACTICAL PROBLEMS

Most practical problems yielding integral equations fail to yield one of the classical types (1.1) - (1.5); we consider here what can be done about this.

For examples which do appear in practice, we cite the following, taken more or less at random from the literature:

1) Population Competition (Downham & Shah (1976); Downham (1974))

$$\underline{f}(x) = \underline{g}(x) + \int_a^b K(\underline{x} - \underline{y})) \, h(\underline{y}) \, \underline{dy}; \quad \underline{f}^T = (f_1, f_2, f_3) \qquad (7.1)$$

that is, a set of three coupled, non-linear Fredholm second kind equations.

2) Quantum Scattering: Close-Coupled Calculations (Horn and Fraser (1975))

$$f(x) = g(x) - \lambda \int_0^\infty K_1 \, (x,y) \int_0^\infty K_2 \, (y,z) \; f \; (z) \; dz \; dy \qquad (7.2)$$

An equation of Fredholm origin but with infinite range and iterated kernel.

3) Problem area as (2), but a different system and an alternative formulation (Chan and Fraser (1973))

$$\left(\frac{d^2}{dx^2} + K^2\right) \underline{f}(x) = \underline{g}(x) + \int_0^\infty K(x,y) \; \underline{f}(y) \; dy \qquad (7.3)$$

$$\underline{f}^T = (f_1, f_2); \quad K = \begin{pmatrix} K_{11} & K_{12} \\ K_{21} & K_{22} \end{pmatrix}.$$

"suitable" boundary conditions are supplied in addition. Equations (3) are integrodifferential, and coupled, with a linear Fredholm kernel.

4) Currents in a superconducting strip (Rhoderick and Wilson (1962))

$$g(x) = \frac{1}{\pi} \int_0^1 \frac{t-x}{(t-x)^2 + a^2} \, f(t) \; dt \qquad (7.4)$$

a genuine, and not uncommon example of a first kind linear Fredholm equation.

5) Flow round a Hydrofoil (Kershaw(1974))

$$f(x) = g(x) + \lambda \int_C f(y) \; K \; (x,y) \; dy \qquad (7.5)$$

subject to $\int_C f(y)\, dy = 0$, C a closed contour.

This is an equation of Fredholm type, but with a subsidiary condition to be satisfied.

These examples illustrate the wide variety of equations encountered in practice; only (4) can be regarded as being of a "standard" type.

Additional difficulties are apparent when we look at the structure of the equations involved. Problems may be:

(i) Linear or non-linear
(ii) Coupled equations or single
(iii) Finite range or infinite
(iv) Well or ill-conditioned
(v) Smooth or non-smooth (Kernels and driving terms)
(vi) Without or with side-conditions

This variety of problem is very difficult to service with a small number of "complete" routines; that is, routines which, on being called, return an approximate solution to the given problem. However, all is not lost. The examples given above, although varied, can be seen to be made up from only three rather standard operators:

a) $\int_a^b dy\, K\,(x,y)$

b) $\dfrac{d^2}{dx^2}$

c) I - the unit operator

provided that we allow K to be non-linear (in the unknown function), matrix valued (for coupled equations) and to cover if necessary infinite intervals.

It is therefore possible in principle to provide a set of routines which perform a discretization of these basic operators, together with a mechanism for

a) Piecing together operators to form a complete equation
b) Solving the equation

It can then be left to the user to piece the bits together and to call on the solution mechanism.

A suitable discretization is given (for example) by the Galerkin method with an orthonormal set of basis functions; this method leads to particularly simple algorithms for adding and multiplying integral operators. Let $\{h_i(x)\}$ be an orthonormal set of functions on $[a,b]$; then the Galerkin discretizations of the operators κ_1 and κ_2:

$$\kappa f_\ell = \int_a^b \kappa_\ell(x,y)\ f(y)\ dy, \quad \ell = 1,2$$

are the N×N matrices K_1, K_2:

$$(K_\ell)_{ij} = \int_a^b h_i(x)\ K_\ell(x,y)\ h_j(y)\ dy \qquad \ell = 1,2 \quad i,j = 1,2,\ldots,N$$

The Galerkin discretization of the operator $\kappa_1 + \kappa_2$ is $K_1 + K_2$ while a rapidly convergent discretization of the operator $\kappa_1\kappa_2$ is K_1K_2. Construction of discretized equations from discretized operators is with this approach reduced to the familiar operations of linear algebra. Provision for the user to "piece the bits together" is a problem in numerical software rather than numerical methods; and a package based on an extended Fast Galerkin algorithm and a suitable extension of the simplified discretization described above is currently under construction (Delves & Hendry (1976)).

With this package, the code required to set up and solve equations (7.2) will have the form

```
    interval  : = (0, infinity);
    nterms    : = 20;
lefthandside  : = unit + lambda * linearfred (K1)
                   * linearfred (K2);
linearsolve (lefthandside, drivingterm (g), sol-
utionvector);
```

For the interested : the language is Algol 68; interval, nterms, lefthandside are global variables. Infinity is a procedure call which sets up a hidden mapping - to deal with the semi-infinite interval. Unit, linearfred, linearsolve and drivingterm are package procedures, and K1, K2, g are user procedures; solutionvector a user-provided vector.

While the package design is not yet completely frozen, (and its performance still not known!) it should be clear from this example that such a package can be simple to use, and gives an economic way of providing for the wide variety of problems encountered in practice.

8. REFERENCES

Atkinson, K.E. (1976) "An Automatic Program for Linear Fredholm Integral Equations of the Second Kind", *ACM Trans. on Math. Software* **2**, pp. 154-171. See also Algorithm 503, Ibid pp. 196-199.

Babolian, E. (1977) "Solution Methods for Fredholm Equations of the First Kind", MSc Thesis, University of Liverpool.

Baker, C.T.H., Fox, L., Mayers, D.F. and Wright K. (1964) "Numerical Solution of Fredholm Integral Equations of the First Kind", *Comput. J.* **7**, pp. 141-147.

Chan, Y.F. and Fraser, P.A. (1973) "S-Wave positron scattering by hydrogen atoms", *J. Phys. B.*, **6**, pp. 2504-2515.

Day, J.T. (1968) "On the Numerical Solution of Volterra Integral Equations", *BIT*, **8**, pp. 134-7.

Delves, L.M. (1977) "A Fast Method for the Solution of Fredholm Integral Equations", Research Report CSS/76/1/2, University of Liverpool; *JIMA* to be published.

Delves, L.M. and Abd-Elal, L.F. (1977) "The Fast Galerkin Algorithm for Fredholm Integral Equations", *Comput. J.* to appear.

Delves L.M. and Hendry, J.A. (1976) "A Package for solving Integral Equations?" Internal Report, Department of Computational and Statistical Science, University of Liverpool (unpublished).

Delves, L.M. and Walsh, J. (Eds.) (1974) "Numerical Solution of Integral Equations", Clarendon Press.

Downham, D.Y. (1974) "Model of Clines", *Adv. in App. Prob.*, **6**, No. 1, pp. 7-10.

Downham, D.Y and Shah, S.M.M. (1976) "The Integral Equation Approach for Models of Clines", Preprint, University of Liverpool, Department of CSS.

El-Gendi, S.E. (1969) "Chebyshev Solution of Differential, Integral and Integro-differential Equations", *Comp. J.*, **12**, pp. 282-287.

Hanson, R. (1971) "A Numerical Method for Solving Fredholm Integral Equations of the First Kind Using Singular Values", *SIAM J. Numer. Anal.*, **8**, pp. 616-622.

Horn, S. and Fraser, P.A. (1975) "Low-energy ortho-positronium scattering by hydrogen atoms", *J. Phys. B.*, **8**, pp. 2472-2475.

Kershaw, D. (1974) "Singular Integral and Boundary Value Problems" ch. 19 (258-266) of "Numerical Solution of Integral Equations", Eds. L.M. Delves and J. Walsh, Oxford University Press.

Logan, J.E. (1976) "The Approximate Solution of Volterra Integral Equations of the Second Kind", PhD Thesis, Department of Mathematics, University of Iowa.

Lyness, J.N. and Kaganove, J.J. (1977) "A Technique for Comparing Automatic Quadrature Routines", *Comput. J.*, **20**, pp. 170-177. See also test results in: EAR reports QR1, QR2 (1975) (Argonne National Laboratory, unpublished) by the same authors.

Miller, G.F. and Symm, G.T. (1975) "Procedure fred 2b", National Physical Laboratory Preprint D5/06/1/A/7/75. Implemented as NAG routine D05AB.

Noble, B. (1969) "Instability when solving Volterra Integral Equations of the Second Kind by Multi-Step Methods", in 'Springer Lecture Notes in Mathematics 109: Conference on the Numerical Solution of Differential Equations'.

Phillips, C. (1977a) "Numerical Solution of Volterra Integral Equations of the Second Kind by Step-by-Step Methods", PhD Thesis, Department of Computational and Statistical Science, University of Liverpool.

Phillips, C. (1977b) "An Improved Algorithm for the Solution of Volterra Integral Equations of the Second Kind by Runge-Kutta Processes", Preprint, Department of Computational and Statistical Science, University of Liverpool; submitted to ACM Trans. Maths. Software.

Phillips, D.L. (1962) "A technique for the Numerical Solution of certain Integral Equations of the first kind", *J. Ass. Comput*

Mach., **9**, pp. 84-96.

Pouzet, D. (1963) "*Methode d'integration numerique des equations integrales et alegro differentielles du type de Volterra de seconde espece. Formules de Runge-Kutta*" in "Symposium on the Numerical Treatment of Ordinary Differential Equations, Integral, and Integro-differential Equations", Birkauser Verlag, Basel: pp. 362-368.

Rhoderick, E.H. and Wilson, E.M. (1962) "Current distribution in thin superconducting films", *Nature,* **194**, 1167-1168.

Riddell, I. and Delves, L.M. (1977) "On Comparing Routines for the Solution of Integral Equations", Research Report CSS/ 77/7/1, University of Liverpool.

Schmaedeke, W.W. (1968) "Approximate solutions of Volterra Integral Equations of the first kind", *J. Math. Anal. and Appl.*, **23**, pp. 604-613.

Spence, A. (1975) "On the Convergence of the Nystrom Method for the Integral Equation Eigenvalue Problem", *Numer. Mathematik,* **25**, pp. 57-66.

Spence, A. (1976) "Error Bounds and Estimates for Eigenvalues of Integral Equations", University of Bath Research Report Math/NA/2.

Thomas, K.S. (1977) "Procedure FE2SR", unpublished, Oxford University. This procedure implements an algorithm described in the author's paper "On the Approximate Solution of Operator Equations, part 2", Submitted to *Numer. Math.*

Tihonov, A.N. (1964) "Solution of nonlinear integral equations of the first kind", *Soviet Maths.*, **5**, pp. 835-838.

Turchin, V.F. Kozlov, V.P. and Malkevich, M.S. (1971) "The Use of Mathematical Statistics Methods in the Solution of Incorrectly Posed Problems", *Soviet Phys. Usp.* **13**, pp. 681-702.

V. 4

PROBLEMS OF ALGORITHM DESIGN FOR PARTIAL DIFFERENTIAL EQUATIONS

J. Walsh

(Department of Mathematics, University of Manchester)

1. INTRODUCTION

In the field of partial differential equations, even more than in ordinary differential equations, the limitations of analytical methods of solution are very quickly apparent. Approximate methods of solution, mainly based on finite differences or on variational principles, have been used in practical work since the days of hand calculation, but their application on a large scale only became possible with the advent of computing machines. Since then, numerical techniques have improved greatly, and large and difficult problems in partial differential equations can be solved on modern computers. However, the development of library routines and packages in this area has been much slower than in other areas. There are good reasons for this. Although in principle one can define methods of solution for broad classes of partial differential equations, for example elliptic equations, in practice the variations within a class, depending on the equation itself, the geometry of the region, and the boundary conditions, make it easier to look at each problem individually. So the programs for solving partial differential equations tend to be designed for very restricted sub-classes, and to lack the generality which is usual in mathematical algorithms in other fields, such as linear algebra. Hence they have appeared to be unsuitable for inclusion in an algorithm library.

Another problem in constructing algorithms is the treatment of errors. A good algorithm should be designed not only to give an approximate solution to a given problem, but also to produce some measure of the error in the result. This is very difficult to achieve in the case of partial differential equations. Even for ordinary differential equations, the problem of error control is not completely solved; most routines only provide estimates for the local error, leaving it to the user to carry out suitable tests on the global error. In partial differential equations the sources of error are more complex, arising from the geometry of the region as well as from the equation and boundary conditions, and it is much more difficult to include automatic checks on the accuracy of the results. Because the problems are usually large, it is difficult to apply even the simple check of repeating the calculation with different parameters. Most routines in current use are therefore constructed with a more limited objective, not to solve the partial differential equation itself, but to carry out certain parts of the calculation which are purely algebraic or geometrical. The process of numerical solution can be resolved into three main stages, first the definition of the region and the discretization of the problem, then the solution of the discrete system (which consists of either algebraic equations or ordinary differential equations), and finally the synthesis of the required results and tests for accuracy. The second stage is the one which is most easily handled by general algorithms, and suitable routines are often available in the library anyway.

These auxiliary algorithms are certainly worth having, but they do not cover the most difficult parts of the calculation. The first stage is usually the most complicated, where the geometrical features of the problem are reduced to algebraic equations, and it is here that average users need the most help. Further, although the intermediate stage (for example the solution of a system of linear equations of band form) can be solved by a standard routine, we may lose thereby any special features of the problem, such as additional structure in the equations. An algorithm with maximum efficiency in computation and storage would combine the first two stages, so that any special properties of the discretized equations could be used in the algebraic solution.

The choice of numerical methods is governed by the basic classification of partial differential equations, which is connected with the idea of "well-posed" problems (Courant and Hilbert, 1962, Ch.III). Let us consider a general second-order equation in the form

$$a\frac{\partial^2 u}{\partial x^2} + 2b\frac{\partial^2 u}{\partial x\partial y} + c\frac{\partial^2 u}{\partial y^2} + g\left(u, \frac{\partial u}{\partial x}, \frac{\partial u}{\partial y}\right) = 0, \qquad (1.1)$$

where the coefficients are not necessarily constant. Then we know that if $b^2 - ac < 0$ the equation is of elliptic type, and it normally occurs in a closed region with boundary conditions specified at all points of the boundary. Any simple form of numerical approximation gives a system of simultaneous algebraic equations, and the solution certainly exists if $a.\frac{\partial g}{\partial u} < 0$. If we try to solve the equation in a step-by-step fashion starting from the boundary, the solution is very unstable. This numerical property reflects a corresponding analytical property, that the problem is not well-posed as an initial-value problem. Consequently an elliptic problem is usually solved numerically by reducing it to a large simultaneous algebraic system. However, in ordinary differential equations inherent instability is no longer regarded as a barrier to step-by-step solution, if we use devices such as multiple shooting or orthonormalization. In a similar way, it may be possible to develop effective ways of controlling instability in elliptic equations, so that some form of step-by-step solution can be used. This approach would be very valuable in overcoming the problem of large storage requirements, which arises with simultaneous methods of solution. Shooting methods have been investigated by Bank and Rose (1975) for elliptic equations of simple type, and it has been shown that excessive instability can be avoided by semi-analytical techniques. The method of nested dissection (George, 1973) is another way of breaking up the large algebraic system, and avoids the simultaneous treatment of the entire region and the boundary conditions.

For the case where (1.1) is elliptic, but we have $a.\frac{\partial g}{\partial u} > 0$, the analytical properties of the equation are rather different. First, there is no guarantee of the existence of a unique solution, and the linearized form of the operator may be singular for certain boundary conditions. Secondly, the inverse operator is not necessarily positive, so that the problem need not be monotone. Thirdly, the variational formulation of the problem does not lead to a positive definite quadratic form. All these properties have their counterparts in the approximating algebraic system. If we wanted to construct a general algorithm for elliptic equations, it would have to take these problems into account, which would make it too complex to solve simple cases efficiently. It is probably better to restrict the class of equations by requiring the condition $a.\frac{\partial g}{\partial u} < 0$, and to treat any other equations as special problems.

Let us consider the effect of the "well-posed" requirement on other classes of partial differential equations. Parabolic equations are characterized by the condition $b^2 = ac$ in (1.1). They often arise from time-dependent problems, and may be expressed alternatively in the following form (in one space dimension)

$$\frac{\partial u}{\partial t} = p\frac{\partial^2 u}{\partial x^2} + q(u, \frac{\partial u}{\partial x}). \tag{1.2}$$

A typical set of boundary conditions would be specified values of u on $t = 0$, and on $x = a$, $x = b$ for $t > 0$. This problem is well posed for forward integration in time if we have $p > 0$, $\frac{\partial q}{\partial u} \leqslant 0$. The significance of these conditions is that the problem is elliptic in the space dimension (so that the steady state is stable), and that small disturbances are damped in the time direction. In the case of a linear problem with linear boundary conditions, the solution consists of descending exponential terms. If we attempt to integrate backwards in time (or if $p < 0$) the solution is extremely unstable, because it is the sum of exponentially increasing terms. It appears that a useful general algorithm can only be based on the stable case, However, it should be noted that the unstable problem is sometimes presented for solution; we may want to find the initial conditions (or the range of initial conditions) which lead to a given solution at time $t = T$. The result is not unique, because any disturbance which dies out in the range $(0,T)$ can be added to it. This problem presents very special difficulties for a general algorithm, and it is best to regard it as an entirely separate case.

The remaining class of problems in the classical theory is that of hyperbolic equations, corresponding to the condition $b^2 - ac > 0$ in (1.1). In this case, the analytical theory centres on the rôle of the characteristics in defining the region in which the solution can be found. As we would expect, the characteristics also play a large part in the numerical solution, but methods based directly on characteristics are very complicated in more than two dimensions. If simpler forms of discretization are used, based on a finite-difference mesh, the essential analytical property of well-posedness must be preserved, and the boundary conditions must be properly specified in relation to the region of integration. This is not a problem that can be solved automatically by the algorithm; it requires preliminary analysis.

The general problem of relating the discrete solution to the continuous equation has been investigated by a number of authors, but the methods are not well developed for automatic computation. The difference correction method (Fox and Goodwin, 1949) can be used very successfully in hand calculations, but it is difficult to apply on a computer for partial differential equations. Let us consider the general equation

$$Lu + g(u) = 0, \qquad (1.3)$$

with appropriate boundary conditions; suppose it is represented by the algebraic system

$$A\underline{u} + \underline{g}(\underline{u}) = 0, \qquad (1.4)$$

where $A\underline{u}$ represents a finite-difference approximation to the differential terms Lu. If the mesh is sufficiently small, we can represent the leading correction terms of the approximation by a sum $C\underline{u}$ say, involving higher differences. The difference correction method uses the iteration

$$A\underline{u}^{(k+1)} + \underline{g}(\underline{u}^{(k+1)}) = -C\underline{u}^{(k)}, \qquad (1.5)$$

where $\underline{u}^{(0)}$ is the solution of (1.4). The conditions for the use of the method, and the order of accuracy of successive iterates, have been analysed by Pereyra (1966), who has also developed an algorithm based on the difference correction for simple elliptic equations.

A new formulation of the basic principle of deferred correction has been given recently by Zadunaisky (1976), who uses a different method for calculating the correction term. Suppose we obtain the first approximation $\underline{u}^{(0)}$ from (1.4) as before, and then interpolate it by a continuous function v say. If v is sufficiently smooth, we can substitute it in the original partial differential equation (1.3) to obtain

$$Lv + g(v) = d, \qquad (1.6)$$

where d is the defect (or residual) function. The equation (1.6) has the exact solution v. We discretize (1.6) in the same way as the original problem and solve an equation of the same form as (1.4) with the addition of the non-homogeneous term d. If the discrete solution is $\underline{v}^{(0)}$, the difference between $\underline{v}^{(0)}$ and v gives an estimate of the difference between $\underline{u}^{(0)}$ and the exact solution u. In the case of monotone problems, this method can be extended to give strict error bounds for the approximate solution $\underline{u}^{(0)}$. The general approach seems promising, but it requires considerably more development before it can be used as the basis for a general algorithm.

2. PARABOLIC EQUATIONS

We turn now to a more detailed consideration of particular problems and methods. In the case of parabolic equations, it has been proposed by many authors that a general algorithm should be constructed by discretizing the space operator, and using an initial-value package to solve the resulting system of ordinary differential equations. An example of this approach is given by Sincovec and Madsen (1975). Let us take a simple parabolic equation in the form

$$\frac{\partial u}{\partial t} = K\nabla^2 u + f(u). \tag{2.1}$$

The space operator on the right is elliptic, and is stable if $K > 0$ and $\frac{\partial f}{\partial u} < 0$, with suitable boundary conditions. It can be represented on a rectangular grid by a set of finite-difference expressions, or on a more general mesh by finite-element expressions. Alternatively it can be discretized in various indirect ways, for example by representing u as a finite sum of Chebyshev polynomials (Dew, 1977). Whichever method is used, the parabolic problem will be reduced to a system of ordinary differential equations of the form

$$\frac{d\underline{u}}{dt} = A\underline{u} + \underline{f}(\underline{u}), \tag{2.2}$$

where $\underline{u}$ is a vector of discrete parameters (usually function values) representing the solution. Let us consider the case of finite-difference approximation and assume that f(u) is linear; then the linear operator on the right-hand side of (2.2) has negative eigenvalues. The complementary function for the system (2.2) is a sum of descending exponentials with a wide range of exponents, and so the differential equations are stiff. However, the stiffness ratio is not exceptionally large, and it is quite possible to solve (2.2) with a properly constructed stiff integration routine.

If we look at the ordinary differential equations in more detail, we see that there are some disadvantages in treating a parabolic equation by this method. The matrix A, which is obtained from the space derivative terms, has a very special form, and if there is more than one space dimension it will generally be of large order. The classical method of solution is to discretize both sides of (2.1), giving equations of the following type (Crank-Nicolson)

$$\underline{u}_{j+1} - \tfrac{1}{2}k[A\underline{u}_{j+1} + \underline{f}(\underline{u}_{j+1})] = \underline{u}_j + \tfrac{1}{2}k[A\underline{u}_j + \underline{f}(\underline{u}_j)], \qquad (2.3)$$

where k is the time-step, and $\underline{u}_j$ is the solution at time t = jk. Often the term f(u) is purely local, so that the Jacobian $\left[\frac{\partial \underline{f}}{\partial \underline{u}}\right]$ is a diagonal matrix F say. Then the operator on the left of (2.3) is of "striped band" form, in two space dimensions. The equations (2.3) can be solved by band matrix methods, but because of the sparsity pattern of the band, it is more efficient to split the operator. If A is separated into difference operators in the x and y directions, A = X + Y say, we can replace (2.3) by the pair of equations

$$\left.\begin{aligned} (I - \tfrac{1}{2}kX - \tfrac{1}{2}kF)\underline{u}_{j+\frac{1}{2}} &= (I + \tfrac{1}{2}kY + \tfrac{1}{2}kF)\underline{u}_j, \\ (I - \tfrac{1}{2}kY - \tfrac{1}{2}kF)\underline{u}_{j+1} &= (I + \tfrac{1}{2}kX + \tfrac{1}{2}kF)\underline{u}_{j+\frac{1}{2}}. \end{aligned}\right\} \qquad (2.4)$$

Assuming that the method is stable (which can be proved in restricted cases), the formulation (2.4) gives a very economical way of handling (2.3), by reducing the linear equations to be solved to triple-diagonal form. If on the other hand (2.2) is solved by a general stiff O.D.E. package, the structure of the space operator will be lost, and there will be a serious loss of efficiency.

Of course it can be argued that using a good O.D.E. package to solve (2.2) will increase efficiency in other ways, by bringing in high-order integration formulae in the time direction. The simple discretizations (2.3) and (2.4) are of low order in time, and therefore require more steps for the same accuracy. However, the use of a high-order method for the time integration is not justified unless we can achieve corresponding accuracy in the space dimensions; indeed the user could be misled if he is asked to set a tolerance in the solution of (2.2) which is independent of the space discretization. The O.D.E. package method completely separates the time and space integrations, whereas in fact the accuracy of the two is related. Ideally the routine should be able to control the space discretization as well as the time step, as the integration proceeds. This is not easy to do automatically by any method, but it is certainly impossible if we use a general package to solve (2.2).

Another way of constructing algorithms for parabolic equations is to treat the space integration as dominant, and to solve a sequence of boundary-value problems in time. For the example (2.1), we may write

$$-\tfrac{1}{2}k[K\nabla^2 u_{j+1}+f(u_{j+1})]+u_{j+1}=\tfrac{1}{2}k[K\nabla^2 u_j+f(u_j)]+u_j, \qquad (2.5)$$

where u_j is now the approximate solution at $t = jk$. From the values of u_j we calculate the expression on the right-hand side, and solve a nonhomogeneous elliptic equation for u_{j+1}. This suggests that an elliptic package can be used as the basis of an algorithm for parabolic equations. The package can be developed from any suitable method, e.g. finite differences, finite elements, Galerkin, or Fast Fourier transform, with some form of correction or error estimation. But we have essentially the same difficulty as before, that the calculations in time and in space have been separated, and that therefore we may get poor results even when the elliptic problem is solved accurately, because of errors in the time direction.

A fully satisfactory algorithm for parabolic problems would have to maintain comparable accuracy in both time and space integrations. It would provide facilities for modifying the space mesh as the integration proceeds, and it would do this in a consistent way so that the diffusion in subregions was continuous. Although some work has been done in this direction, the methods need more development before we can design completely automatic routines.

3. HYPERBOLIC EQUATIONS AND CHARACTERISTICS

In the hyperbolic case, a well-known method of solution in two dimensions is to take a grid of characteristic curves as the basic mesh, and to transform the partial differential equations into differential relations which can be integrated along the characteristics. For non-linear equations the characteristics depend on the solution, and so the grid and the solution are determined together. This ensures that stability is maintained, at the cost of some geometrical complications arising from the non-uniform mesh. An alternative method is to solve the equation on a regular coordinate grid in x and t say, using finite differences or other suitable approximations. We have to ensure that the boundary conditions are specified correctly, so that results are not calculated outside the domain of definition of the solution, and that any stability restrictions on the mesh are satisfied.

Both methods seem to present difficulties in designing a general algorithm for hyperbolic equations. For strongly non-linear problems, with possible discontinuities in the solution, the physical properties of the variables have to be taken into account. The treatment of shock waves, for example, is specific to the type of flow being considered. Such problems require special methods for each type of physical system. If the equations are mildly non-linear, and discontinuities are not expected, a more general approach is possible,using a simple form of discretization on a regular mesh. But there is still the problem of ensuring that the domain of dependence conditions are met.

Let us take as an example the following linear system

$$\frac{\partial \underline{u}}{\partial t} = A\frac{\partial \underline{u}}{\partial x} + \underline{f}(x, t), \qquad (3.1)$$

where $\underline{u}$ and $\underline{f}$ are vectors of n components, which is to be approximated by simple difference formulae on a mesh in x and t. If the problem is of pure initial-value type, with boundary conditions given for all x at t = 0, the solution at any point may be obtained by explicit integration formulae, with suitable restrictions on the time-step to ensure stability. But if the problem is limited in the space direction, which is usually the case in practical work, we have to consider what conditions are needed on the boundaries x = a, x = b say. The requirements for a well-posed problem are discussed by Morton (1977), who also considers the effect on accuracy of different forms of boundary approximation. A suitable formulation can be found in any particular case, but it is difficult to cover all possibilities in a general algorithm. If instead of (3.1) we have a mildly non-linear system, the characteristics are not known initially, and the conditions for the boundaries may change as the integration proceeds. Thus an integration scheme which is satisfactory at the beginning of the calculation may become marginally unstable later on, and it is not easy to detect this automatically.

The method of Sincovec and Madsen (1975), described above for parabolic equations, has also been applied to hyperbolic equations, though the authors recognize the need for caution in this case. Apart from the difficulty about boundary conditions mentioned above, there is the problem of matching the accuracy in space to that in time, just as for the parabolic case. Furthermore, if we reduce the hyperbolic problem to a system of ordinary differential equations, the complementary solution may include both stiff and rapidly oscillating components. It is important to use an integration method that preserves the properties of the solution and is reasonably efficient, and this would generally be different from the method appropriate for parabolic equations. On the whole, it seems better not to separate the space and time integrations.

4. THE PROBLEM OF SINGULARITIES

In the field of elliptic equations, considerable progress has been made with the construction of general packages, and some examples are discussed by Jacobs in Chapter V.5. Many programs are based on finite-element methods, and versions are available which can handle a wide range of problems and regions. For the case of the Laplace operator, special methods may be used as the basis of algorithms, for example the integral equation method (Symm and Pitfield, 1974) and the Fast Fourier transform. All these methods involve different ways of discretizing the problem, but they all assume that the solution is smooth, i.e. that it has at least one or two continuous derivatives.

In practical problems it is very common for an elliptic equation to have singularities in the solution arising from the boundary conditions. Any boundary point at which there is a corner (whether re-entrant or not) gives rise to a discontinuity in some derivative, unless the boundary conditions are very special. If we attempt to approximate the solution by smooth functions, the accuracy obtained will be limited, and methods of error estimation such as those described in Section 1 will not be valid.

A typical example of this difficulty is given by Laplace's equation over an L-shaped region. The problem was first discussed by Kantorovich and Krylov (1958, Ch.7), who noted that the region could be regarded as two overlapping rectangles, in each of which a Fourier series solution could be found. Using Schwartz alternation between the two rectangles, they obtained a rapidly convergent process, but the resulting solution was constrained to be smooth and it did not reflect the singularity at the re-entrant corner. This problem has been considered by many later authors as a test case for new methods. If we take zero boundary conditions on the boundaries adjacent to the corner, the solution in the neighbourhood has the form

$$u = \sum_{1}^{\infty} \alpha_i r^{(2/3)i} \sin \frac{2}{3} i\theta. \tag{4.1}$$

The coefficient α_1 is non-zero, and so the first derivative is discontinuous at $r = 0$. The expression (4.1) is not well represented by the first few terms of a Fourier series, nor can it be approximated by finite differences near the corner. For this particular case, it is fairly easy to obtain an accurate solution by collocation; we take the first few terms of (4.1), and find the coefficients α_i by matching this expression to the remaining boundary conditions. However, such a method is not effective unless the geometry is fairly simple, so that the singular solution can be extended over the whole region.

In considering a general approach to the numerical treatment of singularities, there are two main questions. First, is it always possible to find a local analytical form for the singular solution? In the case of Laplace's equation a large number of particular solutions are known, and we can often find an expression to fit the boundary conditions, but for general elliptic equations the problem is more difficult. A method devised by Fox and Sankar (1969) can be used for certain types of linear equations, and further applications and examples are given by Crank and Furzeland (1976). But the analysis is not simple enough to be carried out automatically, and it does not apply to non-linear problems. For more complicated cases the user would have to simplify the problem in some way, by approximating the equation or the boundary conditions, to obtain a reasonable form for the singular function.

When an appropriate expression has been found, the next question is how to incorporate it into some technique for numerical solution. If the singular function is known explicitly, i.e. if the coefficients of the singular terms are known, we can subtract it out completely to leave a nonsingular problem

(Rosser, 1974). Usually the coefficients are not known, however, and the analytical solution is not necessarily valid over the whole region. We therefore have to apply the singular solution locally, and to match it to some other form of solution in the remainder of the region. This presents considerable difficulties, both in the analysis of accuracy and convergence, and in practical implementation. It is possible to obtain good results with finite differences or finite elements in suitable cases, but the process requires careful programming, and is not suitable for an algorithm. The theory is better developed for finite elements than for finite differences, because the singular functions can be regarded simply as additional elements of the basis. But the method becomes complicated if the singular terms are used over more than one or two elements. An alternative way of matching singular solutions has been suggested by Delves and Hall (1977), who use a number of different analytical functions in subregions, and determine the coefficients by a generalized variational principle which involves the interfaces. This may be more satisfactory than attempting to match an analytical form to a discrete mesh function.

In practical problems, where there are often uncertainties in the data, the effect of singularities may not be the largest source of error, and a detailed treatment is not appropriate. A simple way of limiting the effect of a corner is to reduce the mesh in the neighbourhood, which is fairly easy with finite elements. If the solution is required only at distant points, this is quite sufficient to give good results. But in a general algorithm one would like to have some way of estimating the error due to the singularity, so that the user can be warned about the effect, even if it cannot be removed.

5. DESIGN OF PACKAGES

The problems discussed above illustrate the difficulty of constructing a complete algorithm for any class of partial differential equations. This does not mean that it is not worth trying to write general software, but at present we must recognize that our algorithms are rather limited in scope. The following are examples of some fairly simple routines which can be provided in a satisfactory form, with due cautions to the user

1. Fast Fourier transform method for Poisson's equation in a rectangle.
2. Finite element program of moderate complexity for general

regions, Poisson's or similar equations.
3. Routine for non-linear parabolic equations in one space dimension, by classical discretization without mesh deletions and insertions.
4. Auxiliary routines in linear algebra, particularly for band and sparse matrices.
5. Quadrature routines for finite element programs.

Of course much larger and more complex packages have been produced in special areas, such as structural mechanics or fluid flow. But we are thinking here of general-purpose routines, which are suitable for inclusion in a mathematical library which is not designed for particular applications.

In the field of partial differential equations, it is even more necessary than in other fields to provide full documentation for library routines, explaining the analytical requirements of the method, and drawing attention to possible sources of difficulty. Many of the problems arising in the solution of partial differential equations are associated with the mathematical formulation, and cannot be treated by algorithms. Perhaps the most important advice to give to a user is that he should look carefully at the analytical form of his problem before attempting to put it on the computer. It is true that numerical work is almost always needed at some stage in the analysis, but it is usually expensive and ineffective to try to resolve difficulties in formulation by this means. A further point is that simplified models are often very illuminating in solving a complicated system. For example, it may be advantageous to reduce a problem in three space dimensions to one in two dimensions by a suitable method of averaging, and thus to be able to cover a wider range of cases, and produce more intelligible output. Although improved algorithms will doubtless be developed in the next few years, the need for skill in mathematical modelling will not diminish, and it may in fact become more important when the problems of technique are more easily solved.

6. REFERENCES

Bank, R.E. and Rose, D.J. (1975) "An $O(n^2)$ method for solving constant coefficient boundary value problems in two dimensions", *SIAM J. Num. Anal.* **12**, 529-540.

Courant, R. and Hilbert, D. (1962) "Methods of Mathematical Physics. Vol. II", Interscience, New York.

Crank, J. and Furzeland, R.M. (1976) "The treatment of boundary singularities in axially symmetric problems", Report TR/63, Dept. of Mathematics, Brunel University.

Delves, L.M. and Hall, C.A. (1976) "An implicit matching principle for global element calculations", Report CSS/76/10/3, University of Liverpool.

Dew, P.M. (1977) "A note on the numerical solution of quasi-linear parabolic equations with error estimates", Report 101, Computer Studies, University of Leeds.

Fox, L. and Goodwin, E.T. (1949) "Some new methods for the numerical integration of ordinary differential equations", *Proc. Camb. Phil. Soc.* **45**, 373-388.

Fox, L. and Sankar, R. (1969) "Boundary singularities in linear elliptic differential equations", *J. Inst. Math. Applicns.* **5**, 340-350.

George, J.A. (1973) "Nested dissection of a finite element mesh", *SIAM J. Numer. Anal.* **10**, 345-363.

Kantorovich, L.V. and Krylov, V.I. (1958) "Approximate Methods of Higher Analysis", Noordhoff.

Morton, K.W. (1977) "Initial-value problems by finite difference and other methods", in "The State of the Art in Numerical Analysis", D.A.H. Jacobs (ed), Academic Press, 699-756.

Pereyra, V. (1966) "On improving an approximate solution of a functional equation by deferred corrections", *Num. Math.* **8**, 376-391.

Rosser, J.B. (1974) "Fourier series in the computer age", Report TR/43, Department of Mathematics, Brunel University.

Sincovec, R.F. and Madsen, N.K. (1975) "Software for non-linear partial differential equations", *ACM Trans. on Math. Software* **1**, 232-260.

Symm, G.T. and Pitfield, R.A. (1974) "Solution of Laplace's equation in two dimensions", Report NAC 44, National Physical Laboratory, Teddington.

Zadunaisky, P.E. (1976) "On the estimation of errors propagated in the numerical integration of ordinary differential equations" *Num. Math.* **27**, 21-39.

V. 5

SOME ELLIPTIC BOUNDARY-VALUE PROBLEMS IN THE CEGB

D.A.H. Jacobs

(Central Electricity Research Laboratories, Leatherhead, Surrey)

1. INTRODUCTION

The need to solve elliptic boundary-value problems arises in a very wide range of physical models, from problems in electrostatics (e.g. Poisson's equation) to viscous fluid flow (Navier-Stokes equation), and from stress analysis to heat transfer. During the last decade, significant advances have been made in the methods and techniques available and used. These are in part the result of the availability of computers with large memories and fast processing, both of which are frequently required for complex three dimensional problems. The new solution methods are even more important, providing substantial reductions in computation and core storage. Various techniques have also increased the flexibility of some methods.

Two general numerical discretization methods have evolved for solving elliptic boundary-value problems - finite differences and finite elements. Space precludes an explanation of these techniques and interested readers should refer to one of the many texts on the subjects (e.g. for finite differences, Ames (1969), Smith (1975) for finite elements, Mitchell and Wait (1976) or Zienkiewicz (1967)). Both have their own merits. Finite element meshes employing, for example, triangles in two dimensions and tetrahedra in three dimensions provide a very flexible means of refining the grid locally. Local mesh refinement with finite difference grids is more awkward to employ, particularly with rectangular grids. It tends to destroy the banded structure of the matrix, thereby making solution by the advanced iterative techniques difficult because they may rely on topologically rectangular meshes. However the derivation of finite difference sets of equations is generally easier than that of the finite element analogues.

Other numerical methods of solution also exist. For

example the formulation of the problem in terms of equivalent integral equations. Frequently these also require solution by discretization techniques employing finite differences or finite elements.

With all the methods, a discretized model of the problem is used in which the "continuous" dependent variable is characterized by a finite set of nodal values. The partial differential equation and boundary conditions are replaced by a system of simultaneous algebraic equations involving these nodal values. Frequently the systems are large, and if the governing differential equation is non-linear, so is the system of algebraic equations.

Two general classes of methods exist for solving systems of linear algebraic equations: direct and iterative. They have both been developed to exploit the particular features of systems derived from partial differential equations. Reid (1977) and Fox (1977) contain comprehensive reviews and descriptions of the majority of developments in these areas.

Thus a general package for the solution of certain elliptic boundary-value problems could employ either finite differences or finite elements to produce the system of algebraic equations, and either a direct or an iterative procedure for solving the resultant system.

Within the Generating Board there are a large number of programs which have been developed for solving elliptic boundary-value problems, and the methods employed cover a wide spectrum of those available. The majority form an integral part of larger packages designed to model one particular problem. For example there are programs to solve the neutron-diffusion equation in a nuclear reactor geometry, and several programs to solve the steady-state Navier-Stokes equations in various geometries. In addition, there are several wider "general purpose" packages which have been designed to cover a range of applications. These can be classified into two types.

The first consists of programs developed to cover one area of application. The user supplies the input data which uniquely describe his particular problem. One of these is the BERSAFE (Berkeley Nuclear Laboratories Stress Analysis by Finite Elements) system developed for the static stress analysis of an arbitrary structure. (The BERSAFE suite is available commercially.) The program uses the "frontal" approach (Irons, 1970) of Gaussian elimination to solve the system of algebraic equations derived using finite element techniques.

The second group consists of "numerical algorithm" black-box routines which assist in the solution of particular systems of algebraic equations. These the user incorporates into his own software. They considerably facilitate the use of sophisticated solution procedures without the need for the user to program or even understand the method. Several such routines, generally employing direct methods of solution, are available in most subroutine libraries. A suite of iterative routines based on the strongly implicit procedure (SIP, Stone, 1968) have been developed at the Central Electricity Research Laboratories to cover many of the commonly occurring systems of equations obtained from finite difference models. In addition to providing the routines, techniques and guidance must also be given to users. These assist in the translation of his problem to a form suitable for solution by one of the routines and in the subsequent program design.

The above are only a sample of the software available. They do however illustrate three main categories:

(i) finite differences/finite elements
(ii) direct solution methods/iterative solution methods
(iii) complete programs/black-box routines.

This paper describes two of the software packages developed within the C.E.G.B. BERSAFE, a finite element model using a direct method in a complete program, is described in Section 2. The SIP iterative routines, for finite difference models, are described in Section 3. Thus all six areas categorized above are included. Both sections are necessarily very brief, covering only those aspects of the package necessary to obtain an overall idea of their design and applications. Some of the outstanding needs that can be identified are described in Section 4. Details of some of promising recent developments are also given.

2. THE BERSAFE SUITE

2.1 Introduction

BERSAFE is a suite of programs developed at the Berkeley Nuclear Laboratories by Hellen and co-workers. Commenced in 1968, it was designed to provide a general purpose static stress-analysis package capable of modelling arbitrary structures. The modelling employs finite element techniques and the "frontal" approach to Gaussian elimination is used to solve the system of algebraic equations. The first Phase of the development was completed in 1970. This was superseded by an enhanced version, Phase II, which incorporates additional element types, including those for fracture mechanics, and other features. A third Phase covers non-linear analyses for plasticity and creep.

The basic differential equation solved in Phase II is

$$\tfrac{1}{2}E\ \{\nabla^2\phi + \text{grad div } \phi\} + g = 0$$

where E is the modulus of elasticity, with suitable boundary conditions. Both two and three dimensional problems can be solved

The programs of the suite cover three main areas: data generation, analysis and output presentation. Fig. 1 illustrates the over-all organisation.

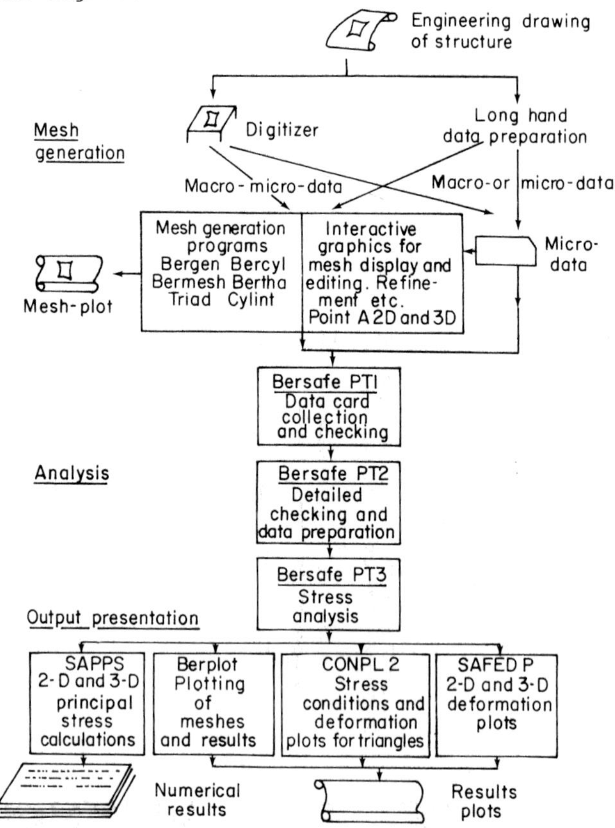

Fig. 1 The BERSAFE suite and some associated programs

2.2 BERSAFE Data Generation

The analysis part of BERSAFE has been designed to accept data input in the form of a sequence of elements. Each element is defined by the numbers of the nodes which make up the element, and the properties of the material comprised in that element. A large number of different types of elements are catered for, each with an identifying code. Thus the constant stress element EP6 is the planar triangular element with 6 degrees of freedom - two at each vertex. More elaborate elements are available (some are shown in Fig. 2a). The corresponding elements denoted by EXn are available in axisymmetric coordi-

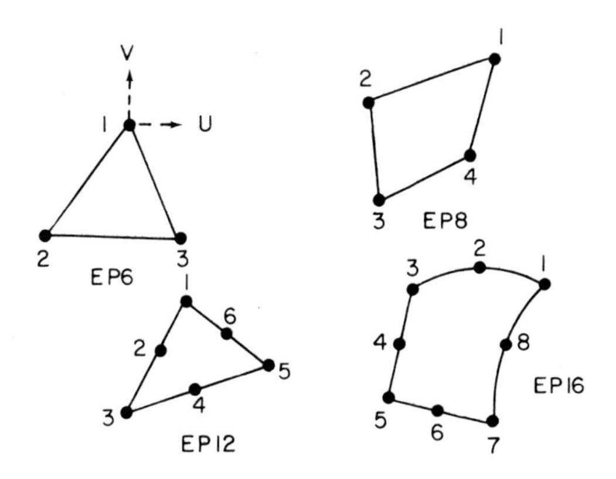

Fig. 2(a) Some of the planar elements available in BERSAFE

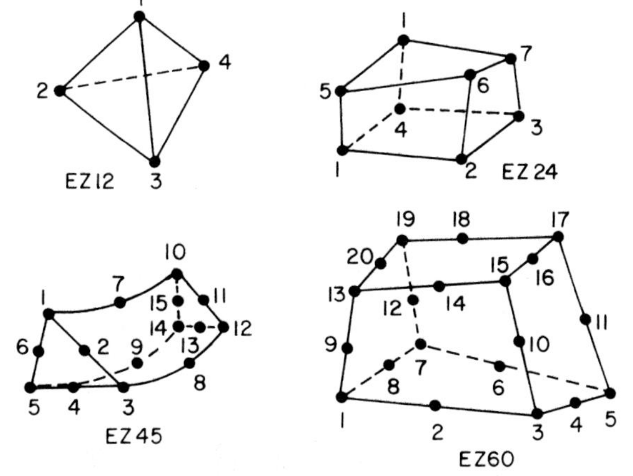

Fig. 2(b) Some of the three-dimensional elements available in BERSAFE

nates (r,z). In three dimensional Cartesian coordinates the simplest volume element is EZ12, and other higher order elements are available (some of which are shown in Fig. 2b). Some of these elements are also available for cylindrical geometry (e.g. EZ45R) and have an R as the final character. There are also many other special element types for treating particular stress problems, e.g. fracture elements.

A complete program such as BERSAFE relies entirely on the data input correctly specifying the problem. This tends to necessitate large quantities of data. To assist the user in preparing this for a particular structure many software aids have been developed within the CEGB. The standard mechanical means of preparing the data involves drawing out the elements required and preparing the data from such a plan. A digitizing table is often used. This tends to be restricted to two-dimensional problems.

BERGEN automatically generates finite element discretizations for a two or three dimensional structure supplied as a sequence of macro-blocks consisting of quadrilaterals in two-dimensions and hexahedra in three-dimensions. It divides each block into a set of basic elements of prescribed type. BERMESH is another automatic mesh generation program, restricted to two-dimensional problems. The input can be macro- or micro-block data in a similar form to BERGEN, or it can be supplied in a form for use by a "data-synthesis" technique which generates the region of interest from only small amounts of input data, or it can be produced from a digitizing table. TRIAD can be used to generate a triangular mesh from node data supplied by the user.

One "geometry" of much interest in engineering applications is that of the intersection of two cylinders. BERCYL can be used to produce such meshes, which tend to be particularly difficult to generate by hand. There is also a program which can generate a three dimensional mesh from layers of two dimensional ones (BERTHA).

Other associated programs have been produced to assist in data preparation, such as mesh refinement. The totality of programs enables meshes to be generated with relative ease for many complex engineering items. The data information produced is compatible across the range of many finite element programs available (see Section 2.5). Interactive programs (POINTA2D and POINTA3D) have been developed at the Headquarters Computing Centre (HQCC) to run on the GEC 4080 computer which front-ends the main frame 370. They incorporate many useful facilities enabling meshes to be displayed, refined and even generated from macro-blocks.

2.3 The "Frontal" Approach

Finite element models generally produce systems of algebraic equations which are positive definite. This means (Wilkinson, 1965) that during a direct solution employing, for example, Gaussian elimination, pivoting is not required. The coefficient matrix is generally very sparse because of the nature of the elements. The avoidance of pivoting means that the sparsity pattern is not rearranged during the elimination.

Band matrix techniques exploit the sparsity, since during the Gaussian elimination of a system of banded equations in which all the non-zeros of the coefficient matrix lie within M rows or columns of the main diagonal, the creation of non-zeros (termed fill-in) only occurs within the semi-bandwidth m about the main diagonal. Jennings (1966) extended this result to the case of a variable bandwidth matrix. For a matrix of order n, with semi-bandwidth m, Gaussian elimination employing band-matrix techniques requires only $O(1/2nm^2)$ computations contrasted with $O(1/3\ n^3)$ for a full matrix. (The storage is also reduced from n^2 to 2 nm.) For typical engineering applications with n = 2000, m might be about 200. The computation is therefore reduced by about 99%. The importance of reducing the semi-bandwidth m is evident from the dependence of the amount of computation on m^2.

Irons (1970) showed that in addition to reducing the computation, band matrix techniques can be used to reduce still further the in-core storage requirements during the elimination. With a semi-bandwidth of m, only a square matrix of order m need reside in core at any time since the elimination of a single variable can involve variables (and therefore the corresponding equation) at most m rows (or columns) away. The term "frontal" approach is used because the elimination proceeds along a "front" which traverses across the problem. The in-core storage required is then only m^2. (All of the above figures can be halved, or almost halved in the case of storage for symmetric matrices.)

BANDOPT developed by Donovan at HQCC is a program which seeks a revised node and element ordering to reduce the semi-bandwidth of the matrix. The routine is based on an idea of Collins (1973).

2.4 Analysis and output

When the mesh and ordering has been obtained a plot can be displayed if required before running the analysis routines. These comprise three parts:
Part 1 collects and checks all the data on each card individually. Any errors are identified for the user to correct. The checked input is written onto a file for subsequent processing.
Part 2 reads this data file and checks for other possible errors e.g. misplaced cards. It also sets up the arrays in the correct form and order for the subsequent solution routine. The point at which each nodal value (or degree of freedom) can be eliminated during the frontal solution is determined so that it can be interlaced with the generation of the algebraic equations. The semi-bandwidth is also determined.
Part 3 generates the algebraic equations in turn, and performs the frontal Gaussian elimination. For complex element types, the generation of the equations can require considerable computation. The forward elimination is followed by the back substitution which determines the nodal deflections and hence the element stresses. Output files are created for subsequent plotting and output.

Solutions can be displayed graphically using BERPLOT and the boundaries of strain and yield zones illustrated. Programs for the selective printing of output are also provided. Only very brief descriptions of the programs have been given here. The purpose of this review has been to illustrate one particular type of package. It is supplemented by many user facilities. In the development to date considerable emphasis has been placed on the "input side". This has been required to make the powerful analysis programs easily available as the data preparation thus becomes the most tedious (in terms of man-effort) and also the most error prone area. The interactive graphical display programs currently being implemented will be a major aid to data preparation. They are also supplemented by the development of interactive display software.

2.5 Associated software

Apart from the directly associated "service" software for data preparation and output retrieval, numerous other packages have been developed to interface to various parts of the BERSAFE suite. Indeed the availability of such a large suite, of supporting software for BERSAFE and its related programs encourages the designers of other systems to make use of as much of it as possible. Many of the structures for which BERSAFE is used to analyse the stress are also subjected to non-uniform thermal conditions, for example components of boilers and reactors. A program FLHE has been developed by Fullard at BNL which analyses both the steady-state and transient thermal conditions within an arbitrary system. It again uses finite element techniques and interfaces to the BERSAFE suite in a natural manner, not only making use of the same grid if appropriate, but also employing the same auxiliary software for mesh generation, etc.

For analysing the dynamic behaviour of structures BERDYNE developed by Leech and Spires at HQCC can be used. It too is compatible with the auxiliary software. Hutton (BNL) is currently developing programs for the analysis of fluid flows by solving the Navier-Stokes equations using the finite element method and the frontal solution. For such developments, the existing suite of software provides a very useful armoury of data generation and output display programs.

3. ITERATIVE BLACK-BOX ROUTINES

3.1 Iterative methods and the strongly implicit procedure

Iterative methods for solving systems of algebraic equations have been comprehensively reviewed by Fox (1977).

The basic form of most of these is a factorization of the coefficient matrix A as the sum of two matrices

$$A = B + C$$

where the matrix B is "easily" inverted, typically in O(n) computations, where n is the order of the matrix. The iteration procedure is then based on the repeated use of the up-date equation

$$B\,w^{k+1} = b - C\,w^{k}$$

Since the matrix B was chosen to be easily inverted, the system can be solved very economically. The solution is approached asymptotically as iterations proceed.

The classical methods, including successive-over-relaxation (SOR) have been complemented by more robust and more economic techniques. Amongst these is the alternating-direction-implicit (ADI) schemes, and more recently the strongly implicit procedure (SIP) devised by Stone (1968). A feature in the evolution has been the increased use of implicit soluction: SOR treats each equation separately; the line methods such as ADI and S line-OR solve implicitly (simultaneously) the equations for the the nodes (degrees of freedom) on one line.

The SIP is an approximate direct method in which new values for all the nodes are derived simultaneously. However, the LU factorization which normally takes $O(1/3\ n^3)$ computations is not sought for the coefficient matrix A, but for a matrix B which is "close" to A and has the important feature that the factorization is calculated very simply using $O(n)$ computations. This is achieved by restricting the matrices L and U to have non-zero elements only on the main and on the other diagonals which have the same positions as those of the actual coefficient matrix A in the lower and upper triangular parts respectively. The product of two such matrices

$$LU = B$$

has non-zero diagonals in the same positions as those of the matrix A plus a number of additional ones. The scheme was formulated for problems for which the coefficient matrix A has a very well-defined banded structure, such as that derived from a finite difference model for Poisson's equation and boundary conditions using a topologically rectangular grid in a closed rectangle - this would yield the "standard" five diagonal system of equations. The additional coefficients introduced in the matrix B are partially cancelled by examination of the equivalent algebraic equation. This cancellation involves the use of an acceleration parameter to establish rapid convergence.

The experiments conducted by Stone (1968) and the successful use of the method for many "real" problems has proved it to be a very robust and economic iterative technique. It does not demand diagonal dominance. Possibly the most exciting advance over previous iterative schemes is the insensitivity to the choice of acceleration parameters. Unlike SOR for example, for which careful optimization of the relaxation parameter is required for most problems, a standard set of acceleration parameters automatically generated for each node layout suffices for all but the most "awkward" problems.

Three major disadvantages are evident when SIP is compared to classical iterative schemes. Auxiliary workspace normally amounting to about half of that required to store the difference equations' coefficients is required, although this is still very small compared to that required by direct solution methods. To preserve the structured diagonal form of the coefficient matrix when solving problems in non-rectangular regions, it is necessary to circumscribe the region by a rectangle of grid lines and to include all the nodes within this rectangle. Although this increases storage, some of the extra computation can be avoided. The third and most serious disadvantage is the need to have a topologically rectangular mesh if the structure of the coefficient matrix is to be maintained. Thus local mesh refinement can be awkward. (Means are being developed at CERL to overcome this problem.)

It was the robustness and the avoidance of parameter optimization which led to the provision of SIP routines within the CEGB for solving systems of algebraic equations derived from finite difference models of elliptic problems. The complexity of the method compared with SOR which was at that time in wide use, necessitated the production of black box routines for various difference equations in both two and three dimensions. These were produced in 1971/72, and have been used by a number of people on a wide variety of problems. They now form a constituent part of several applications packages available to CEGB.

3.2 The Iterative solution Routines

The original suite of routines developed by Jacobs (CERL) required as input the coefficients and source terms of the algebraic equations, and an initial approximation. The convergence criterion on the maximum residual of the algebraic equations specified the accuracy required, and the routine sought a satisfactory solution.

However, several limitations of this apparently simple structure must be overcome.

The SIP was developed principally for linear systems, however being iterative, the solution of non-linear systems is a simple extension. The repeated calls to a solution routine with new coefficient values necessitate additional logic and "flagging" to ensure the use of the complete set of acceleration parameters, otherwise the rate of convergence suffers. The same remarks are true when seeking the solution of a time dependent problem cast in terms of the solution of a sequence of elliptic systems.

Another difficulty is the determination of a solution to problems for which the solution is not unique in so far as an arbitrary constant can be added to the whole solution (e.g. Poisson's equation with Neumann boundary conditions). For these problems a consistency condition is required to be satisfied by the algebraic equations. However, without the necessary care, the convergence can be impaired by the truncation of large numbers.

These minor complications can all be overcome by additional parameters and the necessary logic. The first set of routines covered these features.

However, the greatest difficulty is associated with the use of deferred correction (Fox and Mayers, 1968). For example, the five point molecule solution routine can be used to solve systems of equations which are predominantly of five point form, but a few of the equations involve additional terms, at say, the boundary nodes. Such extensions which are needed for more complex problems required "software surgery", which is undesirable. These difficulties have been overcome in the new suite of improved routines under current development.

The revised packages reflect these two levels of requirements. There are now two routines for each type of basic iterative molecule considered. The basic routine for which the input is the current residual and the coefficients of the appropriate algebraic equations to be used in the iteration. The routine performs one iteration of the SIP to determine the up-date vector predicted. The user must supply the calling software which controls the iterative procedure, calculating the residual vector prior to each call of the routine, and combining the up-date vector determined with the previous estimated solution. The calling software must also check for convergence. This basic routine can be used very easily for problems employing deferred correction.

They also form the kernel of the standard "driver" routines which are designed to control the iterative procedure for all "normal" applications. The input includes the coefficients of the system of equations, the source terms, an initial approximation to the solution and a convergence criterion. The routines organize the iteration: calculate the current residual, call the basic routine, up-date the approximation; and repeat this cycle till convergence. For the ease of incorporation by existing users, the call statements for the new "driver" routines are very similar to the original versions. They are, however, more efficient, particularly in the reduced core required.

The systems of equations for which routines have been written are

1. The two dimensional five-point molecule (e.g. Poisson's equation, second order equation).

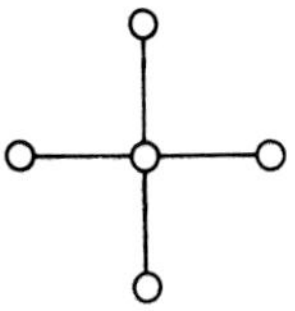

2. The two dimensional nine-point molecule (e.g. high accuracy Laplace solutions, second order equations with $\frac{\partial^2}{\partial x \partial y}$ term).

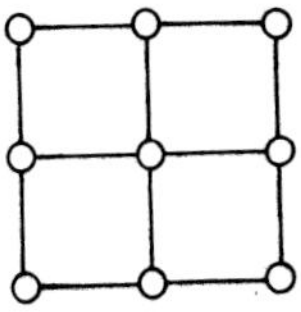

3. The two dimensional thirteen-point molecule (e.g. fourth order equations such as the Biharmonic).

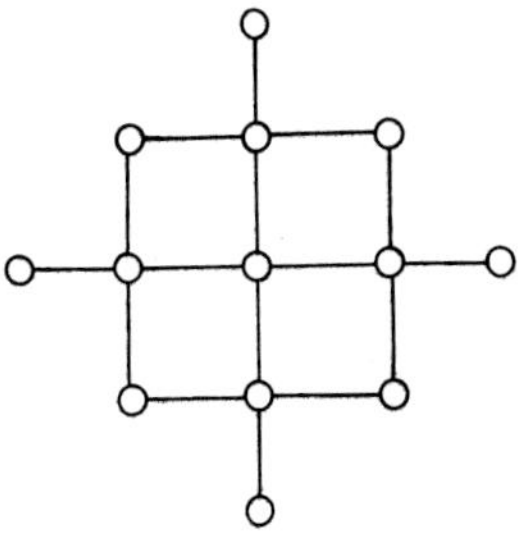

4. The three dimensional seven-point molecule (e.g. three dimensional second order equations).

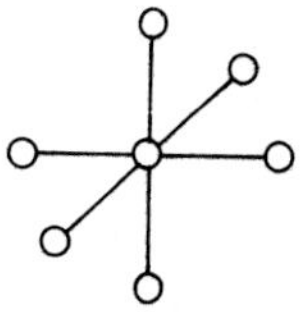

All the routines were developed as generalizations of specific demands from various users. Several additional packages have been provided for particular, but often occurring, problems:

(i) a general Poisson solver for rectangular regions using a rectangular Cartesian grid (input: grid, source terms, boundary conditions, initial approximation; output: converged solution).

(ii) a general Poisson solver for a circular or annular region with (r,θ) coordinates (input: (r,θ) grid, source terms, boundary conditions, initial approximation; output: converged solution).

(iii) as (ii) but for a three dimensional cylindrical region using (r, θ, z) coordinates is under development.

3.3 Aids to Incorporating and Using the Software

The principal aid to users must of course be the program users guide (in the CERL this is contained in the written document called the "Program"). However, this must be fairly brief and cannot include detailed explanations on, for example, the derivation of the finite difference equations for both internal and boundary nodes. However, the routines are frequently used by people untrained in finite difference methods.

Within the CEGB an intensive 1 week course has been run every year for the last 4 years covering the solution of ode's, pde's and optimization. The course is designed specifically for engineers with little or no relevant knowledge, and course members are taught, amongst other topics, how to use finite differences and finite elements, and how to use the iterative and direct methods. Guidance on the use of the SIP routines is included.

Several additional techniques (which are also included in the course) enable wider ranges of problems to be solved by standard iterative methods in general, and by employing the SIP routines in particular. These techniques include the following.

(i) Methods for solving problems in infinite regions exterior to a particular body by a method of region inversion proposed by Dorny (1968). Examples of such problems include the determination of electric fields around isolated bodies.

(ii) Deferred correction methods for a wide range of problems including:

a) problems within regions for which the boundaries are not parallel to coordinate lines and on which gradient boundary conditions are prescribed.

b) problems for which the elliptic pde has a cross derivative term whose magnitude is small compared to the other second order terms.

c) solving steady-state viscous fluid flow problems (non-turbulent) using a second order accurate central difference for the residual calculation, but basing the iteration on the more stable "upwind-difference". (Jacobs 1974).

For all of these, the residual is determined using the system of algebraic equations to which a solution is required, but the iterative SIP up-date is determined using coefficients of a simpler, or more stable system of equations.

(iii) Methods for selective mesh refinement, which tend to

upset the regular banded structure of the coefficient matrix, and therefore preclude the use of SIP in its standard form. However, the gains obtained by using the SIP are generally sufficiently great to enable more nodes to be used, or for the problem solution to be split into two parts. First determining the solution on a coarse grid, and secondly a refined subregion calculation using appropriate coarse node values for boundary conditions. (A good initial approximation is also then available.) Provided the boundaries of the subregion are chosen where the coarse grid is still satisfactory, the technique provides good results. Program complexity is eased if the two regions have the same shape with only the nodal separations changing, thus permitting the use of the same code.

To summarize, the provision of numerical software must be supplemented by techniques for its application to a variety of possible uses.

4. NUMERICAL SOFTWARE - OUTSTANDING NEEDS

4.1 Direct Methods

Undoubtedly one of the most urgent requirements for improving the numerical software currently available is a substantial reduction in the computations required to solve systems of equations by direct means. This would permit the solution to a wider range of problems than is currently possible, particularly non-linear ones.

There is also a substantial class of problems which are currently solved by using iterative techniques because of their size, for which the associated systems of algebraic equations are tending to become ill-conditioned. For these a highly efficient (both in computation and core store) direct method would be invaluable.

Two methods are available which go part of the way towards achieving these goals.

The first are those employing cyclic reduction and Fast Fourier transform (Hockney, 1970, Buzbee, Golub and Nielson 1970) see Fox (1977) for review). They make use of the special symmetry exhibited by the difference equations for certain applications, such as Poisson's equation. The capacitance matrix technique enables problems in non-square regions to be solved. The amount of computation required is reduced to $O(n \log_2 n)$ for a matrix of order n representing a two-dimensional problem; the core storage is also substantially reduced compared with other direct methods.The current limitations of the method are to the solution of Poisson's type equations with uniform mesh spacing. If these could be overcome, the method would have an even greater impact. Currently, for problems for which they are applicable, these methods are by far the most efficient available, even rivalling iterative techniques. Combined with deferred correction, the methods can also be used very effectively for an extended range of problems.

The second recent development is that of nested dissection and associated methods (George 1973). Nested dissection is a systematic way of ordering the nodes and equations of a finite element or finite difference representation to reduce the computation and core store requirements. To achieve these benefits sparse matrix techniques (see Reid 1977) must be used. George (1973) has shown that for two-dimensional problems, the orderings minimize the computation during Gaussian elimination. For a problem in a topologically rectangular region the number of computations is reduced from $O(1/2\ nm^2)$ to $O(1/3\ nm)$. For three dimensional problems, a similar reduction is achieved, but the ordering has not been proved to be the best possible There is hope that significant reductions can be achieved. Various developments are currently being researched (George 1977a).

The application of nested dissection relies on certain geometrical properties of the grid. George (1977b) has extended the ideas to automatic algorithms implementing the minimal degree algorithm which is based on selecting the order for nodal elimination which results in the minimum computation (and fill-in) during the Gaussian elimination.

Some very exciting developments are currently taking place in this field.

4.2 Iterative Methods

The advancement of direct methods has not detracted from the research into and the development of new iterative methods. For very large problems, and for problems with widely differing values of the coefficients, the majority of iterative methods behave poorly. Considerable effort has been directed at understanding the strongly implicit procedure which, although very powerful, is seen as being based on rather heuristic methods. However little of substance has been obtained.

For problems which involve a selection of the following features, the rate of convergence of all iterative schemes can become very slow, and faster methods would be highly desirable.

a) severe non-linearities
b) large numbers of nodes
c) Neumann or periodic boundary conditions (the latter associate values on different boundaries)
d) linked equations.

Fast convergence is required not only for economic reasons, but also to enable acceptable error bounds to be obtained.

The "Incomplete Cholesky-Conjugate Gradient method" (Meijerink & Van der Vorst (1976)) is claimed to provide faster convergence than SIP. It uses an incomplete Cholesky factorization analogous to the incomplete banded LU factorization used in Stone's SIP. The factorization matrix L is restricted to have non-zero elements in only those positions with non-zero elements in the original matrix. Since the resultant factorization is only approximate, the conjugate-gradient method is incorporated to obtain the iterative up-dates to the dependent variable.

Other "new" iterative schemes occur with surprising frequency in the literature. An area which is worthy of some detailed objective research is the production of test problems so that new methods can be compared, at least partially.

4.3 General Problems

Currently means exist for solving the majority of linear elliptic partial differential equations in both two and three dimensions. These could use either finite difference or finite element discretizations. Their solution might require large amounts of both computation and core store and the resolution permitted by current limitations of computers may be inadequate. Ill-conditioning of the system of equations can be a severe problem. It can arise because of limitations on numbers of nodes imposed by either run-time or core store restrictions. This in turn forces the use of grids too coarse and non-uniform, consequently the problem is not represented sufficiently accurately. Ill-conditioning makes solution by iterative techniques unreliable and expensive because of slow convergence; solutions obtained by direct means can also be of dubious accuracy.

The solution of non-linear problems causes great problems. A perennial problem is that of solving the Navier-Stokes equation. The gross non-linearity brings its own problems in addition to the difficulties already mentioned associated with both direct and iterative methods. For many problems there is also the potential requirement for automatic mesh refinement. The Navier-Stokes equation is perhaps an extreme example with which to illustrate the difficulties, but for all non-linear equations improved techniques are urgently required. Currently linearization and various forms of deferred correction are used. An intuitive "rule-of-thumb" is that the "better" the linearization and the "greater" the "implicitness" of the schemes used, the faster the rate of convergence. However with strongly linked variables, considerable under relaxation is often required.

In conclusion, the solution of complex problems involves the separate consideration of each problem and the search for the most appropriate method. If considerably more efficient solution methods were available, both iterative and direct, much of the individual investigation would be eliminated. Both the generation of new methods, and the application of existing methods to ever more complex and large problems, must be pursued with the common goal of solving a greater range of problems economically.

4.4 General Requirements of Software

One aspect of software which is much discussed is the escalation in development costs. The production of truely "modular" programs producing not only sub-packages of "reusable" code for other related systems, but also containing well defined interfaces on which subsequent enhancements can be incorporated, cannot be over emphasized. The software design should model the problem structure, and the individual modules, macro-and micro-, should be self-contained and cohesive with the minimum number of connections in and out. Large general purpose packages should always be designed to facilitate the incorporation of replacement modules developed by users with special requirements, as well as general enhancements. Such enhancements are less easily incorporated in existing software packages, not built using modular techniques!

The modular design of numerical software, particularly that required by software libraries, enables users to easily make use of different packages as they emerge, and also to compare different techniques where appropriate. Since advances in both the direct and iterative solution of systems of algebraic equations are bound to be made in the future, adherence to modularity will permit many existing "production" packages to employ the new methods.

Finally, the trend to larger and more powerful computers is an excellent excuse to accept small losses in efficiency caused by modular programming and the production of useful reusable code. This is in marked contrast to the environment some years ago when the last byte, even bit, was squeezed out of software.

5. CONCLUSIONS

Within the Generating Board two different types of general purpose packages for solving problems involving elliptic boundary value equations are provided.

There are packages for solving a particular equation, or class of equations, and which only require input data to specify the particular problem. Such packages, although built around a particular numerical technique, generally necessitate considerable software for assisting in the preparation of the large amounts of data required.

Many other problems involving the solution of partial differential equations are (currently) judged too complex or too variable to permit the development of one "all singing, all dancing" package which would be sufficiently economic and robust to be useful. For these types of problems, the provision of general purpose routines (e.g. direct and iterative solution routines) which the user incorporates into his own program provide a means of making available the latest techniques to users, permitting him to concentrate on the modelling and physical, engineering, chemical aspects or whatever are appropriate, of the problem. The incorporation of such standard software packages also facilitates the change of solution routine, either to determine the best one suited to his problem, or when new routines become available.

That both types of package are widely used in the CEGB demonstrates the need for both. However, all requirements have not been satisfied. More robust and economic means of solving a wide range of problems are required, and there are still many complex problems for which no economic means is available.

6. ACKNOWLEDGEMENTS

I would like to thank the many people within the CEGB who have provided me with details of many packages, programs, methods, and developments. I would particularly like to mention Dr. A.J. Donovan and Mr. M.J. Kellaway for many valuable discussions, and Mr. K. Fullard, Mr. T.K. Hellen and Mr. G.W. Marshall.

The work was carried out at the CEGB, and this is published by their permission.

7. REFERENCES

Ames, W.F., 1969 Numerical Methods for Partial Differential Equations , Nelson Press, London.

Buzbee, B.L., Golub, G.H. and Nielson, C.W., 1970 "On direct methods for solving Poisson's equations", *SIAM J. Numer. Anal.* **7**, 627-656.

Collins, R.J., 1973 "Bandwidth reduction by automatic renumbering", *Int. J. Numer. Meth. in Enger.* **6**, 345-356.

Dorny, C.N., 1968 "Finite-difference approximations of the exterior problem for Poisson's equation" *J. Comp. Phys.* **2**, 363-380.

Fox, L., 1977 "Finite-difference methods for elliptic boundary-value problems" in Jacobs, 1977.

Fox, L. and Mayers, D.F., 1968 "Computing Methods for Scientists and Engineers", Clarendon Press, Oxford.

George, J.A., 1973 "Nested dissection of a regular finite-element mesh", *SIAM J. Numer. Anal.* **10**, 345-363.

George, J.A., 1977a "An automatic one-way dissection algorithm for irregular finite element problems" Proc. of 1977 Biennial Conference on Numerical Analysis.

George, J.A. 1977b "A minimal storage implementation of the minimum degree algorithm",University of Waterloo, Department of Computer Science Research Report CS-77-09.

Hockney, R.W., 1970 "The potential calculation and some applications", *Meth. Comp. Phys.* **9**, 135-211.

Irons, B.M., 1970 "A frontal solution programm for finite element analysis", *Int. J. Numer. Meth. Engng.* **2**, 5-32.

Jacobs, D.A.H., 1974 "A corrected upwind differencing scheme using the strongly implicit procedure" in 'Numerical Methods in Fluid Dynamics', eds. C.A. Brebbia and J.J. Connor, Pentech Press.

Jacobs, D.A.H., ed. 1977, 'The State-of-the-Art in Numerical Analysis', Academic Press, London.

Jennings, A., 1966 "A compact storage scheme for the solution of symmetric simultaneous equations", *Comp. J.* **9**, 281-285.

Meijerink, J.A. and Van der Vorst, H.A. (1976). "An Iterative Solution Method for Linear Systems of which the Coefficient Matrix is Symmetric M-matrix". *Math. Comp.* **31**, 148-162.

Mitchell, A.R. and Wait, R., 1976 "The Finite Element Method in Partial Differential Equations", Wiley, New York.

Reid, J.K. 1977 "Sparse matrices", in Jacobs, 1977.

Smith, G.D., 1975 "Numerical Solution of Partial Differential Equations", Oxford, London.

Stone, H.L., 1968 "Iterative solution of implicit approximations of multi-dimensional partial differential equations", *SIAM J. Numer. Anal.* **7**, 104-111.

Wilkinson, J.H., 1965 "The Algebraic Eigenvalue Problem", Oxford University Press.

Zienkiewicz, O.C., 1967 "The Finite Element Method in Structural and Continuum Mechanics", McGraw-Hill, London.

PART VI

OPTIMIZATION

VI. 1

MATHEMATICAL PROGRAMMING SYSTEMS

E.M.L. Beale

(Scientific Control Systems Limited and Scicon Computer Services Limited)

1. INTRODUCTION

If this paper were entitled "Mathematical Programming", it would be expected to cover the whole field of constrained optimization, and possibly unconstrained optimization as well. This could hardly be covered adequately in a single paper. And Greenberg (1978) provides a particular reason for not trying, since this book contains the proceedings of a two-week international conference on the design and implementation of optimization software held in 1977. Mathematical Programming Systems refer specifically to suites of computer programs for solving linear, mixed integer and some non-linear programming problems with large numbers of variables and constraints. Examples of such systems are: MPSX/MIP/370, written for the IBM 370 range, XDLA, written for the ICL 1900 range, and SCICONIC, written for the Univac 1100 range. When a purely linear programming model is applied to a practical problem, changes in circumstances may make it necessary to add some integer variables or non-linear constraints before the model can represent the new situation adequately. Since these models nearly always have a large volume of input and output data, it is then very useful to have a unified system for solving different types of mathematical programming problem, with common input and output formats.

People often wish that mathematical programming systems could be easily transported from one type of computer to another and easily modified to try out new algorithmic ideas. It is tempting to think that these things are difficult simply because the developers of mathematical programming systems have no real interest in portability, and are too concerned with performance, in terms of both speed and problem size. But the difficulties are really more fundamental, being caused by the facts that:

(a) the systems contain a large amount of code, mostly concerned with data-handling rather than algorithms;
(b) the systems must cope with large volumes of data, and their efficient transfer to and from the backing store of the computer;
(c) complicated data structures are needed to handle sparse matrices efficiently, and this technology is fundamental to a mathematical programming system.

Section 2 of this paper amplifies these remarks by analysing the structure of a typical mathematical programming system. But it is perhaps worth adding that some mathematical programming research does not depend on very efficient methods for handling large sparse problems, and the programs developed by Land and Powell (1973) have been widely used for such work.

Section 3 is concerned with algorithmic aspects of mathematical programming. Sections 3.1, 3.2 and 3.3 discuss procedures for linear, integer and non-linear programming, respectivel

Users of a mathematical programming system normally communicate with it through Matrix Generator and Report Writer Programs. A Matrix Generator summarizes the data, and presents the coefficients in the mathematical formulation of the model in the right format for the system. And the Report Writer summarizes the solution given by the system in practical terms. IBM formats have become a defacto industry standard for defining both these input coefficients and the output data for the rows and columns in the optimum solution. This standardization breaks down over features that were not in the IBM system when they were first introduced into other systems, but the necessary translation to another system with the same features is not difficult. So one could take the view that Matrix Generators and Report Writers were not an integral part of Mathematical Programming Systems. But this seems a wrong view for a volume on applications of software, so some remarks are made about Matrix Generators and Report Writers in Section 4.

2. THE STRUCTURE OF A MATHEMATICAL PROGRAMMING SYSTEM

The structure of a Mathematical Programming System can be illustrated by considering the functions, and approximate lengths, of the major routines in a typical system, SCICONIC. SCICONIC is written partly in Assembler and partly in FORTRAN. The count of FORTRAN statements includes comments, and 1000

FORTRAN statements is equivalent to about 2000 lines of Assembler. When overlaid, the instructions and local data occupy 8000 36 bit words on a Univac 1100 series machine.

SCICONIC uses the following types of file:

An Input File: Punched Cards, or more usually card images on magnetic storage.

A Problem File: This contains Problems, Linear Programming Bases, Saved Solutions, and Saved Ranges. Here a Problem means a complete description of the rows, columns and nonzero coefficients. The rows and columns are identified by 8-character names, although the nonzero coefficients in each column are identified by their row sequence numbers and their numerical values. A Problem may have a set of alternative objective functions, alternative sets of upper and lower bounds on the variables, and alternative sets of right hand sides on the constraints, so it may represent a large number of specific problems.

A Work File: This contains the necessary information about the specific problem being solved, held in a compact format. For example the rows and columns are identified only by sequence numbers, and if any variable has a finite nonzero lower bound, the problem is expressed in terms of the amount by which the variable exceeds this lower bound, so that the lower bound need not appear on this file.

There are other work files containing the elementary transformations defining the factors of the inverse of the current linear programming basis. These are held entirely within the core storage of the computer unless the problem is large.

A Solution File: This contains information about a solution (possibly a trial solution) in unedited form, with 14 words per row or column containing the information given on a line of the printed solution.

The primary purpose of each routine can be expressed in terms of transferring information to one of these files.

CONVERT takes information from an Input File, checks it, and transfers it to a Problem File. It contains about 2500 FORTRAN statements.

REVISE takes information from an Input File, checks it, and uses it to modify an existing Problem File. There are many options, some of which can be used interactively. Users with good matrix generators rarely need this routine, but others have many different types of requirement. This section of the code contains about 5000 FORTRAN statements.

SETUP sets up a Work File and a Solution File from information on the Problem File. The Solution File contains the row and column names, the bounds on the variables and other relevant input data, though it obviously does not include the optimum values of the primal or dual variables at this stage. The routine contains an option called PRESOLVE, which is equivalent to the REDUCE option on other systems. This option fixes the value of some decision variables when this can be done by a preliminary analysis, and eliminates unnecessary rows, using methods similar to those described by Brearley, Mitra and Williams (1975). This section of the code contains about 4000 FORTRAN statements.

These various Input routines may seem mathematically trivial: many system developers have run into severe trouble by taking this attitude.

We now consider the main algorithmic routines, which take information from the Work File and output information to the Solution File. In SCICONIC they are all written in Assembler.

PRIMAL and INVERT form the algorithmic heart of the system. Together they enable it to solve a linear programming problem. These routines contain about 6000 lines of Assembler.

The parametric routines, which allow parametric variation in either the objective function or the right hand sides of constraints, contain a further 2000 lines of Assembler. This includes the special version of parametrics used in Integer Programming to solve the linear programming subproblems created by the Branch and Bound process.

Convex Interpolation, used to solve convex problems containing general non-linear functions of single arguments, contains a further 1500 lines of Assembler.

GLOBAL implements integer variables, special ordered sets and linked ordered sets, as described by Beale and Forrest (1977), for the solution of integer and nonconvex non-linear programming problems. This section uses many of the routines discussed earlier, and contains a further 5000 lines of Assembler.

PRINTSOLN edits and prints the information on the Solution File. This routine contains about 1500 lines of Assembler.

If a solution is sent to the solution file, it overwrites any solution already there. So there is a short routine SAVESOLUTION that saves the current solution on the Problem File, and another routine POSIT that transfers a saved solution back to the Solution File.

Report Writer programs can access solutions by using four utility routines from the SCICONIC System. POSIT transfers a solution to the Solution File, STROW moves to the Start of the Rows section, STCOL moves to the start of the Columns section, and NEXVEC reads the information in the next line of the solution output. These and other general I/O editor utilities take about 1000 lines of Assembler.

There are many other routines, and we consider the more important of them.

RANGING should perhaps be considered an algorithmic routine. In effect it does the first step of parametric programming on each element in the objective function and on each right hand side element. It therefore takes information from the Work File and outputs information to a Range File which is similar in format to the Solution File. It contains about 1000 lines of Assembler.

PICTURE and GENPIC provide schematic representations of a matrix, which sometimes help to find errors in a model. They take information from the Problem File and output information to the Line Printer. They contain about 3000 FORTRAN statements.

BASISOUT and BASISIN are in effect Getoff and Restart facilities for Linear Programming, and GLOBOUT and GLOBIN are the corresponding facilities for Integer Programming, since they store the information about the Branch and Bound tree. These routines together contain about 1000 FORTRAN statements.

PUNCHMATRIX and PUNCHBASIS copy information from a Problem File to an Input File. Together they contain about 300 FORTRAN statements.

LISTROWS, LISTCOLS and SUMMARY are other output routines describing data on a Problem File. Together they contain about 1000 FORTRAN statements.

Finally, we consider how these routines are called. Some routines automatically call others; for example PRIMAL calls INVERT when appropriate without any user intervention. But the over-all strategy must be set by the user. In early Mathematical Programming Systems, such as LP/90/94, the routines were called in a predetermined sequence; and this was achieved by feeding the computer with a sequence of Agendum Cards containing the names of each successive routine required, together with the values of any relevant parameters. This is not always convenient. For example the next step after PRIMAL may depend on whether the problem is feasible or infeasible. So SCICONIC uses a sequence of cards called a Run Stream, which may be considered as a generalization of Agendum Cards into a subset of FORTRAN, which executed interpretively. The information in the Run Stream is stored in the computer, so it can include jumps, either forwards or backwards, and IF statements. The Run Stream can refer to User-Defined variables, which are essentially local variables to the Run Stream itself, and System-State variables which are common to the Run Stream and the other SCICONIC routines. The Run Stream Analysis program contains about 4000 lines of Assembler.

3. ALGORITHMIC ASPECTS

3.1 Linear Programming

We have already noted that the linear programming module is the algorithmic heart of a mathematical programming system. Many systems have implemented the important developments in methods for exploiting random sparseness in linear programming problems that were published in the early 1970's and summarized in Section 12 of Gill and Murray (1977). Specifically, most systems use the concept of "super-sparsity" due to Kalan (1971), which means that each distinct nonzero value is stored once as a double-precision number, and each occurrence of this value is identified by a half-word containing a pointer to this value. They also represent the basis in factorized form, and update the factors at each iteration of the simplex method. The most popular scheme is that of Forrest and Tomlin (1972), since this works conveniently for both in-core and out-of-core problems. Theoretically, this scheme needs a panic-inve procedure to make it numerically stable when the algorithm creates an excessive off-diagonal element in a row transformation. But we have not found it necessary to provide this panic invert in SCICONIC. This may be partly because, like many other systems, SCICONIC uses the pivotal column and row selection methods developed by Harris (1973). Goldfarb and Reid (1977) shed more light on Harris's column selection algorithm. They show that the criterion she approximates can be calculated exactly, and that this may save iterations, but that it is doubtful if this saving is enough to compensate for the extra work.

Further work is desirable on providing mathematical programming systems with special facilities for problems with special structures. Very efficient programs for network flow problems have been developed using the logic of the primal simplex method, updating the basis using list-processing techniques and carrying out all arithmetic as addition and subtraction of single precision fixed-point numbers. The Dantzig-Wolfe decomposition algorithm could usefully be reconsidered for some types of structured linear programming problem, in spite of its erratic performance when tried in the 1960's. The work of Ho (1977) suggests that decomposition is computationally effective as long as each basis contains only a few proposals from each subproblem.

3.2 Integer Programming

Integer programming problems are solved by Branch and Bound methods in all commercial mathematical programming systems. The facilities available in different systems are broadly similar, and are summarized by Beale (1977). The detailed differences between different systems are summarized by Land and Powell (1977).

These integer programming facilities have been developed primarily to solve problems arising in operational research. These typically have more continuous variables than integer variables, although they may have several hundred integer variables. Codes that are still under development have become much more efficient for these problems, being about ten times faster on the same hardware as they were five years ago. Two factors are about equally important in this improvement: more efficient handling of the underlying linear programming problem for reasons outlined in Section 3.1, and more efficient implementation of the Branch and Bound process, including more effective heuristic search strategies.

We can expect further developments over the next five years. These may bring improvements in the solution of purely combinatorial optimization problems, where the methods have been less successful. These will probably be achieved by combining ideas from other approaches to integer programming such as enumeration, cutting planes and Lagrangian methods with Branch and Bound.

3.3 Non-linear Programming

The last ten years have seen dramatic developments in hill-climbing numerical methods, methods for finding local optimum solutions to unconstrained problems, and in the adaptation of these methods to constrained problems. These developments are summarized by Brodlie (1977), Gill and Murray (1977), Fletcher (1977) and in various contributions to Greenberg (1978) But the only non-linear programming facilities offered by mathematical programming systems are some form of separable programming, based on Miller (1963), and some form of iterative linear programming, based on Griffith and Stewart (1961). SCICONIC has gone further than other systems by providing automatic interpolation in the representation of non-linear function of single arguments; so, although the program uses piecewise linear approximations to these functions, the pieces are made arbitrarily small in the relevant places within the pricing routine of the linear programming algorithm and without user intervention. We have also extended the Griffith and Stewart method by using a sophisticated conjugate gradient algorithm to solve the underlying unconstrained optimization problem. But that is all.

This is disappointing, but it is not clear to what extent mathematical programming systems are failing to meet a real need. People certainly have small-scale non-linear optimization problems, but these can conveniently be solved without using a mathematical programming system. Large non-linear problems can often be conveniently expressed in separable form. And when the real world is more complicated it may still not be worthwhile developing a good model for this extra complication and collecting the necessary data. Futhermore, large non-linear models often contain integer variables. We must then use global optimization methods, which must be based on global information about the objective and constraint functions. Hill-climbing methods are then fundamentally inappropriate.

When the objective and constraint functions can easily be expressed as sums of non-linear functions of single arguments, if necessary by adding extra variables and equations, the necessary global information can be expressed by using the separable formulation, special ordered sets, and automatic interpolation as described by Beale and Forrest (1976). Branch and bound methods will be invoked if necessary to produce a global optimum. The newer concept of linked ordered sets, described by Beale and Forrest (1977), extends this approach to functions containing sums of products of functions of single arguments. Linked ordered sets always need Branch and Bound, but this can be combined with the Branch and Bound operations used to analyse any integer variables.

A practical advantage of this approach is that the problem is expressed formally as a linear programming problem, with restrictions on the combinations of variables that are allowed to be nonzero. This simplifies the task of unifying the input and output formats with those used in ordinary linear programming. In general there is a linear programming variable associated with each possible value of the argument of each non-linear function, so there are families of variables containing an infinite number of members. But this difficulty is overcome by defining each such family as a single pseudo-variable, with the entries in each row of the model defined as a function of the entry in a particular special row, known as the reference row.

4. MATRIX GENERATORS AND REPORT WRITERS

Applications of Mathematical Programming Systems start with the identification of a problem and end with the implementation of the conclusion or, if we limit ourselves to those parts of the operation that use the computer, they start with the development of the mathematical model and end with the production of a report on the solution.

It does not seem appropriate to try to summarize the art of model-building, but the views expressed by Beale, Beare and Bryan-Tatham (1974) on methods of documenting a model and developing a Matrix Generator program from it seem worth repeating. They point out that the key to being able to develop comprehensible large-scale models, and computer programs for matrix generation and report writing, is the compactness of an algebraic formulation that uses subscripts and summation signs. By including summation over one or more subscripts, one can represent

an equation containing many variables in a single line of algebra
And, by including other subscripts not covered by summation signs, one can extend the same single line of algebra to represent several similar equations. It is therefore natural to start the development of a computer system to solve any class of mathematical programming problems with an algebraic formulation. But it is quite hard to write such a formulation of a complex mathematical programming model in a way that someone other than the originator can understand it. The following order of presentation of the model was recommended to mitigate this difficulty.

1.	Subscripts	Use lower-case letters.
2.	Sets	(Which may be needed to give a precise definition of ranges of summation, or conditions under which a constraint, variable or constant is defined). Use Capital letters.
3.	Constants	Use Capital letters.
4.	Variables	Use lower-case letters not used as subscripts.
5.	Constraints (including the objective function).	

If there are too few different letters to give suitably mnemonic symbols, then capital letters can be used as "literal subscripts to create additional composite letters. These literal subscript can be used to make the names of constants correspond to their FORTRAN names.

This may seem a perverse approach, since with a very simple model one can go straight to the heart of the matter, which is the constraints, and then explain them by defining the constants and variables. The subscripts are then defined implicitly in the course of the other definitions. But the recommended approach has the following advantages.

(a) In a model of any complexity, the constraints are hard to appreciate until the terms in them have been defined.

(b) An explicit definition of the subscripts at the outset conveys a great deal of information about the scope of the model. Once they have been defined, they need not be repeated in the definitions of the constants and variables using them.

(c) Sets may not be needed in simple models, and will generally be developed at a late stage in the formulation. But they are often the only compact way of defining domains of summation or the conditions for the existence of a constraint. It seems expedient to list them immediately after the subscripts, with which they are closely associated.

(d) The constants, variables and constraints are normally developed in parallel. But it is expedient to list the constants first, since they implicitly define the model for the user who trusts his mathematical colleagues to implement them properly. For example, if a capacity is defined as a constant, then the user may take it for granted that the recommended solution will not violate this capacity: he may even have difficulty in distinguishing between the definition of this constant and the capacity constraint implementing it.

(e) It is useful to make a sharp distinction between the constants, or assumed quantities, and the decision variables to be determined by the model, particularly as this may not be obvious from the context. For example the level of production may be treated as a piece of data, or left to be determined by the model. So capital letters are always used for constants and lower case letters are always used for variables and subscripts.

(f) It is natural to use combinations of letters to provide mnemonic names for both constants and variables, as in programming languages. But it seems best to maintain the mathematical convention that muliplication is implied when two alphabetic symbols are written side by side and not as subscripts. So when a combination of letters is used to represent a single quantity, all letters after the first are written as subscripts, in capital letters to distinguish them from subscripts that can take numerical values, which are written as lower case letters.

Given this mathematical formulation, the tasks of the matrix generator program are essentially to read in the data defining the constants, to check them for consistency and to print them in convenient formats so that the user can check them for accuracy, and then to create an input file for the mathematical programming system defining the row (or constraint) names and types, the nonzero elements for each column (or decision variable) in turn, and finally the constant terms in each row and any lower or upper bounds on the variables. The task of writing a computer program to generate this input file, given the formulation and the data suitably laid out in the computer, is reasonably straightforward using FORTRAN or any other high-level language. But for this very reason the task can itself be automated. Scicon's Matrix Generator Generator, called MGG for short, takes the algebraic formulation as input, together with FORTRAN statements or FORTRAN functions defining the numerical values of the coefficients, and produces a FORTRAN Matrix Generator program. This MGG program is very popular with our own staff and our clients. Unfortunately it uses a number of special Univac software features and so is not transportable.

This part of MGG is specific to mathematical programming. But other parts of both this and the Report Writer Generator, or RWG, are essentially general-purpose report writing systems. This aspect of mathematical programming has more in common with Commercial Data Processing than with Scientific Computing. Mathematical programming applications systems in production use may also need the facilities of a Database Management System. We, and other developers of mathematical programming software, have made a start in these directions. For example RWG takes care of the tasks of aligning column headings with the numerical data in the columns, and of extending tables sensibly when the columns or rows do not fit on a single page. And the associated database system Scipio provides automatic paging facilities so that the user need not be concerned with the problems of keeping all arrays in core. Other proprietary database management systems for use with mathematical programming systems have been developed. But progress towards a really satisfactory system is likely to come only slowly by a process of trial and error. Different organizations have very different requirements, and perhaps the only thing they will really agree about is that this is not primarily numerical software.

5. REFERENCES

Beale, E.M.L. (1977) "Integer Programming" in The State of the Art in Numerical Analysis, Ed. D.A.H. Jacobs, pp 409-488 Academic Press. London, New York and San Francisco.

Beale, E.M.L., Beare, G. C. and Bryan-Tatham, P. (1974) "The DOAE Reinforcement and Redeployment Study: A Case Study in Mathematical Programming", in Mathematical Programming in Theory and Practice,Ed. P. L. Hammer and G. Zoutendijk, pp 417-442 (North Holland. Amsterdam).

Beale, E.M.L. and Forrest, J.J.H. (1976) "Global optimization using special ordered sets", *Mathematical Programming* **10**, pp 52-69.

Beale, E.M.L. and Forrest, J.J.H. (1977) "Global optimization as an extension of Integer Programming", in Towards Global Optimisation 2, Ed. L.C.W. Dixon and G.P. Szegö, North Holland. Amsterdam.

Brearley, A., Mitra, G. and Williams, H.P. (1975) "Analysis of mathematical programming problems prior to applying the simplex method", *Mathematical Programming* **8**, pp 54-83.

Brodlie, K.W. (1977) "Unconstrained minimization", in The State of the Art in Numerical Analysis, Ed. D.A.H. Jacobs, pp 229-268, Academic Press, London, New York and San Francisco.

Fletcher, R. (1977) "Methods for solving non-linearly constrained optimization problems", in The State of the Art in Numerical Analysis, Ed. D.A.H. Jacobs, pp 365-407, Academic Press, London, New York and San Francisco.

Forrest, J.J.H. and Tomlin, J.A. (1972) "Updating triangular factors of the basis to maintain sparsity in the product form simplex method", *Mathematical Programming* **2**, pp 263-278.

Gill, P.E. and Murray, W. (1977) "Linearly-constrained problems including linear and quadratic programming", in The State of the Art in Numerical Analysis, Ed. D.A.H. Jacobs, pp 313-363, Academic Press. London, New York and San Francisco.

Goldfarb, D. and Reid, J.K. (1977) "A practicable steepest edge simplex algorithm", *Mathematical Programming* **13**, pp 361-371.

Greenberg, H.J. (Editor) (1978) Proceedings of the NATO Advanced Study Institute on the design and implementation of optimization software . SOGESTA. Urbino. Italy. 20th June - 2nd July 1977.

Griffith, R.E. and Stewart, R.A. (1961) "A nonlinear programming technique for the optimization of continuous processing systems", *Management Science* **7** pp 379-392.

Harris, P.M.J. (1973) "Pivot Selection Methods of the Devex LP Code", *Mathematical Programming* **5**, pp 1-28.

Ho, J.K. (1977) "Nested decomposition of a dynamic energy model", *Management Science* **23**, pp 1022-1026.

Kalan, J.E. (1971) "Aspects of large-scale in-core linear programming", Proceedings of the 1971 Annual Conference of the

ACM. Chicago Ill. August 3-5 1971, pp 304-313.

Land, A. and Powell, S. (1973) Fortran Codes for Mathematical Programming: Linear, Quadratic and Discrete, John Wiley and Sons.

Land, A.H. and Powell, S. (1977) "Computer Codes for Problems of Integer Programming", Paper presented at DO77, held at the University of British Columbia. August 1977.

Miller, C.E. (1963) "The Simplex method for local separable programming", in Recent Advances in Mathematical Programming, Ed. R.L. Graves and P. Wolfe, pp 89-100, McGraw Hill, New York.

VI. 2

OPTIMIZATION BY NON-LINEAR SCALING

W.C. Davidon

(Haverford College, Haverford, PA 19041, USA)

1. INTRODUCTION

Optimization algorithms usually give a more rapidly converging sequence of estimates x_k for the minimizer of an objective function when the components of the vector x are suitably scaled. This is a special case of the more general scaling of the objective function $f : X \to \mathbb{R}$ by a linear map $S : W \to X$ whose domain W is a convex neighbourhood of 0 in some Euclidean space. An initial scaling $S_0 : W \to X$ is often chosen based on prior knowledge of the objective function or past experience with similar ones. Some optimization algorithms then use successive function and derivative evaluations to update these to scalings $S_k : W \to X$ which improve the conditioning of the Hessian of the composition $fS_k : W \to \mathbb{R}$ near its minimizer. For example, each iteration of conjugate gradient algorithms in effect scales the domain of the objective function just in the direction of the preceding step, while each iteration of variable metric algorithms updates more general linear or affine scalings $S_k : w \to x_k + J_k w$. The optimization algorithms considered here update still more general collinear scalings

$$S_k : w \mapsto x_k + \frac{J_k w}{h_k \cdot w + 1}$$

to make the Hessian of the composite $fS_k : W \to \mathbb{R}$ more nearly constant as well as better conditioned.

There is an invertible collinear scaling $S_k : W \to X$ which makes $fS_k : X \to \mathbb{R}$ quadratic iff $f : X \to \mathbb{R}$ is a rational function of the form

$$f : x \mapsto \frac{x \cdot Ax + 2b \cdot x + \gamma}{(a \cdot x + \alpha)^2} .$$

The use of these collinear scalings is essentially equivalent to using the corresponding rational approximations for the objective function. Other non-quadratic approximating functions have been suggested by Fried (1971), Jacobson and Oxman (1972), Charalambous (1973), Davison and Wong (1974), Kowalik and Ramakrishnan (1976), and others. Their approximating functions have the property that the Newton step from any point is in the direction of the minimizer. Most of their approximating functions $f : X \to \mathbb{R}$ have quadratic compositions of $sf : X \to \mathbb{R}$ with some non-linear scaling $s : \mathbb{R} \to \mathbb{R}$ of their range, rather than the non-linear scalings $S : W \to X$ of the domain X considered in this paper.

2. AN ALGORITHM SCHEMA

These algorithms call a subroutine for calculating the value f_k and n-dimensional gradient vector f'_k of a differentiable objective function $f : X \to \mathbb{R}$ at any point x_k of its n-dimensional convex domain X. Input quantities used by the algorithm are:

- x_0, an $n \times 1$ column vector estimating a minimizer of $f : X \to \mathbb{R}$;
- J_0, an $n \times m$ matrix whose columns estimate conjugate displacements from x_0 which each increase the objective function by half. These columns span the m-dimensional space of displacements consistent with any linear constraints, so that n-m is the number of linearly independent constraints. When there are no constraints, $m = n$, J_0 is invertible, and $(J_0 J_0^T)^{-1}$ is the Hessian of a quadratic approximation to the objective function.
- h_0, an $m \times 1$ column vector typically equal to 0. Non-zero h_0 can be used to estimate asymmetric behaviour of the objective function about x_0, or to specify a hyperplane in $\mathbb{R}^n$ where a singularity in an extension of $f : X \to \mathbb{R}$ is expected.

Step 0: Call f_0 and f'_0 at x_0. Then for $k = 0, 1, 2, \ldots$

Step 1: Set $w_k = - J_k f'_k$.

Either stop, or else find a point $x_{k+1} = x_k + \Delta_k x$, typically with $\Delta_k x = J_k\, w_k/(h_k \cdot w_k + 1)$, where $f_k > f_{k+1}$, $0 > f'_k \cdot \Delta_k x$, and $(f_k - f_{k+1})^2 > (f'_k \cdot \Delta_k x)(f'_{k+1} \cdot \Delta_k x)$.

Set $\rho_k = ((f_k - f_{k+1})^2 - (f'_k \cdot \Delta_k x)(f'_{k+1} \cdot \Delta_k x))^{\frac{1}{2}}$,

$\gamma_k = - f'_k \cdot \Delta_k x/(f_k - f_{k+1} + \rho_k)$, and

$r_k = J_k^T (\gamma_k f'_{k+1} - f'_k) - \gamma_k (f'_{k+1} \cdot \Delta_k x) h_k$.

Step 2: Choose an $m \times 1$ column vector v_k with $v_k \cdot v_k = 2\rho_k \neq r_k \cdot v_k$, and with $(v_k - r_k) \cdot v_j = 0$ for as many j in the sequence $k-1, k-2, \ldots$ as possible.
Set $u_k = (v_k - r_k)/(v_k - r_k) \cdot v_k$.

Step 3: Set $h_{k+1} = \gamma_k h_k + (1 - \gamma_k - \gamma_k h_k \cdot v_k)$ and

$J_{k+1} = \gamma_k J_k + (\Delta_k x - \gamma_k J_k v_k)\, u_k^T - \Delta_k x\, h_{k+1}^T$.

Return to step 1 for the next iteration.

Comments: The calculation stops in step 1 when no significant decrease in function value can be obtained. There is an x_{k+1} satisfying the three conditions of step 1 whenever $f'_k \neq 0$, the objective function has a lower bound, and rounding is negligible. If such an x_{k+1} is not obtained directly with $\Delta_k x = J_k\, w_k/(h_k \cdot w_k + 1)$, then one or two doublings or halvings of this step usually gives an acceptable x_{k+1}. There is no need for a one-dimensional optimization.

The only divisions are by $f_k - f_{k+1} + \rho_k$ in step 1 and by $(v_k - r_k) \cdot v_k$ in step 2. Since $f_k > f_{k+1}$ and $\rho_k > 0$, the denominator $f_k - f_{k+1} + \rho_k$ is the sum of two positive terms, and since v_k is chosen so that $v_k \cdot v_k \neq r_k \cdot v_k$, the denominator $(v_k - r_k) \cdot v_k$ is also non-zero.

Different choices for the $m \times 1$ column vector v_k in step 2 give different updates for h_{k+1} and J_{k+1}, and hence for the collinear map $S_{k+1} : w \mapsto x_{k+1} + J_{k+1}\, w/(h_{k+1} \cdot w + 1)$. Were only affine rather than the more general collinear scalings considered, then $\gamma_k = 1$ and $h_k = 0$ for all k. In this case the algorithm reduces to a variable-metric one, and the different choices for the vector v_k give different updates equivalent to those in the Broyden-Fletcher family of updates for $H_k = J_k J_k^T$ (Broyden, 1970 and Fletcher, 1970). Then if the m×1 column vector s_k is defined by $\Delta_k x = J_k s_k$, a v_k in the direction of s_k gives the BFGS update, a v_k in the direction of r_k gives the DFP update, and a v_k for which $s_k - v_k$ and $v_k - r_k$ are multiples of $s_k - r_k$ gives the symmetric rank one update.

If in the one-dimensional case the step $\Delta_k x = J_k w_k / (h_k \cdot w_k + 1)$ is always accepted, then the algorithm can be simplified considerably. The kth iteration can be started with just f_k and f'_k at x_k and a step $\Delta_k x$, and it then proceeds as follows.

Set $x_{k+1} = x_k + \Delta_k x$. Call f_{k+1} and f'_{k+1} at x_{k+1}.
Set $\rho_k = ((f_k - f_{k+1})^2 - (f'_k \cdot \Delta_k x)(f'_{k+1} \cdot \Delta_k x))^{\frac{1}{2}}$,
$\gamma_k = - f'_k \cdot \Delta_k x / (f_k - f_{k+1} + \rho_k)$, and

$$\Delta_{k+1} x = \frac{\Delta_k x}{\dfrac{1}{\gamma_k^{\;3}} \dfrac{f'_k}{f'_{k+1}} - 1} .$$

3. PROPERTIES OF THE ALGORITHM

Theorem: If the objective function $f: X \to \mathbb{R}$ is of the form

$$f: x \mapsto \frac{x \cdot Ax + 2b \cdot x + \gamma}{(a \cdot x + \alpha)^2},$$

and if k iterations of the algorithm are completed with $(v_j - r_j) \cdot v_i = 0$ for all $k > j > i$, then

$$fS_k(w) = f_k + f'_k \cdot J_k w + \tfrac{1}{2} w \cdot w$$

for all vectors w in the subspace spanned by $\{v_j : k > j\}$ for which $S_k(w) = x_k + J_k w/(h_k \cdot w + 1)$ is in X. The minimum value $f_k - \frac{1}{2} w_k \cdot w_k$ of the restriction of $f:X \to \mathbb{R}$ to the affine subspace in X spanned by $\{x_k : k \geq j\}$ is at $S_k(w_k)$, where w_k is the orthogonal projection of $-J_k^T f'_k$ into the vector subspace in W spanned by $\{v_j : k > j\}$.

The proof of this theorem is a straightforward exercise in linear algebra and mathematical induction. Less work is needed if homogeneous coordinates are introduced. Thus when analysing the kth iteration, it is convenient to replace the objective function $f:X \to \mathbb{R}$ by a degree-0 homogeneous function with an n+1 dimensional domain whose value at $\lambda(x-x_k, 1)$ equals f(x). The n+1 dimensional gradient of this homogeneous function has the value $(f'(x), -f'(x) \cdot (x-x_k))/\lambda$ at $\lambda(x-x_k, 1)$.

It is an immediate corollary of this theorem that when the domain of the objective function $f:x \mapsto (x \cdot Ax + 2b \cdot x + \gamma)/(a \cdot x + \alpha)^2$ is n-dimensional, then if rounding is negligible the minimum of f can be located in at most n iterations. Numerical experience strongly suggests that the algorithm has superlinear convergence, though this has not yet been proven except in the one-dimensional case. Petter Bjorstad (1977) has established r-quadratic convergence for the one-dimensional version of this algorithm, but his methods do not appear to generalize readily to higher dimensional domains.

These algorithms appear to warrant further development because of these general features:
1) Function as well as gradient values are used quantitatively.
2) The rational functions used for approximating the objective function have a second order singularity along a hyperplane and so may give a better fit to exponentials, penalty functions, or other functions which increase more rapidly than quadratics near some hyperplane.
3) These algorithms are invariant not only under affine transformations, as are Newton-Raphson and variable-metric algorithms, but also under projective transformations as well. Thus these have the largest group of automorphisms of any optimization algorithm which has been proposed.
4) These algorithms share the duality property of projective geometries under interchange of points and hyperplanes. This duality has so far found more applications in linear programming

than in unconstrained optimization.

4. CONCLUDING REMARKS

Though there has not yet been sufficient numerical testing of these algorithms on which to base accurate comparisons with others, they have reduced computation times for some standard test problems to about half that needed by variable-metric algorithms. It seems probable however that other developments have the potential for much more substantial reductions in calculation time than the use of collinear rather than affine scalings of the objective function. Two specific suggestions are:
1) Different modes of function evaluations. Instead of always evaluating the objective function and its derivatives to maximum accuracy, much time can probably be saved in many problems by making only rough approximations for these until a neighbourhood of the minimum is reached. Calculation time for evaluating the objective function can be reduced at the cost of accuracy by using a coarse grid for numerical integrations, using fixed knot locations in spline interpolations, or by judicious sampling of data in curve-fitting and least-squares problems.
2) Selective evaluations of directional derivatives. Experience with a variety of objective functions suggests that when searching for a minimizer in a domain of many dimensions, major reductions often result from steps in a certain subspace of few dimensions, particularly during the early stages of the search. Thus when differences of function values are used to estimate derivatives, time could be saved in each iteration by evaluating these differences for only a few directions other than those of recent steps.

5. REFERENCES

Bjørstad, P. (1977) unpublished.

Broyden, C.G. (1970) "The convergence of a class of double-rank minimization algorithms", *J. Inst. Math. Appl.* **6**, 76-90.

Charalambous, C. (1973) "Unconstrained optimization based on homogeneous models", *Math. Programming* **5**, 189-198.

Davison, E.J. and Wong, P. (1974) "A robust conjugate gradient algorithm which minimizes L-functions", Control Systems Report 7313, University of Toronto.

Fletcher, R. (1970) "A new approach to variable metric algorithms", *Comp. J.* **13**, 317-322.

Fried, I. (1971) "N-step conjugate gradient minimization scheme for non quadratic functions", *AIAA J.* **9**, 2286-2287.

Jacobson, D.H. and Oxman, W. (1972) "An algorithm that minimizes homogeneous functions of n variables in n+2 iterations and rapidly minimizes general functions", *J. Math. Anal. Appl.* **38**, 535-552.

Kowalik, J.S. and Ramakrishnan, K.G. (1976) "A numerically stable optimization method based on a homogeneous function", *Math. Programming* **11**, 50-66.

VI. 3

OPTIMIZATION IN PRACTICE

S.E. Hersom

(Numerical Optimisation Centre, Hatfield, Hertfordshire)

1. INTRODUCTION

This paper is based on the experience of the Numerical Optimisation Centre at the Hatfield Polytechnic which, in 1967, manned its research and consultancy activities in optimization with full-time personnel. The Centre had to develop its algorithms to meet the requirements of the problems posed by our clients and has now an operational set of programs which we call the OPTIMA package (NOC, 1975). A succinct description is given for each of the routines in this package. In the second half of the paper a small selection of the many problems to which these routines have been applied are briefly described so as to illustrate features of general interest.

2. THE ROUTINES

2.1 OPVM

This is a variable metric algorithm for minimizing a function when there are no constraints on the variables. Fig. 1 shows the essential steps in the process. An approximation to the inverse Hessian of the function, $H^{(k)}$, is available at the kth step of the iteration and $g^{(k)}$ is the gradient of the function at $x^{(k)}$. A search is made along $-H^{(k)}g^{(k)}$ until an acceptable point is found, then $H^{(k)}$ is updated and the process repeated. If no acceptable point is found, the process terminates.

The search does not necessarily find the minimum along the line, but uses the criterion shown. The updating for H is based on one of the Broyden Fletcher Shanno family modified to take into account any cubic component along the search direction (Biggs, 1973).

$$\text{Min } F(X) \quad \text{No constraints}$$

$$g(X) = \partial F(X)/\partial X$$

$$H \simeq [\partial^2 F/\partial X_j \partial X_i]^{-1}$$

Initially (usually) $H^{(0)} = I$

Line search from $X^{(k)}$ along $p = -H^{(k)} g^{(k)}$

If no acceptable point, terminate.

Else

$$\delta = X^{(k+1)} - X^{(k)}$$

$$\gamma = g^{(k+1)} - g^{(k)}$$

$$H^{(k+1)} = H^{(k)} + \Delta H(\delta, \gamma, H^{(k)})$$

and repeat.

Acceptability: find α such that

$$0 < \varepsilon_1 < \frac{F(X + \alpha p) - F(X)}{p^T g(X)} < \varepsilon_2 < 1$$

Typically $\varepsilon_1 \simeq 0.001$

$\varepsilon_2 \simeq 0.75$

Fig. 1. Variable Metric Algorithm - OPVM

2.2 OPLS

Many optimization problems, especially those arising from system identification, parameter estimation or just curve-fitting, require the minimization of a function which can be expressed as the sum of squared terms. It is not unreasonable to assume that we can take advantage of this form of the objective function in order to obtain a better estimate of the second degree terms than we can with the variable metric updating method. Algebraically, as we see from Fig. 2, the Hessian matrix can be approximated by the $J^T J$ matrix, which depends only on differential coefficients of the first order. The approximation can break down if the individual terms and their second derivatives are not very small at the solution, and when this happens the algorithm becomes very inefficient. We can express precisely in mathematical terms when this condition is present, but it is not a criterion which can be readily built into an algorithm. When presented with a least squares problem we therefore first try OPLS and if the performance is not satisfactory, we switch to OPVM.

$$\text{Min } F(X) = \sum_i S_i^2 \quad \text{No constraints}$$

$$J = [\partial S_i/\partial X_j]$$

$$g = 2S^T J$$

$$G = [\partial^2 F/\partial X_i \partial X_j]$$

$$= 2J^T J + 2\sum_k S_k [\partial^2 S_k/\partial X_i \partial X_j]$$

$$\simeq 2J^T J$$

$$H \simeq \tfrac{1}{2}(J^T J)^{-1}$$

$$\text{Search Direction is } -Hg \simeq (J^T J)^{-1} S^T J.$$

Fig. 2. Least squares algorithm - OPLS

There are, in fact, very few problems arising from practical projects, apart from these curve-fitting type problems which are constrained. We find, therefore, that the main use of OPVM is as a backstop for OPLS.

2.3 OPRQP

This is our constrained optimization algorithm and is by far the one most commonly used. It is a development by my colleague Dr. Bartholomew-Biggs (1972, 1975) who followed an earlier idea by Dr. Walter Murray (1969). It is basically a penalty function technique, but it works with a quadratic approximation to the penalty function and calculates approximate Lagrange multipliers corresponding to the constraints.

It also uses the concept of a set of "active" constraints. All equality constraints are always regarded as active as are, at any stage of the iterations, all violated inequality constraints. If, however, a previously active constraint is no longer violated but the sign of its Lagrange multiplier indicates that the penalty function would be decreased if the constraint boundary were crossed, then this constraint is also included in the active set. The penalty function formulation is given in Fig. 3(a).

$$\text{Min } F(X) \quad \text{s.t.} \quad e_i(X) = 0$$
$$b_j(X) \geqslant 0$$

Use $P(X,r) = F(X) + \frac{1}{r} \Sigma [g_i(X)]^2$

g_i - "active" constraints

$$f = \nabla F(x), \quad A = [\partial g_i/\partial X_j], \quad L = \nabla^2 F + \frac{2}{r} \sum_i g_i \nabla^2 g_i,$$

$$P(X + p,r) = f + \frac{2}{r} A^T g + (L + \frac{2}{r} A^T A)p$$

We require $p = p^*$ such that $P(X + p^*,r) = 0$

but $(L + \frac{2}{r} A^T A)$ tends to become ill-conditioned as $r \to 0$.

Fig. 3(a). Penalty function formulation

The direct method for solving $\nabla p = 0$ involves the matrix $(L + \frac{2}{T} A^T A)$ which tends to become ill-conditioned as $r \to 0$. This problem can be shown to be equivalent to the quadratic programming problem shown in Figure 3(b) and leads to the algorithm outlined in Fig. 3(c). The matrices occurring in this algorithm need not be ill-conditioned and it is straightforward to make use of updated approximation to L^{-1} as progress is made.

Problem equivalent to:

$$\min_p \frac{1}{r} p^T L p + f^T p$$

$$\text{such that } Ap = -\frac{r}{2} \lambda - g$$

$$\text{where } (\frac{r}{2} I + AL^{-1} A^T) \lambda = AL^{-1} f - g.$$

Solution is $p^* = L^{-1}(A^T \lambda - f)$

Fig. 3(b). Quadratic Program Equivalent

$$B \simeq L^{-1}$$

$$\text{Solve} \quad \left(\frac{r}{2} I + ABA^T\right) \lambda = ABf - g$$

$$p^* = B(A^T\lambda - f)$$

$$\text{Try} \quad X^{(k+1)} = X^{(k)} + p^*$$

Either accept $X^{(k+1)}$

or use p* as search direction

Calculate g, f, A at next point

$$r = 2\alpha \sqrt{\frac{g^T g}{T}} \qquad 0 < \alpha < 1$$

$$(\text{usually } r^{(k+1)} \simeq \alpha r^{(k)})$$

Update B

Repeat.

Fig. 3(c). Constrained quadratic programming algorithm - OPRQP

The routine uses the solution of the quadratic programming problem, p = p*, in a tentative manner. It tries it for acceptability but if it does not give an acceptable point it uses the direction p* as a search direction and looks for an acceptable point in a way similar to the previous routines.

The vector λ tends to the vector of Lagrange multipliers associated with the binding constraints and so can be used for determining the active set for the next iteration. The penalty parameter, r, is adjusted according to progress made and, hopefully, the calculation proceeds such that the sequence of points, $x^{(k)}$ progressively approaches the constraints from an "infeasible" point, that is one at which at least one constraint is violated, but causes the minimum increase in the objective function.

2.4 OPND

All the previous algorithms expect to have access not only to the value of the objective function and of all the constraints but also the values of all their first derivatives with respect to the optimization variables.

In practice, an analytical solution is only rarely available and so resort is made to finite differences. The OPND routines are ancillary routines which perform this purpose. They usually work with central differences, although one-sided differences are used if the derivative step enters a region where it is impossible to calculate the function or a constraint.

A comment is that even when analytical derivatives are available, the coding for their calculation is the one most likely to have an error and so we often use OPND just to check this part of the program.

2.5 OPSEN

For unconstrained problems, we have the concept of a quadratic approximation to the objective function. This provides us with a means to gain some appreciation of what would happen if, in implementing the results, the precise values found for the optimization variables were not used. This is important bearing in mind that most of the problems we are dealing with arise from parameter estimation projects. The OPSEN routine executes a sensitivity calculation at the optimum point as shown in Fig. 4.

(1) Create quadratic approximation to F(X) at optimum point.

(2) Calculate eigen-values/vectors.

(3) Take step $d/\lambda^{\frac{1}{2}}$ along each eigenvector,

d - given by user

λ - eigenvalue

and calculate value of F(X) and its approximation at these points.

Fig. 4. Sensitivity routine - OPSEN

By using this routine for varying step-lengths, d, the user gains an appreciation of the size of the region for which the quadratic approximation is valid and, within that region, of the shape of the objective function. It is not unusual to find that the Hessian of the quadratic surface is ill-conditioned in that the ratio of maximum to minimum eigenvalue is very large. One particular case gave a ratio of approximately 20 000:1 which meant that the lengths of the longest and shortest axes had a ratio of over 140 : 1.

Depending on the object of the whole project we can draw one of two complementary conclusions.

(a) If the requirement is to find any function which fits the data then errors in the direction of the shortest axis are serious, but almost any point along the longest axis would suffice.

(b) Conversely, if the object is to find the values of the parameters, then along the shortest axis they are found with high precision, but they have only very low precision in the direction of the longest axis.

For constrained problems, we can, in effect, work with the projection of the objective function on to the tangent planes of the constraints which are active at the solution and this projected function can be used instead of the actual objective function, in a manner similar to the unconstrained objective function. In practice, however, there may be no guarantee that, in varying the values given to the optimization variables, one or more of the constraints might not be violated. An alternative approach is to divide these active constraints into two classes, "hard" and "soft", that is, into those which must not be violated under any circumstances and those which might be allowed small violations. We can project the objective function into the planes of the hard constraints and repeat the calculation as before, but we must expect to find a change to the first order both in the objective function and in the violation of the soft constraints.

One of the by-products of running OPSEN, is to reveal the difficulties under which the optimization routines have been working. If the variables could have been scaled and undergone a suitable transformation the objective function would have been spherical at the solution and the convergence could have been more readily obtained. It does not mean, of course, that such a scaled transformation would have benefited the progress of the algorithm far away from the optimum point. The opposite might well apply. Nevertheless we have found that even simple scaling of the variables can make a significant improvement to the rate at which an optimization routine converges to its solution.

Another feature is that when the small steps are made in the OPSEN routine, function values are found which are slightly less than those at the optimum point. This is almost always due to the rounding-errors or other features which prevent the function from being calculated with sufficient precision to give a smooth set of values in the neighbourhood of the solution. Rounding errors could be reduced by working with multiple-precision numbers, but it is not easy to see how far one should go in this direction. In all the optimization routines we are trying to drive a vector to zero. The elements of this vector are the components of the gradient of the function, or of its projected gradient or something equivalent, and, in floating point arithmetic, zero is any number with no significant digits and can be of any magnitude. In any case, other cases of "roughness" are often present. It may be that the function uses tabulated data which have to be interpolated. On the other hand, it might use a subprogram which is itself iterative and so the smoothness of the results depend upon the stopping criterion of the inner iterative process. Another cause might be that the function evaluation depends on the integration of equations in which the integration interval is of variable length and so minor variations occur due to variations in the number of integration steps or to the variable length of the last step.

None of these numerical difficulties has yet been found to be insuperable, but they have certainly retarded progress.

3. APPLICATIONS

I would now like to mention some of the applications to which we have applied these algorithms. I shall not produce a catologue of projects, but I would like to mention a few which illustrate features of general interest.

One problem was the scheduling of a cutting machine which could be set up to produce material suitable for one or two orders at the same time. For any one set-up, therefore, amounts of material, proportional to the time of running the machine, would go towards satisfying these orders, Fig. 5. By choosing values for a suitable vector of these times, some of the orders could be precisely satisfied, but the others would be over supplied. The problem was to decide which set-ups to arrange on the machine and for how long the machine should run for each of them to obtain optimum productivity. The straightforward solution was to calculate all possible set-ups of the machine for the group of orders being dealt with and make the length of run for each set-up an optimization variable. The objective function, the productivity, could be readily calculated since it was a comparatively simple non-linear expression of the financial return from selling the prime material, the scrap, overrun, etc. divided by the return from selling all the material used at the prime rate.

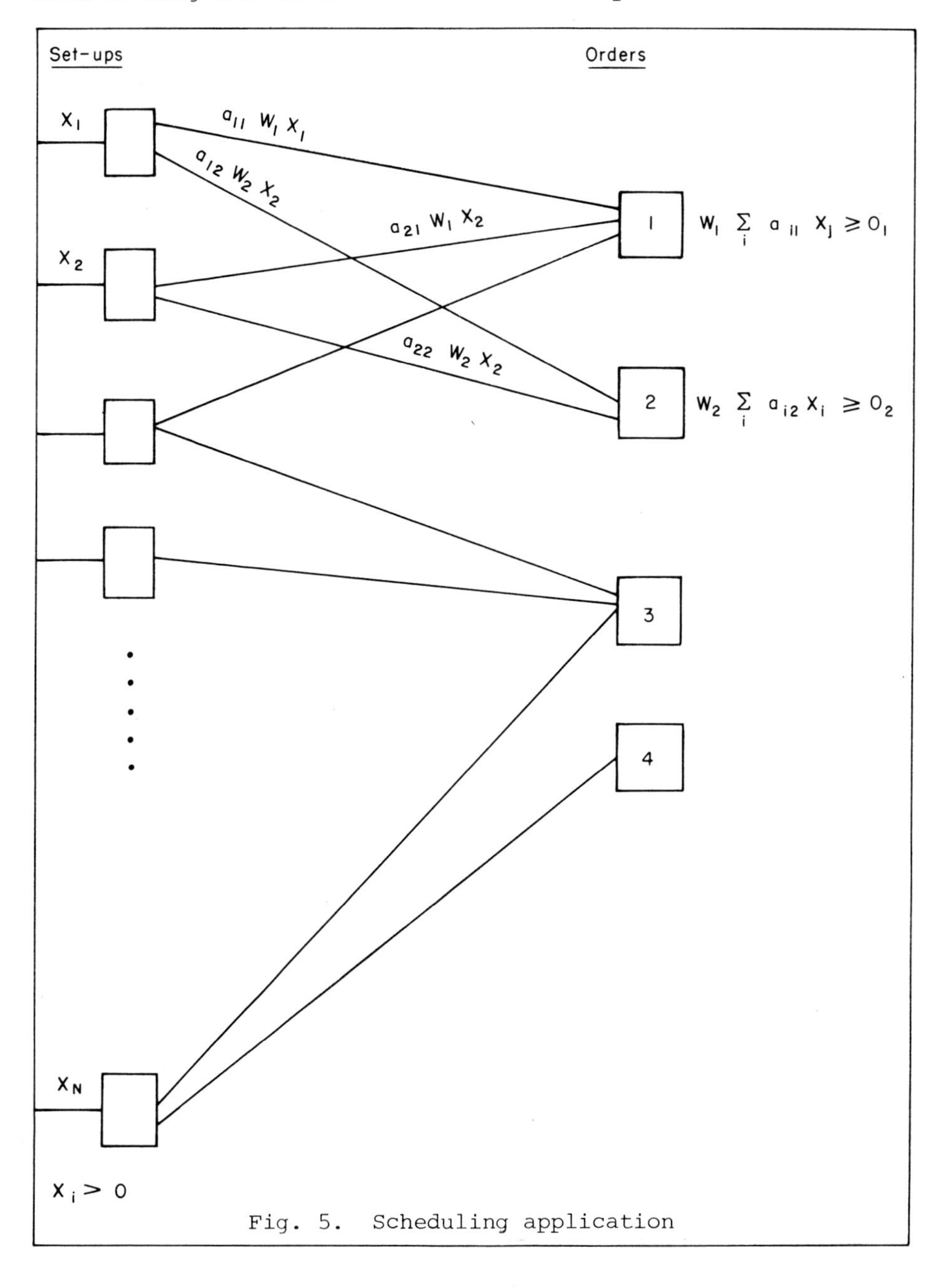

Fig. 5. Scheduling application

For a typical order pattern there could be well over 100 possible set-ups and since the time of running is approximately proportional to the square or even the cube of this number this approach was completely impractical. Yet, at the solution, we knew that only a few of the set-ups would be required, that is, of all the optimization variables only a few, about 10, would be non-zero at the solution.

With this particular problem we produced results by first of all obtaining a heuristic solution, using few variables, which satisfied all the orders, but which was not optimum. We then optimized this using the same set of variables, i.e. the same set-ups but we found values of the variables which gave greater productivity while still satisfying all the orders. This solution is the same as if we had used all the variables but had constrained all the unused variables to be zero. It was therefore possible to calculate the Lagrange multipliers associated with each of these constraints and to list those with positive marginal cost, that is, those which it would be beneficial to include in the set. The next set of variables consisted of those with non-zero values from the previous iteration together with those with the greatest marginal cost of being kept out previously. The optimization was then repeated. This is closely analogous to Linear Programming practice. It was not possible to include, always, all those with a positive marginal cost since the total number of variables could still be too high, so an arbitrary upper limit was imposed. The final solution was when an optimum had been obtained and no Lagrange multiplier associated with a zero valued variable indicated a positive marginal cost.

We have met another class of problems, mainly in the structural engineering field, which have the common feature that the, what might be called natural, vector of optimization variables, the p's, had a large number of elements, but there existed another vector, d of considerably smaller dimensions which could be said to define the state of the system. Taking p as the vector for optimization, although there were only simple non-negative constraints to be met, had the disadvantage that it was large and that there might also be sub-spaces which gave zero change to the objective function F.

We can take d as the vector of optimization variables but then we find that one of two conditions normally occurs.

Either

(1) There is an infinite set of p's which can give rise to a particular d. If this occurs at the optimum value for d then this optimum has an infinity of solutions, but this is not usually the case.

or (2) No value of p can be found to produce the given d and also satisfy the non-negativity constraint.

In the second case we can set up another constrained program which, for a given d determines a p which in some sense produces the best solution and some measure of how far the given d is from a feasible point. For example, we can introduce slack variables such that all the elements of p are non-negative, and calculate a function of these slack variables which is the required measure (Fig. 6). This measure is used as the constraint for the optimization program.

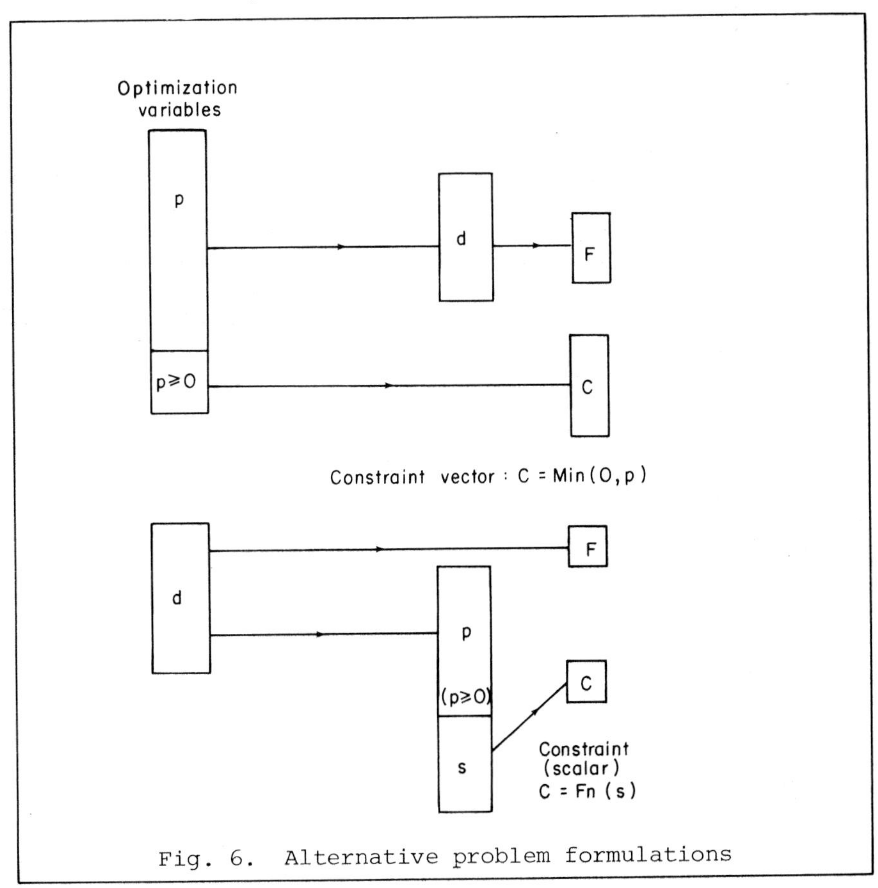

Fig. 6. Alternative problem formulations

In summary, the process is that the optimization algorithm produces a trial value for the vector d and the user's program calculates the value of the objective function and attempts to solve for the p's. If non-zero slack variables are required, it returns to the optimization program a function of these slack variables as the violation of a constraint as well, of course, as the value of the objective function. The final solution normally occurs at a boundary between the region where there is an infinite number of non-negative p's which can supply the d and the region in which there is none. On this boundary we can expect the number of positive (non-zero) elements of p to be equal to the dimension of d.

This process is, of course, only useful when the time saved by reducing the number of variables substantially exceeds the extra time involved in calculating p and s from d.

Finally, and I think it appropriate in this space age, I shall end with a space problem. Satellites are being launched with great frequency. I understand Russia alone launched 100 last year. A satellite must be launched so as to give maximum pay-load, and part of that pay-load is the fuel to keep it on-station or to put it into a required sequence of orbits. Some time after the launch, these station-keeping and similar manoeuvres must be carried out with the least fuel consumption since the life of the vehicle is just as long as there is fuel remaining. The problem is therefore, given a particular state of the vehicle, that is its position and velocity, how to fire one or more thrust motors to acquire a desired orbit.

In some configurations, the control can be specified simply as the times of starting and stopping these thrust motors, the directions of the thrusts being prescribed. The total number of control variables is not large and they can be made into optimization variables in a problem where the fuel used is the objective function and the desired orbit can be expressed as a set of equality constraints.

In other configurations, the direction of thrust may be continuously variable and is specified typically by two angles in space. One way would be to take each pair of angles at closely spaced points in time as the optimization variables, but we find we have a thousand or more variables in the problem and this imposes prohibitive computational costs.

The classical solution to this type of problem is to apply Pontryagin's Maximum Principle from which we derive the set of equations as shown, somewhat simplified, in Fig. 7(a). This results in a two-point boundary problem in which we have to determine the initial value of the adjoint variables, the p's, so that the final values of X also satisfy a set of end conditions. The values of the angles at intermediate points can be determined analytically from the maximization of the Hamiltonian H. The attainment of the objective function is implicit in the solution of these equations.

$X_0(t)$ - Fuel used

$X_i(t)$ - other state variables $i = 1, \ldots, N$

$\tau(t)$ - thrust (= 0 or 1)

$P_i(t)$ - adjoint variables

$u(t)$ - control variables (angles of thrust)

$X(t_0)$ - given

$$\left.\begin{array}{l} \dot{X}_i(t) - \tau(t)\, f_i(X(t), u(t)) \\ \dot{p}_i(t) = -\sum_j p_j\, \partial f_j/\partial X_i \\ \max_u \left(H = \sum_j p_j f_j\right) \end{array}\right\} \text{for all } t$$

$$\text{i.e.} \quad \Sigma p_j\, \partial f_j/\partial u = 0$$

$$\tau = Fn(p, f)$$

Determine $p(t_0)$ such that:

$$\psi(X(t_f)) = 0 \text{ (Final orbit)}$$

and $\exists\, \lambda$ such that

$$p_i(t_f) + \lambda^T\, \partial\psi/\partial X_i = 0$$

Fig. 7(a). Trajectory optimization - classical approach

This is well known to be a numerically difficult problem. Further difficulties can arise when there are discrete actions to be taken, for example, the switching on and off of a thrust motor [$\tau(t) = 0$ or 1]. This can be theoretically achieved by the use of "switching functions" derived from the adjoint variables, but rounding errors can introduce a large degree of imprecision in these actions, and in some circumstances the theory also breaks down (Bell and Jacobson, 1975).

Our hybrid approach was to retain the differential equations for the state and adjoint variables, to bring back the objective function explicitly and to apply an optimization algorithm which would supply the initial values of the adjoint variables and any discrete quantities, such as the times of starting and stopping the thrust motors, so that a complete trajectory could be integrated in a manner parallel to the classical method. At the end of the trajectory calculation the fuel consumption was returned to the optimization routine and any measures of the misalignment of the final orbit with that required. These measures went in as the constraints to the optimization process.

$$\underset{p(t_0),\ \theta}{\text{Min}}\ X_0(t_f)$$

$$\text{s.t. } \psi(t_f) = 0$$

$X_0(t_f)$, $\psi(t_f)$ obtained from integrating

$$\dot{X}_i(t) = \tau(t)\, f_i(X(t), u(t))$$

$$\dot{p}_i(t) = -\sum_p p_j \partial f_j / \partial X_i$$

$$\sum_j p_j\, \partial f_j / \partial u = 0$$

θ = vector: times for switching motors and other actions, other discrete parameters

$$\left[\frac{d^k}{dt^k}\left(\sum_j p_j\, \partial f_j / \partial u\right)\right]_{t=t_0} = 0 \qquad k=1,2,\ \ldots$$

Obtain $p(t_0)$ from $u(t_0), \dot{u}(t_0), \ldots$

Fig. 7(b). Trajectory optimization - a hybrid approach

To improve convergence, the initial values of the adjoint variables were not, as such, in the set of optimization variables since employing these leads to an ill-conditioned problem. However, the maximization of H at $t = t_0$ leads to a set of equations connecting $u(t_0)$ and $p(t_0)$. By differentiating these equations a sufficient number of times it is possible to express the p's as functions of $u(t_0)$, $\dot{u}(t_0)$ and so on, and these can be used as optimization variables in place of the $p(t_0)$. Since these angles and their derivatives are physical quantities, if the physical problem is not ill-conditioned, then the numerical problem will not be.

4. FUTURE

For the future I see two particular lines of development taking place. One is the development of Global Minimization techniques. All algorithms I have mentioned are essentially local minimization routines, and, especially with the constrained problems, can stick at a local optimum which is far from the best position. Stochastic methods show the most promising approach for solving these global minimization problems (Archetti (1975), Dixon (1977), Gomulka (1978) and Törn(1978)). However, a small amount of practical knowledge applied to the formulation of the problem in the first place can often circumvent the difficulty and as such, can be more valuable than a large amount of algorithmic expertise.

The other line is the possibility of having cheap computers which can execute parallel computation. For most problems the time of execution is almost directly proportional to the time taken to carry out the gradient calculations. Since the calculation of one element of the gradient vector does not depend on that of another, all elements could be calculated simultaneously. For problems with large dimensions, the time would be reduced by a factor approximately equal to the number of processors available so that factors of 10 or more are already within sight, and hence problems of larger dimensions could be envisaged.

5. CONCLUSION

Numerical optimization routines can be regarded as established techniques. They are not perfect and no doubt further developments will be forthcoming. However, as with many applied computational techniques, the most significant advance usually occurs when the practical problem is formulated appropriately.

6. REFERENCES

Archetti, F. (1975) "A Sampling Technique for Global Optimisation in Biggs (1975).

Bell, D.J. and Jacobson, D.H. (1975) "Singular Optimal Control Problems", Academic Press.

Biggs, M.C. (1972) "Constrained Minimisation using Recursive Equality Quadratic Programming", in "Numerical Methods for Non-linear Optimisation", ed., F. Lootsma, Academic Press.

Biggs, M.C. (1973) "A Note on Minimization Algorithms which make use of Non-quadratic Properties of the Objective Function", *J. Inst. Maths. Applics*, **12**, pp. 337-338.

Biggs, M.C. (1975) "Constrained Minimization using Recursive Quadratic Programming: Some Alternative Subproblem Formulations", in "Towards Global Optimisation", Eds., L.C.W. Dixon and G.P. Szegö, North Holland.

Dixon, L.C.W. (1977) "Global Optima without Convexity", Technical Report No. 85, Numerical Optimisation Centre, The Hatfield Polytechnic.

Gomulka, J. (1978) "Numerical Experience with Törn's Clustering Algorithm and two Implementations of Branin's Methods", in Törn (1978).

Murray, W. (1969) "An Algorithm for Constrained Minimization", in "Optimisation", ed., R. Fletcher, Academic Press.

NOC, (1975) OPTIMA - Routines for Optimisation Problems, Numerical Optimisation Centre, The Hatfield Polytechnic.

Törn, A. (1978) "A Search Clustering Approach to the Global Optimisation Problem" in "Towards Global Optimisation II", eds., L.C.W. Dixon and G.P. Szegö, North Holland.

SUBJECT INDEX

P

Q

R